Recent Developments in Mechanical Engineering

Volume I

Recent Developments in Mechanical Engineering
Volume I

Edited by **Bilroy Lockhart**

NY RESEARCH PRESS
New York

Published by NY Research Press,
23 West, 55th Street, Suite 816,
New York, NY 10019, USA
www.nyresearchpress.com

Recent Developments in Mechanical Engineering: Volume I
Edited by Bilroy Lockhart

International Standard Book Number: 978-1-63238-389-1 (Hardback)

Printed in China.

Contents

Preface

With the age of industrialization, the origin of the smallest of objects that we use in our daily existence can now be traced to a machine. The subject of Mechanical Engineering was derived from the need to design and manufacture everything from small individual parts, in order to reduce human effort. It involves simultaneous production and usage of heat and mechanical power for production purposes.

Among one of the oldest and broadest engineering disciplines, Mechanical Engineering involves everything from stability, maintenance design and manufacturing of mechanical systems. Utilizing the basic principles of engineering, physics and material science, Mechanical engineering is a diverse subject. From creating inkjet printer nozzles to spacecraft and multipurpose turbines, all these processes are premised on Mechanical Engineering.

An understanding of subjects such as structural analysis, thermodynamics, electricity, kinematics and mechanics is crucial to excel in this discipline of engineering. A good knowledge of control, instrumentation, heat transfer and design will always come handy in this field. Today, Mechanical Engineering finds it's applications in areas such as heating and cooling systems, watercraft, aircraft, weapons, medical devices, robotics and others.

I am grateful to all the researchers who've contributed their expertise and findings in this book. I would also like to thank the publishing house for finding me worthy of doing this project and giving me this incredible opportunity.

Editor

Study of Stress Sources and Critical Stress Combinations for the Input Shaft of a Longitudinally Mounted Four Speed Automotive Automatic Transmission Model

N. A. Abdel-Halim
Automotive Engineering Department, Faculty of Engineering, Helwan University, Cairo, Egypt

ABSTRACT

In the present paper; two models of the input shaft for a Longitudinal Mounted four Speed Automotive Automatic Transmission for *the first time* were introduced to describe the input shaft critical loads. In the first model; the DC (Direct Clutch) connects two gears together. This gives no change in the set torque (the set output torque equal to the set input torque). In the second model; the ODB (Over Drive Brake) fixes one element of the planetary set. This is resulting, the gear set gives reduction ratio (the set output torque is not equal to the set input torque). So, the transmission input shaft is worked under two different working operating conditions of torque. Also, it is loaded by a two vertical loads which are coming from the turbine and planetary set loads respectively. They are shown that there are three critical combinations of forces (contact force, shear force, and normal force) applied on the input shaft. The critical forces can be possibility exist three types of cracks for the input shaft cross section they are: transverse (torsion stress), longitudinal (bending stress), and vertical (shear stress). *The three cracks are studied* in this article. The article considers three stress factors: shearing torsion stress, shear stress, and bending stress.

Keywords: Longitudinally Mounted Automotive Automatic Transmission; Input Shaft Stresses; Torque Converter Torque Ratio; Torque Converter Weight; Simple Planetary Gear Set Weight

1. Introduction

In study of stress sources and critical stress combinations of the input shaft for the present longitudinally mounted four speed automotive automatic transmissions, which they are resulting in failure. It is common to focus on the main loads acting on the spline part of the input shaft which are forces due to the turbine torque and turbine weight, and also focus on the loads acting on the solid part of the input shaft which are forces due to the planetary set torque (the maximum set torque is occurred at the brake set is applied) and the planetary set weight respectively [1].

Turbine forces are transmitted to the input shaft through the spline part. For the simplest calculations it is assumed that the turbine transmits force and torque to the spline part of the shaft at the middle of its width, and study the stressed-strained is based on the corresponding cross-section. Although, actually, turbine force of interaction between the turbine hub and the spline shaft part is distributed along the length of the turbine hub. Also, the planetary set force is distributed along the bearing of the set.

Forces on the input shaft for the mentioned longitudinal mounted four speed automatic transmission induce stresses which are of torsional, bending, shearing, and axial nature. In studies of stressed-strained states of the input shaft resulting in fatigue failure, it is common to forces on the input shaft induce stresses which are of a torsional, bending, shearing, and axial nature. focus primarily on torsion phenomena. In so doing, torsional shear stresses are considered for the most part. The effect of axial forces is usually not significant and can be omitted.

2. Basic Concepts

The longitudinally four speed automatic transmission **Figure 1** input shaft is flowed the torque converter turbine torque to a simple planetary gear set. The turbine torque is varied according to the variation of the pump torque. The turbine torque variation has two categories. The first one is given at low torque converter speed ratio $\left(n = \omega_T / \omega_P\right)$. Where, the ratio of the turbine and pump torques is given by the Equation [2]:

$$R_T = T_T/T_P = 1/g\left(U+\omega_{up} - nUR_t^2 - \omega_{ut}R_t\right)\Big/\left(1/g\left(U+\omega_{up}-\omega_{ur}R_p\right)\right)$$

The second stage is occurred when the torque converter lock-up clutch is activated (at the design point), the turbine and pump are connected to the engine output. The angular velocity and torque of turbine are expressed as follows [3]:

$$\omega_T = \omega_P = \omega_E,\ T_T = T_p = T_E .$$

The resistant torque of the planetary gear set for the turbine torque has two values. One value occurs when the DC clutch is applied **Figure 2** and its value is:

$$T_{I/P} = T_T \times i_{\text{DC}} \times SF .$$

The other value is occurred when the ODB brake is applied **Figure 3** and the input shaft resistance torque is become:

$$T_{I/P} = T_T \times i_{\text{ODB}} \times SF .$$

The turbine torque (T_T) transmits through a spline part on the input shaft **Figure 4** effect on the whole shaft until the carrier bearing. The spline part of the shaft affected manly torsion moment (no bending).

Diagrams of bending moment M_b, caused by turbine weight (W_1) and first planetary gear set (W_2), and shear force (F_s), are shown in **Figure 5**.

Hence, the diagrams of contact forces from torsion $\left(F_t^c\right)$ are shown in **Figure 6(a)**. Diagram of torsional moment τ_t^c, caused by contact forces from torsion is shown in **Figure 7**. For any tooth since F_t^c and τ_t^c are invariants relative to tooth position (see **Figures 6(b)** and **7**).

The bending and shear stresses value changes with the distance from neutral axis are shown in **Figures 8** and **9**.

Since the stressed state of any *spline tooth* under torsion is the same, dangerous stress combinations for any

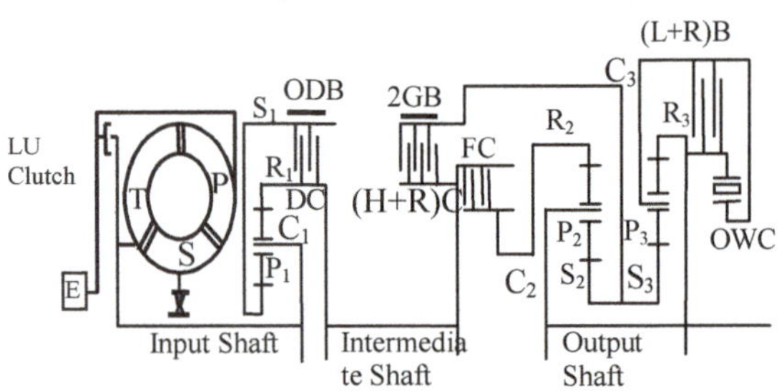

Figure 1. A longitudinally mounted four speed automotive automatic transmission model [3,4].

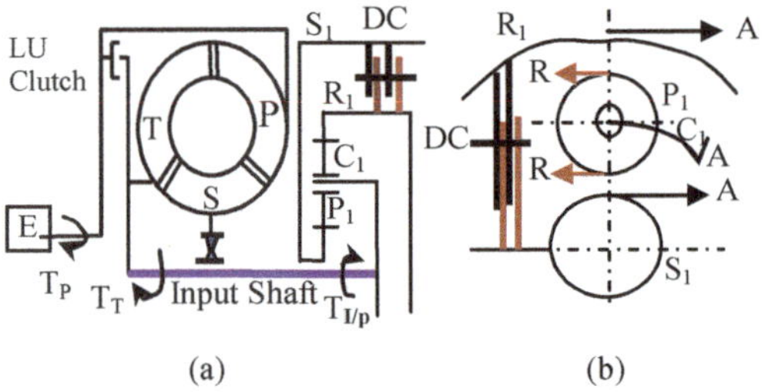

Figure 2. Input shaft model with dc clutch (a) Side view; (b) Plan view.

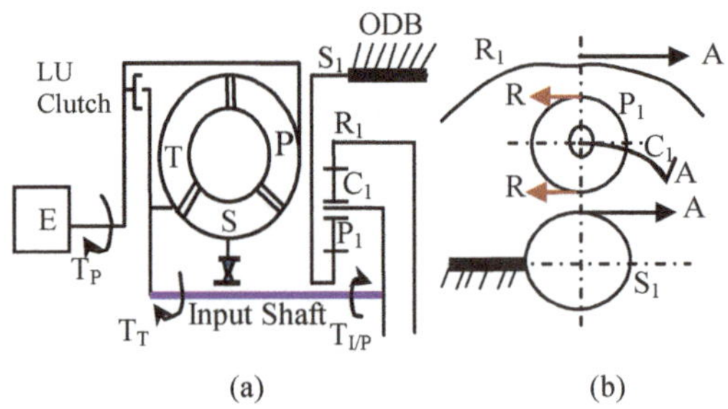

Figure 3. Input shaft model with ODB brake (a) Side view; (b) Plan view.

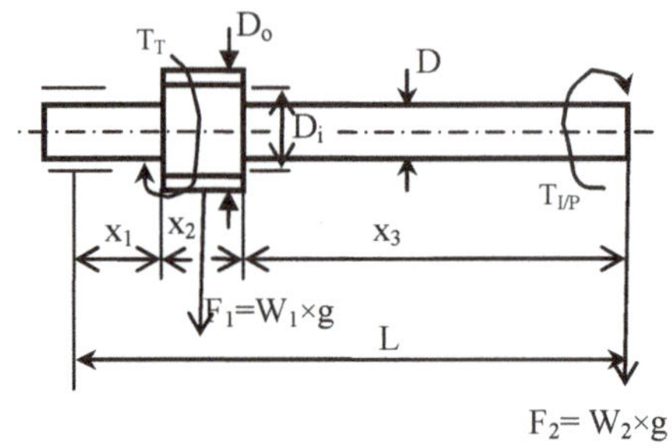

Figure 4. Input shaft design model.

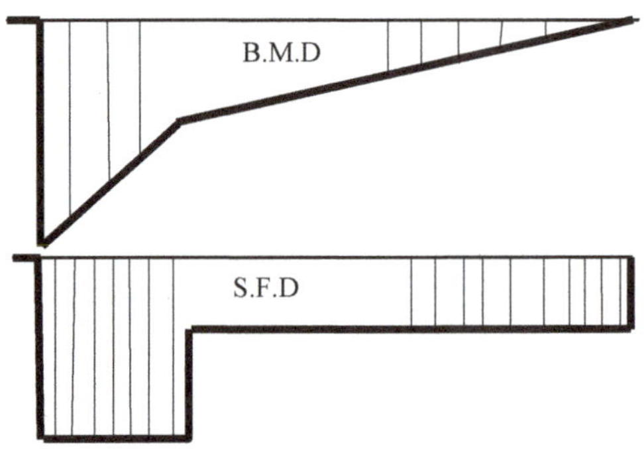

Figure 5. Bending moment and shear force diagrams.

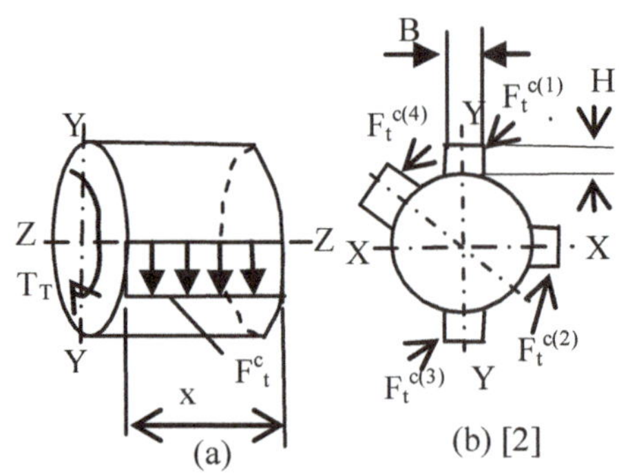

Figure 6. Diagram of contact forces caused by torsion.

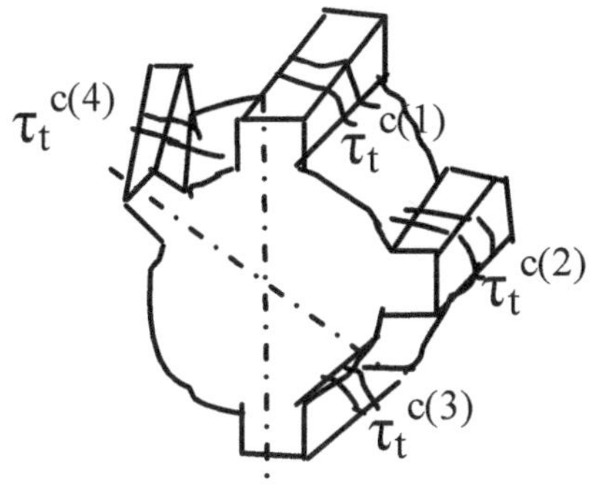

Figure 7. Torsional moment diagram [5].

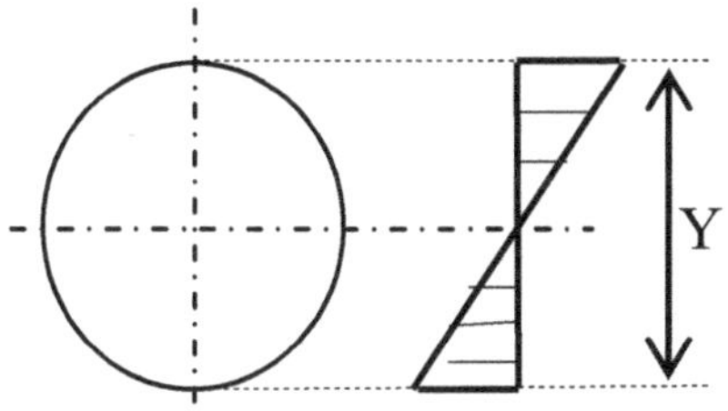

Figure 8. Bending stress diagram.

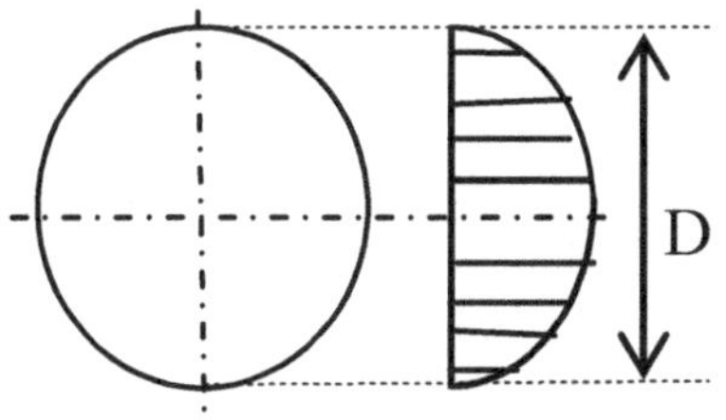

Figure 9. Shearing stress diagram.

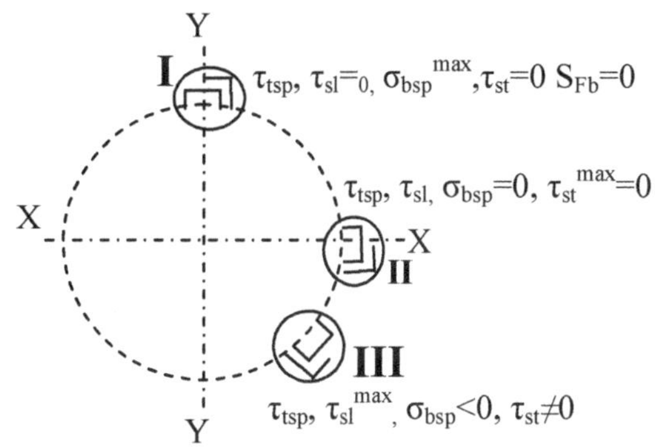

Figure 10. Three critical positions with different stress combinations of spline tooth [5].

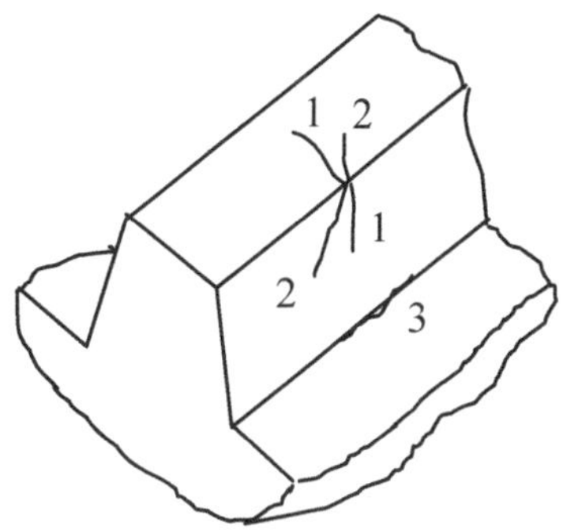

Figure 11. Three possible types of tooth cracks [5].

$$W_{\tau sp} = 2\times(I_X + I_Y)/D_i$$
$$= 2\times((\pi/64)\times D_i^4 + (\pi/64)\times D_i^4)/D_i,$$
$$W_{\tau sp} = (\pi/16)\times D_i^3 = 0.2D_i^3$$

tooth will be defined by bending. In the general case (torsion plus bending), the engineer should consider the spline tooth in three positions with different stress combinations (see **Figure 10**). In each of these positions, the engineer should analyze the possibility of appearance of three types of cracks shown in **Figure 11**: 1) transverse, 2) inclined, 3) longitudinal.

3. Torsion Stress

3.1. Spline Part Stress (Inclined Crack)

The Model of longitudinally mounted four speed automotive automatic transmission input shaft has spline part. This part is transmitting the torque converter turbine torque to the first planetary transmission gear set through its carrier. A stress analysis was first performed related to the torque converter size Equation (1) which can cause crack number (2) **Figure 11** [5]:

$$\tau_{tsp} = (T_{T\max}/W_{\tau sp})\times K_\tau,$$
$$F_{t\max}^c \times (D_m/2)\times n_{sp} = T_{T\max} \quad (1)$$

$W_{\tau sp}$= Polar moment of inertia (J) over distance of stress point from the neutral layer (y).

3.2. Solid Part Stress (Inclined Crack)

The same stress analysis (Equation (2) and **Figure 12**) is performed for the solid part of the input shaft which has different dimension. This difference is appearing into torsional section modulus ($W_{\tau sl}$) and the applied force value (F_t).

$$\tau_{tsl} = (T_{T\max}/W_{\tau sl})\times K_\tau,$$
$$T_{T\max} = F_t \times (D/2) \quad (2)$$
$$W_{\tau sl} = (\pi/16)\times D^3 = 0.2D^3$$

4. Shear Stress

4.1. Spline Part Stress

4.1.1. Longitudinal Crack

In practice, when the spline tooth of the input shaft is loaded by twist force, the shearing force at a section always parallel to the tangential force (circumference force or contact force from torsion), $F_{t\max}^c$. The resultant shearing stress at section [Equation (3)] assumes much importance in check the longitudinal input shaft crack

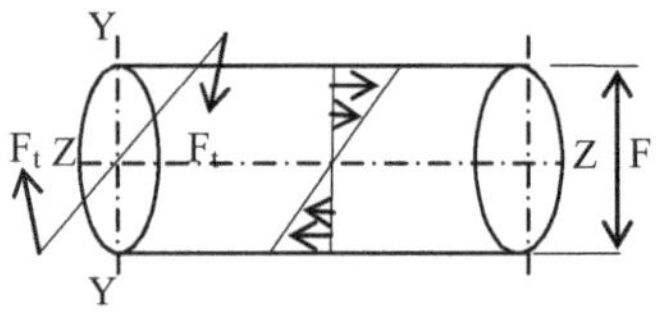

Figure 12. Torsional stress of solid shaft.

number (3) **Figure 11**.

$$\tau_{sl} = F^{c}_{t\max} / (B * X_2) \quad (3)$$

4.1.2. Transverse Crack

Also, there is direct shearing force **Figure 5** at a section parallel to the perpendicular force which coming from the turbine weight (W_1) and the planetary set (W_2). The resultant direct shearing stress [Equation (4)] **Figure 9** at section assumes much importance in check the transverse input shaft crack number (1) **Figure 11**.

$$\begin{aligned} \tau_{st} &= F_s / \left(\pi D_i^2 + B \times H \times n_{sp}\right) \\ F_s &= F_1 + F_2 = W_1 \times g + W_2 \times g \end{aligned} \quad (4)$$

4.2. Solid Part Stress

The solid part of the transmission input shaft is affected by direct shearing force resulting from the planetary gear set weight (W_2). Equation (5) shows the solid shearing stress which is suitable to the perpendicular crack of (Z-Z) direction **Figure 12**.

$$\tau_{ss} = F_s / \left(\pi D^2\right) \quad (5)$$

5. Bending Stress

5.1. Spline Part Stress

The level of the bending stress [Equation (6)] through the spline part of the input shaft is measured by two forces F_1 and F_2 which are resulted from two weights W_1 and W_2 multiply by their distances starting from the forces applied points to the input shaft bearing support point (at the left point of the shaft). Also, the value of the modulus of shaft section effects on the stress [5].

$$\begin{aligned} \sigma_{bsp} &= M_{bsp} \times \left(K_b / W_{bsp}\right) \\ M_{bsp} &= F_1 \times x_2 + F_2 \times \left(x_2 + x_3\right) \\ W_{bsp} &= I_{xsp} / y_{sp} = \left((\pi/64) \times D_i^4\right) / D_i / 2 \\ &= (\pi/32) \times D_i^3 = 0.1 \times D_i^3 \end{aligned} \quad (6)$$

5.2. Solid Part Stress

The bending stress [Equation (7)] through the solid part of the shaft slightly different from the spline part where forces distances and modulus of section values are not the same.

$$\begin{aligned} \sigma_{bs} &= M_{bs} \times \left(K_b / W_{bs}\right) \\ M_{bs} &= F_1 \times \left(x_1 + x_2/2\right) + F_2 \times L \\ W_{bs} &= I_{xs} / y_s = \left((\pi/64) \times D^4\right) / (D/2) \\ &= (\pi/32) \times D^3 = 0.1 \times D^3 \end{aligned} \quad (7)$$

6. Compound Stress

The reliable study of power transmitting shafts stress sources and critical stress combinations are supported on three steps of study. First; torsion stress characteristics were studied for the given shaft for its different cross sections. Second; shear stresses also for the different cross sections were calculated. Third; bending stresses were established. Finally; compound stresses into the spline and solid parts were defined by Equations (8)-(11). Equations (8) and (10) are represented compound torsion stresses while Equations (9) and (11) are showed compound bending stresses [5].

Spline Part Stress

$$\tau_{sp\Sigma} = \sqrt{\left(\tau_{tps}\right)^2 + \left(\tau_{sl}\right)^2} \quad (8)$$

$$\sigma_{eqsp} = \sqrt{\left(\sigma_{bsp}\right)^2 + 4 \times \left(\sigma_{sp\Sigma}\right)^2} \quad (9)$$

Solid Part Stress

$$\tau_{s\Sigma} = \sqrt{\left(\tau_{tsl}\right)^2 + \left(\tau_{ss}\right)^2} \quad (10)$$

$$\sigma_{eqs} = \sqrt{\left(\sigma_{bs}\right)^2 + 4 \times \left(\sigma_{s\Sigma}\right)^2} \quad (11)$$

7. Conclusions

1) The input shaft models were established having two different shapes of cross sections. Firstly; a circular cross section has splines around it. The splines are loaded by the torque converter turbine tortional force and weight. Secondly; a circular cross section hasn't spline. It is loaded by the first planetary gear set weight and gear set torional force.

2) A combined action of compound torsion and bending has been considered in studying the stressed-strained state of splined and solid parts of the input shaft.

3) Generally, there are three critical of forces for the spline part tooth. For each the three forces, three types of cracks, transverse, vertical, and longitudinal, may be occurred.

Also, the above three forces affect on the solid part of the input shaft. They are causing three cracks, axially, vertically and transversally [6].

REFERENCES

[1] H. Heisler, "Advanced Vehicle Technology," 2nd Edition,

2002, pp. 117-123.

[2] P. M. Heldt, "Torque Converters or Transmissions," 5th Edition, 1955, pp. 272-277.

[3] E. O. A. Abd Elmaksoud, E. M. Rabeih, N. A. Abdel-Halim and S. M. El Demerdash, "Investigation of Self Excited Torsional Vibrations for Different Configurations of Automotive Automatic Transmission Systems during the Engagement Period," *Ain Shams Journal of Mechanical Engineering* (*ASJME*), Vol. 2 No. 1, 2010, ISSN: 1687-8612,

[4] Y. Zhang and X. Chen, "Dynamic Modeling and Simulation of a Dual-Clutch Automated Lay-Shaft Transmission," *Journal of Mechanical Design*, Vol. 127, No. 2, 2005, pp. 302-207.

[5] B. P. Volfson, "Stress Sources and Critical Stress Combinations for Splined Shaft," *Journal of Mechanical Design*, Vol. 104, No. 3, 1982, pp. 551-556.

[6] J.-D. Wu and J.-H. Lin, "Experimental Investigation of Active Vibration Control for Gear Set Shaft," *International Journal of Vehicle Noise and Vibration*, Vol. 2, No. 1, 2006, pp. 1-16.

Nomenclature

E Engine
LU Locking clutch
ODB, 2GB, $(L + R)B$ Brakes
OWC One Way Clutch
DC, $(H + R)C$, FC Clutches
P_1, P_2, P_3 Planetary gears
S_1, S_2, S_3 Sun gears
R_1, R_2, R_3 Ring gears
C_1, C_2, C_3 Carriers
M_1, M_2, M_3 Moments
A Action
R Reaction
D Input shaft diameter
L Input shaft length
W_1 Turbine weight
W_2 Planetary set1 weight
F_1, F_2 Normal forces
G Acceleration of gravity
U The velocity of the part on which the jet acts
ω_{up} The tangential component of the relative velocity to the exit fluid of the impeller.
ω_{ut} The tangential component of the relative flow velocity to the exit fluid of the runner.
ω_{ur} The tangential component of the relative flow velocity to the exit fluid of the reactor (stator).
ω_T Angular velocity of the turbine.
ω_P Angular velocity of the pump.
ω_E Angular velocity of the engine.
N The ratio of turbine (runner) to pump (impeller) speeds (rpm).
R_T Turbine over pump torques ratio.
R_t Radius of the exit fluid from the runner.
R_P Radius of the exit fluid from the reactor.
T_T Turbine torque.
$T_{T\max}$ Maximum turbine torque.
T_P Pump torque.
T_E Engine torque.
$T_{I/P}$ Designed resistant torque of the planetary gear set.
i_{DC} Planetary set reduction ratio with DC clutch.
i_{ODB} Planetary set reduction ratio with ODB brake.
SF Safety Factor.
F_t^c Contact force from torsion
τ_t^c Torsional moment.
τ_{tsp} Torsional stress through the spline part.
τ_{tsl} Torsional stress through the solid part.
τ_{sl} Longitudinal shear stress through the spline part.
τ_{st} Transverse shear stress through the spline part.
τ_{ss} Shear stress through the solid part.
$\tau_{sp\Sigma}$ Compound torsional stress of spline part.
$\tau_{s\Sigma}$ Compound torsional stress of solid part.
σ_{eqsp} Compound bending stress of spline part.
σ_{eqs} Compound bending stress of solid part.
F_s Shear force.
F_t Tangintial force from torsional.
$W_{\tau sp}$ Torsional section modulus of the shaft spline part.
$W_{\tau sl}$ Torsional section modulus of the shaft solid part.
W_{bsp} Bending section modulus of the shaft spline part.
W_{bs} Bending section modulus of the shaft solid part.
K_τ Stress safety factor.
D_o Outer diameter of the shaft spline part.
D_m Mean diameter of the shaft spline part.
D_i Inner diameter of the shaft spline part.
D Diameter of the shaft solid part.
σ_{bsp} Bending stress through the spline part.
σ_{bs} Bending stress through the solid part.
M_{bsp} Bending moment through the spline part.
M_{bs} Bending moment through the solid part.
n_{sp} Number of splines.
I_X Moment of inertia around X axis
I_Y Moment of inertia around Y axis
I_{xsp} Moment of inertia around X axis of spline part.
y_s Distance from the section axis to the stress layer.
B Width of spline tooth (m)
H Height of spline tooth (m)
X_1, X_2, X_3, L Distances (m)

2

Vibrating Drivers for Transportation

Ben Zion Sandler, Vladimir Chapsky
Mechanical Engineering Department, Ben Gurion University, Beer Sheva, Israel

ABSTRACT

The movement of objects by means of vibrations is a widely known idea, used for bodies transportation in automated industry based on vibrating bases, on which the transported elements are placed. We consider an inversed idea: vibration is applied to the movable element. Surface, on which this displacement must be realized, is unmovable. The asymmetry of the friction forces in the different phases of the vibration is the cause of motion in this case. A distinctive feature of the proposed device is a slope of the plane of vibration of the inertial mass, which leads to increasing of the asymmetry of friction. In this paper, we consider an example of application of the device to the lateral vehicle parking. The idea is numerically estimated and tested with a laboratory prototype. The movement along a straight line of the trolley with sloped vibrating mechanism under influence of asymmetric friction forces has been estimated and practically simulated with the laboratory prototype.

Keywords: Inertial Drivers; Asymmetric Friction Force; Parking Mechanism

1. Introduction

In this paper, the author addresses a powerful means for developing new technical concepts, finding novel solutions and solving certain technical problems. It involves a method of thinking that produces results over several different cognitive levels: from the simplest design level in up to the highest conceptual level used to formulate new technical ideas. The method has already been partially described in this author's book [1].

The movement of objects by means of vibration is the widely known idea, used for transportation different parts and bodies in automated industry based on the use of vibrating bases, on which the transported elements are placed. This vibration is mainly realized by electromagnets and variable voltage. In this paper we consider the inversed idea: the vibration is applied to the moved element while the surface, on which this displacement must be realized, is unmovable.

The most known use of vibration of the object to move it himself is the vibro-hammers, used in the building and mining industries. These designs are mainly intended for movement in the vertical direction. Here, except the inertia forces, the force of weight acts in the vertical direction and leads to the asymmetry of the acting forces. Another situation occurs in devices intended for horizontal movement. In them the difference is caused by the asymmetry of the friction forces, acting on the base at different periods of the movement of the inertial mass, mounted on the base.

Example of such device is the mechanisms described in the patent USP No. 2639777 [2]. It intended for vehicles parking. In the method described in this patent the difference of friction between the wheels and the road in different phases of the rotation of the inertial mass (**Figure 1(a)**) is provided by an additional pressure, caused by the force of inertia of the rotating masses. This leads to the displacement of the car (**Figure 1(b)**). But calculations showed low efficiency of this design.

Search the efficiency of this method has led to a method of inertial driver, described in this article. In the new proposed method, the plane of rotation of the inertial mass is inclined to the line of moving direction. Calculations and simulations show that the proposed method is much more efficient than previously known, and can actually be used in practice. In addition, there are prospects for the use of this type of inertial drives to simply move the mobile objects.

2. Description of the Proposed Mechanism

2.1. Construction

The inclined inertial device is shown in **Figures 2(a)** and **(b)**. It consists of two identical engage gears (1), driven by motor (2). The gears are equipped with two symmet-

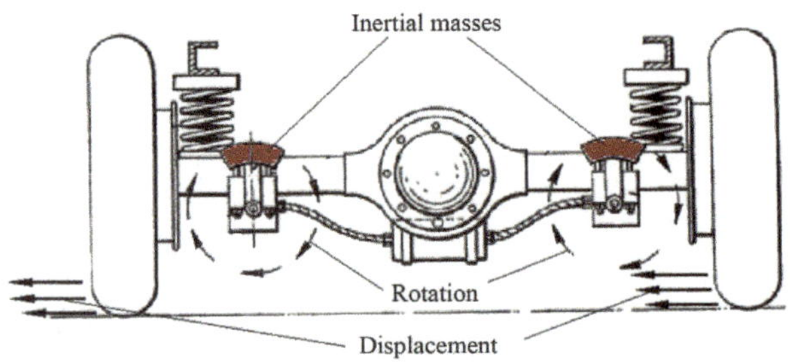

(a)

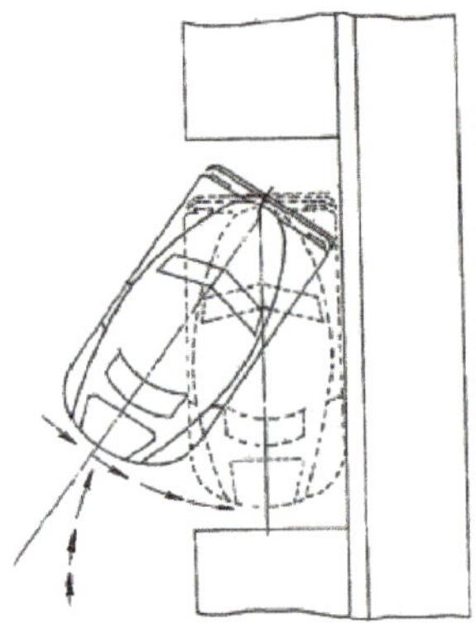

(b)

Figure 1. Parking of the car. Figures from USP No. 2639777.

rically fixed masses (3). The gears are assembled in a bearing common base (6) by means of two axes 4 and ball bearings (5). The assembly is mounted on a platform (7) at an inclination angle α. The platform slides on the surface (8) in the direction X. The transmission of rotation from the motor (2) to gears (1) is provided by two pulleys (9) and belt (10). Two force vectors $F_v(t)$ and $F_h(t)$ are created as a result of masses (3) rotation, which influences the normal force creating the friction between surface (8) and the devices support and the driving force moving the body the device is fastened on.

2.2. The Working Principle

We consider the operation of proposed vibration driver mechanism on example of **car parking** (**Figure 3**).

Laboratory model that demonstrates this movement is shown in the following video [3].

In following example the proposed mechanism consists from two rotating gear wheels mounted on rear axis of the car at some inclination angle as shown schematically in **Figure 4**.

The working principle is based on difference between friction forces, acting in two half-periods of wheels rotation. The rotation of the inertial masses, fixed on the disks, in the opposite direction leads to inertia forces, acting on the car, in the direction perpendicular to the line connecting the centers of disks ($F(t)$ in **Figure 4**). The components of inertial forces, acting along the line, connecting the centers of disks, are equals and opposite in direction, so they cancel each other out. If the rotation of the discs is with a constant angular velocity $\omega = 2\pi f$, then the function of inertia forces versus time is of a harmonic function with frequency *f*, as shown in **Figure 4**.

$$F(t) = F_0 \operatorname{Sin}(2\pi f t) \tag{1}$$

where F_0 is magnitude of the inertial force, created by two rotating masses:

$$F_0 = 2m(2\pi f)^2 r \tag{2}$$

m, mass of one inertial body; *f*, the frequency of rotation of the discs in Hertz units (or rotations per second); *r*, the distance between the centers of mass of the inertial bodies and the centers of the rotating disks.

The vertical component of force $F(t)$ changes the pressure on the car axle, and consequently changes the force of friction in the different phases of the rotation of the discs. The horizontal component of harmonic force $F(t)$ together with the different friction forces in the different phases of the discs rotation, leads to the movement of the car.

2.3. Calculation of the Angle of Car Displacement

In common case the movement of the car consists of three parts in each period of disc rotation: 1) *sliding*, the motion with friction, when the horizontal component of force is greater than the friction force and the vertical force component is less than the weight of the moving object; 2) *hovering* above the road, when the vertical component of force is greater than the weight of the moving object (which leads to its flying); 3) *deceleration*, motion in the presence of the friction force after touching the road.

Let's assume some restrictions on the physical model for considered here numerical examples (the estimated quantities are approximate):

- The friction index does not depend on the motion speed;
- Vehicle mass is much greater than the inertial mass of rotating bodies;
- The magnitude of the loading force, acting on bearing of rotating rings does not exceed 10,000 Newton;
- The rotation speed of the rings does not exceed 1200 rpm.

As shown in **Figure 3**, the car's back part is mowed inside the free place due to the vibrating driver. The same way, after the device is rotated for 180 degrees, it can be used for bringing the car out of the parking place by moving it reverse.

For the numerical example of the calculation of turning the car, we will take the geometric and physical pa-

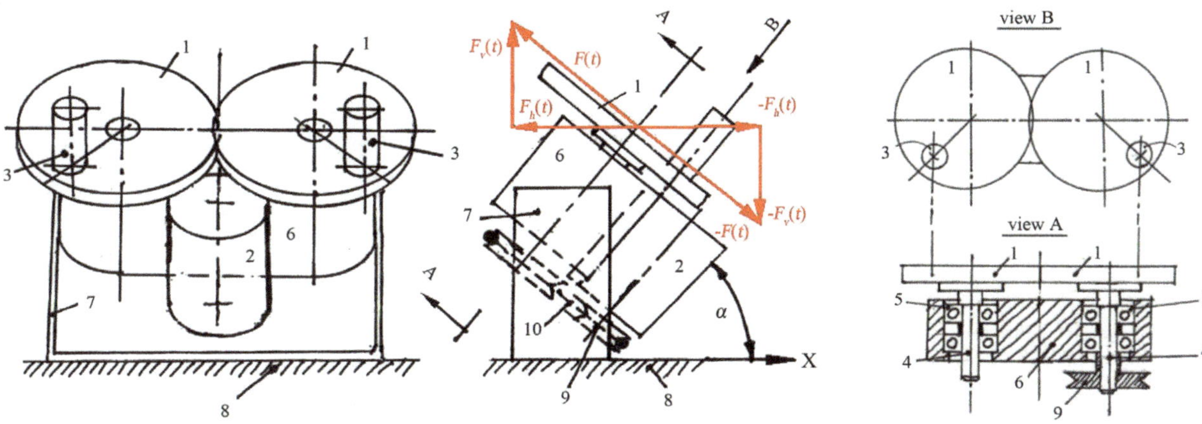

(a) (b)

Figure 2. General view of the vibrations generating device.

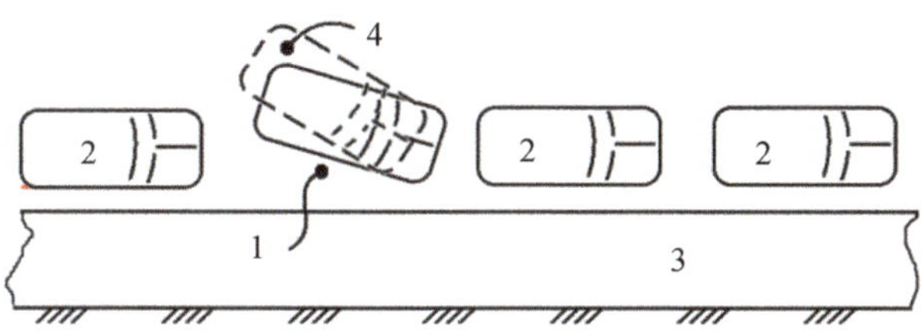

Figure 3. Lay-out of the parking place 1 in a street near the side walk 3. Car 2 finishes the parking process by moving its back part 4.

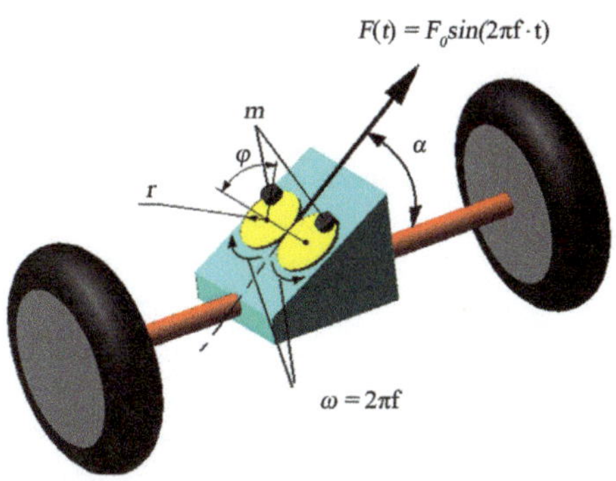

Figure 4. Schematic diagram of the back car's axe with inclined inertial device, mounted on it.

rameters of the vehicle and the inertial unit close to the real. Here are the following data:

- mass of the car is $M = 1000$ kg ;
- distance between rear and front axles is $L = 2.5$ meters;
- moment of inertia of the car relative to center of the front suspension $J = 2600\ \text{kg}\cdot\text{m}^2$;
- pressure on the rear axle due to gravitation weight of the car is $P = 5000$ N (Newton);
- radius of rotation disc is $r = 0.1$ meter;
- mass of the inertial body $m = 3$ kg ;
- rotation speed is 1200 rpm $(f = 20\ \text{rps}, T = 0.05\ \text{sec})$;
- inclination angles are $\alpha_1 = 45^\circ, \alpha_2 = 30^\circ, \alpha_3 = 15^\circ$;
- index of friction is $k = 0.8$;
- the permissible load on the bearings of the rotation discs assumed as $F_b = 5000$ N .

The purpose of the calculation is to estimate the angular displacement of the car relative the center of front suspension for all given α during one cycle, one second, and ten seconds.

For definiteness, the initial conditions assume as:

- The discs are already spinning with angular velocity $\omega = \text{rad / sec}$
- At the time $t = 0$ the position of the inertial body $\varphi = 0$.
- The car is not moving.

1) ***Sliding.*** For calculation of the first part of displacement, when car moves with friction, the time, when car starts to move and time when car starts to hover should be defined. Also, we need to define the equation of motion in horizontal direction.

Let $F_v(t)$ be the time function of the vertical component of the inertial force $F(t)$:

$$F_v(t) = F(t)\text{Sin}(\alpha) = F_0\,\text{Sin}(\alpha)\text{Sin}\left(\frac{2\pi}{T}t\right) \tag{3}$$

where $T = \dfrac{1}{f}$ is the period of disc rotation. F_0, magnitude of the inertial force (Equation 3);

$F_0 = 9474.82$ N . (The load on the bearings $F_b = F_0/2 = 4737.41$ N is less than permissible.

At the time, when the vertical component of inertial force equals or greater than the pressure force on the rear car axle, it breaks away from the road.

$$F_v(t)\big|t = t_{\text{break}} = F_v(t_{\text{break}}) = P, \tag{4}$$

The break time t_{break} was calculated from Equations (3) and (4). The graphs of function $F_v(t)$ for different inclination angles and the break time t_{break} were calculated using the Wolfram's Mathematica 8 software, shown in **Figure 5**. As shown in **Figure 5**, for $\alpha_1 = 45^\circ$ t_{break} = 0.0067 seconds. As can be seen from **Figure 5**, in the current example only for α_1 the hovering takes place. Thus, further we consider the current example separately for α_1, and then for α_2 and α_3.

Determination of time t_{start} when back part of the car starts to move transversal with friction, sliding.

The condition for horizontal moving is the horizontal component of the inertial force becomes greater than friction force. The following equation expresses this condition:

$$F_0 \operatorname{Cos}(\alpha)\operatorname{Sin}\left(\frac{2\pi}{T}t\right) \geq k\left(P - F_0 \operatorname{Sin}(\alpha)\operatorname{Sin}\left(\frac{2\pi}{T}t\right)\right) \quad (5)$$

where $F_0 \operatorname{Cos}(\alpha)$, the magnitude of horizontal component of inertial force.

From Equation (5) implies that:

$$\operatorname{Sin}\left(\frac{2\pi}{T}t\right) \geq \frac{kP}{F_0\left[\operatorname{Cos}(\alpha) + k\operatorname{Sin}\alpha\right]},$$

and

$$t_{\text{start}} = \frac{T}{2\pi}\operatorname{ArcSin}\left(\frac{kP}{F_0\left[\operatorname{Cos}(\alpha) + k\operatorname{Sin}(\alpha)\right]}\right) \quad (6)$$

Equation (6) is true for all $\alpha_1, \alpha_2,$ and α_3. Calculation gives the following:

$$t_{\text{start1}} = 0.00269 \text{ sec},$$
$$t_{\text{start2}} = 0.00271 \text{ sec},$$
$$t_{\text{start3}} = 0.00292 \text{ sec}.$$

Let $F_h(t)$ be the horizontal component of the inertial force $F(t)$, $F_{loc}(t)$ be the locomotive power, defined as difference of horizontal component of the inertial force and force of friction $F_{fr}(t)$:

$$F_{loc}(t) = F_h(t) - F_{fr}(t),$$

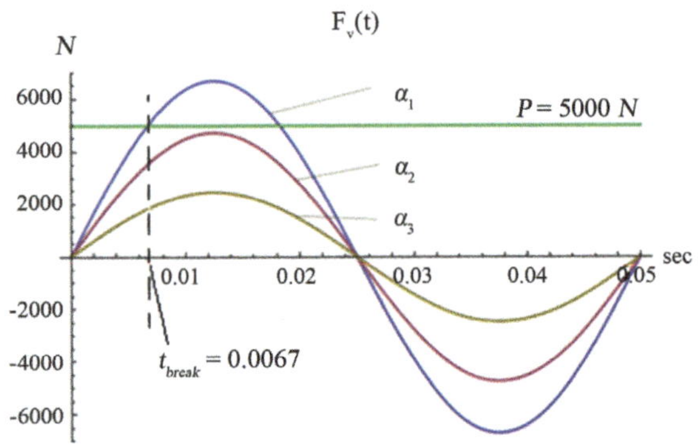

Figure 5. Vertical component of the inertial force versus time for different inclination angles of the plane of rotation discs.

where:

$$F_h(t) = F(t)\operatorname{Cos}(\alpha) = F_0 \operatorname{Cos}(\alpha)\operatorname{Sin}\left(\frac{2\pi}{T}t\right), \quad (7)$$

$$F_{fr}(t) = k\left(P - F_v(t)\right) \quad (8)$$

The moments of the corresponding forces relative to center of the front axle are:

$M_v(t) = F_v(t)\cdot L$, the moment of vertical component of inertial force,

$M_h(t) = F_h(t)\cdot L$, the moment of horizontal component of inertial force,

$M_{fr}(t) = F_{fr}(t)\cdot L$, the moment of friction force,

$M_{loc}(t) = F_{loc}(t)\cdot L$, the moment of locomotive power,

where L, the distance between rear and front axles.

The angular acceleration of the car relative to center of front axle, caused by moment of locomotive force during first part of motion (*sliding*):

$$A_s(t) = \frac{M_{loc}(t)}{J} \quad (9)$$

Graphs of function $A_s(t)$ versus time for different inclination angles of discs plane are shown in **Figure 6**.

The angular velocity $V_s(t)$ of the car relative to center of front axle as function of time is calculated by integration of the function $A_s(t)$ over time and is represented in **Figure 7**.

The area under velocity plot from t = t_{start1} = 0.00269 sec to $t = t_{\text{break1}} = 0.0067$ sec. numerically equals to angular displacement (φ_s) of the car during one cycle. The result of calculation gives φ_s and the angular velocity at the time of breaking V_{break}:

$$\varphi_{s1} = 0.000013977 \text{ rad}, V_{\text{break}} = 0.0102 \text{ rad/sec} \quad (10)$$

2) *Hovering*. The second part of the motion is the movement of a car under the action of gravitational and inertial forces, without touching the road with an initial velocity.

The vertical component of the initial angular velocity equals to zero. Equation of the second part of motion (hovering) in vertical direction looks as follows:

$$A_{vhov}(t) = \frac{F_v(t) - P}{J}L,$$
$$\varphi_v(t) = \int_{t_{\text{break}}}^{t}\int A_{vhov}(t)\,dt \quad (11)$$

where $A_{vhov}(t)$, vertical component of angular acceleration during hovering; $\varphi_v(t)$, angular displacement in vertical direction during hovering as function of time. The numerical representation of Equation (11) and graph of function $A_{vhov}(t)$ for inclination angle $\alpha_1 = 45^\circ$ is shown in Equation (12) and **Figure 8**.

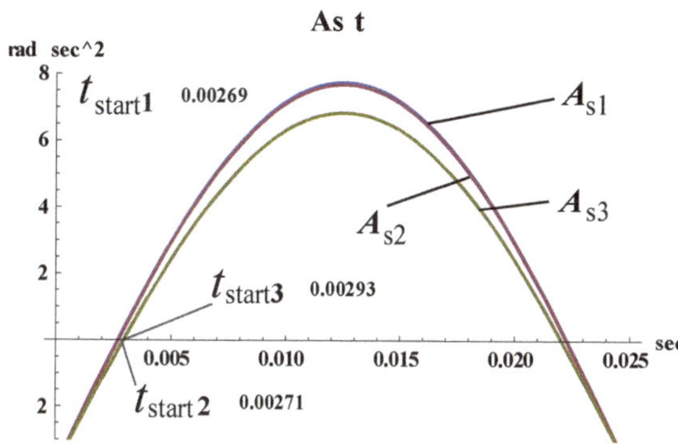

Figure 6. Horizontal angular acceleration of the car during sliding and friction on the road as a function of time.

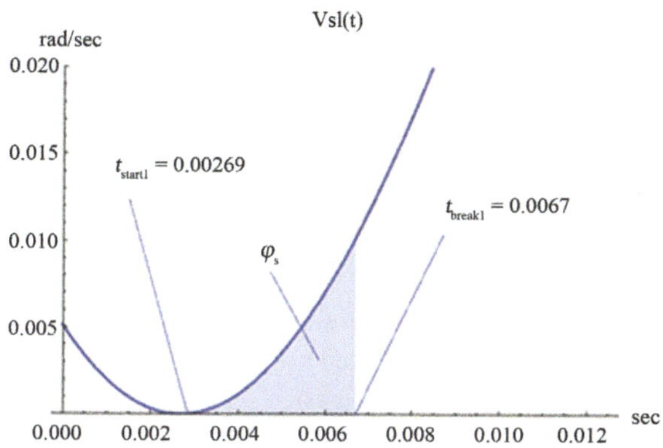

Figure 7. The horizontal angular velocity of the car for inclination angle α_1 during sliding.

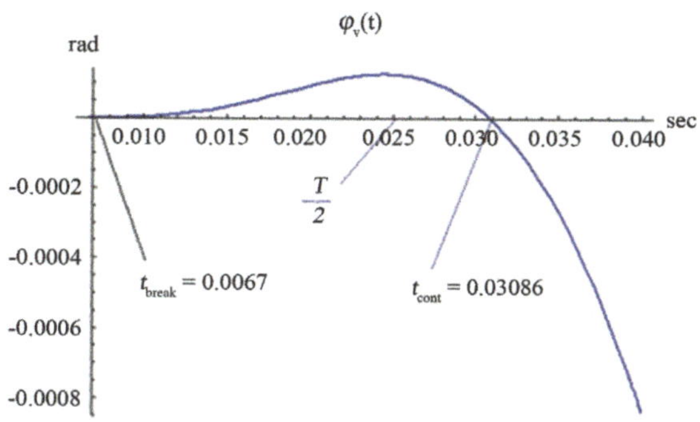

Figure 8. Vertical angular displacement $\varphi_v(t)$ during hovering as function of time.

$$\varphi_{vhov}(t) = -0.00003236 + 0.066354t$$
$$-2.40386t^2 - 0.000408\,\mathrm{Sin}\left(\frac{2\pi}{T}t\right) \quad (12)$$

Obviously only positive values of the vertical displacement make sense. **Figure 8** shows that contact with the road exists in the second half of the period T of disks rotation. It means that after the time of $t = T/2$ inertial force acting opposite the movement of the car and slows it.

The horizontal component of the initial angular velocity equals to the angular velocity at the time of breaking (V_{break}). Equations of the second part of motion (hovering) in horizontal direction looks as follows:

$$A_{hhov}(t) = \frac{F_h(t)}{J}L,$$
$$V_{hhov}(t) = V_{\mathrm{break}} + \int_{t_{\mathrm{break}}}^{t} A_{hhov}(t)\,\mathrm{d}t, \quad (13)$$
$$\varphi_{hhov}(t) = \int_{t_{\mathrm{break}}}^{t} V_{hhov}(t)\,\mathrm{d}t.$$

Or in the numeric representation:

$$V_{vhov}(t) = 0.0443162 - 0.051264\,\mathrm{Cos}\left(\frac{2\pi}{T}t\right) \quad (14)$$

where $A_{hhov}(t)$, horizontal component of angular acceleration during hovering, $V_{hhov}(t)$, horizontal component of angular velocity during hovering, $\varphi_{hhov}(t)$, angular displacement in horizontal direction during hovering.

Figure 9 shows the horizontal component of angular velocity $V_{hhov}(t)$ of the car relative to center of front axle as function of time. The area under velocity plot from $t = t_{\mathrm{break1}} = 0.0067$ sec to $t = t_{cont} = 0.03086$ sec . Numerically equals to the angular displacement (φ_{hhov}) of the car during one hover. The result of calculation gives:

$$\varphi_{hhov} = 0.001649 \text{ rad}.$$

The horizontal component of angular velocity at the time of contact with road is:

$$V_{cont} = 0.0822946 \text{ rad / sec}. \quad (15)$$

3) ***Deceleration.*** We assume that the collision with the road was of the nature of the inelastic and there was no rebound. Thus, the third part of movement is the horizontal sliding with presence of friction force with initial angular velocity V_{cont} under action of the moment of horizontal component of inertial force $M_h(t)$. The

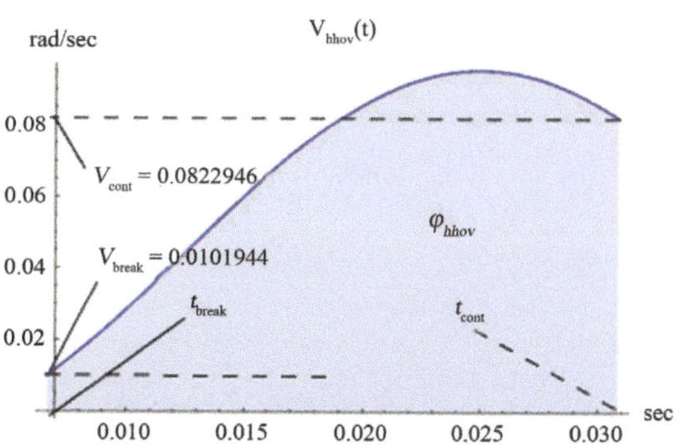

Figure 9. Angular velocity of the car in horizontal hovering, as a function of time (without friction).

equations of motion for third part are the same, as for the first part, but with other initial conditions. There are as follows:

$$A_d(t)=\frac{M_{loc}(t)}{J},$$
$$V_d(t)=V_{cont}+\int_{t_{cont}}^{t}A_d(t)\mathrm{d}t, \qquad (16)$$
$$\varphi_d(t)=\int_{t_{cont}}^{t}V_d(t)\mathrm{d}t,$$

where $A_d(t),V_d(t),\varphi_d(t)$, are the angular acceleration, velocity, and displacement during the third part of movement (deceleration). Numerical representation of the angular velocity $V_d(t)$ is follows:

$$V_d(t)=0.13263-3.84615t-0.0922753\,\mathrm{Cos}\left(\frac{2\pi}{T}t\right) \qquad (17)$$

Graph of this function is shown in **Figure 10**. The area under the graph of $V_d(t)$ numerically equals to angular displacement of the car during part of deceleration:

$$\varphi_d(t)=0.0002532\text{ rad} \qquad (18)$$

Finally, the absence of a reverse movement should be checked. For this purpose, the model must satisfy the following condition: the value of horizontal force less than the value of friction force, $F_h(t)\le F_{fr}(t)$, for $t>t_{stop}$, or, according to Equation (5):

$$\left|F_0\,\mathrm{Cos}(\alpha)\mathrm{Sin}\left(\frac{2\pi}{T}t\right)\Bigg|_{t_{stop}}^{T}\right| \le \left|k\left(P-F_0\,\mathrm{Sin}(\alpha)\mathrm{Sin}\left(\frac{2\pi}{T}t\right)\right)\Bigg|_{t_{stop}}^{T}\right| \qquad (19)$$

Graphs of left and right sides of Equation (19) are shown in **Figure 11**.

As clear from **Figure 11**, the condition of Equation (19) is satisfied. Reverse movement during deceleration is absent.

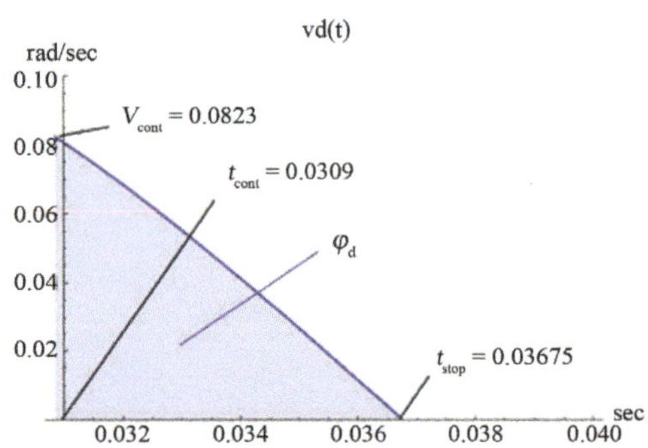

Figure 10. Angular velocity of car's movement during horizontal deceleration, as a function of time.

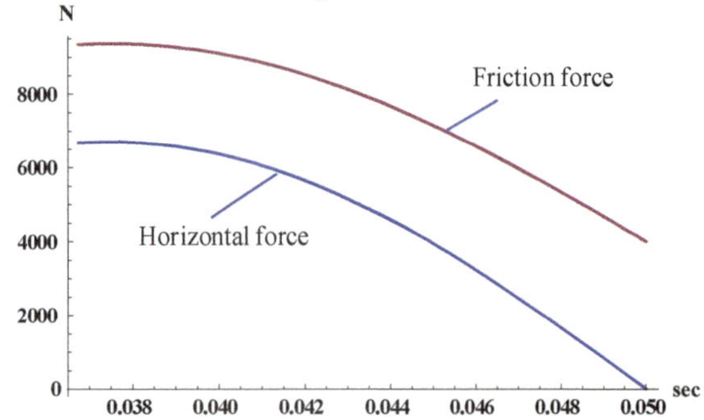

Figure 11. Comparison of the force values during deceleration.

Full rotation in one cycle is the sum of three parts of angular displacement:

$$\Phi_{\text{cycle1}}=\varphi_{\text{cycle1}}+\varphi_{hov}(t)+\varphi_d(t)=0.001916\text{ rad} \qquad (20)$$

The full rotation in ***one second*** is:

$$\Phi_{\text{sec1}}=f\cdot\Phi_{\text{cycle1}}=0.038323\text{ rad} \qquad (21)$$

The full rotation in ***fifteen seconds*** is:

$\Phi_{15\_1}=15\cdot\Phi_{\text{sec1}}=0.575\text{ rad}$, or in degrees:

$$\Phi_{15\_1D}=32.9^\circ \qquad (22)$$

The calculations of the angular displacements for inclination angles $\alpha_2=30^\circ$ and $\alpha_3=15^\circ$ were performed in the same way as for Parts 1 and 3 of previous calculation, but only for continuous time range from t_{start} to t_{stop}. The corresponding start times, which have been calculated previously by Equation (6), are:

$t_{\text{start2}}=0.00271\text{ sec}$, $t_{\text{start3}}=0.00292\text{ sec}$.

By integration of function of angular acceleration of the car Equation (9) the function of angular velocity for α_2 and α_3 should be obtained. The results of calculations are follows:

$$V_{s2}(t)=0.0969367-3.84615t-0.0918\,\mathrm{Cos}\left(\frac{2\pi}{T}t\right)$$
$$V_{s3}(t)=0.09906089-3.84615t-0.08504\,\mathrm{Cos}\left(\frac{2\pi}{T}t\right) \qquad (23)$$

The graphs of listed functions are shown in **Figure 12**.

The angular displacements numerically equal to arias under graphs $V_s(t)$, and should be calculated by integration in time range from t_{start} to t_{stop}.

$$\begin{aligned}
\Phi_{cycle2} &= 0.00173 \text{ rad} \\
\Phi_{cycle3} &= 0.00147 \text{ rad} \\
\Phi_{sec2} &= 0.0346 \text{ rad} \\
\Phi_{sec3} &= 0.0293 \text{ rad} \\
\Phi_{15_2D} &= 29.74^\circ \\
\Phi_{15_3D} &= 25.18^\circ
\end{aligned} \tag{24}$$

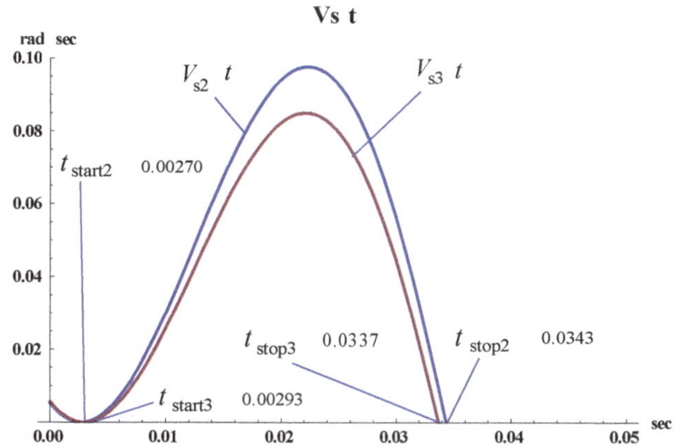

Figure 12. Function of angular velocity during sliding for inclination angles $\alpha_2 = 30^\circ$ and $\alpha_3 = 15^\circ$.

The results of calculations showed that the proposed model of the inertial drive with inclination angle of discs plane 30° - 45° is acceptable for use for car parking.

2.4. Some Possible Improvements of the Proposed Devise

The above example shows the opportunity to apply the proposed inertial drive for parking a small car. Some additional design solutions can enhance the capabilities of the devices, described above.

A design in which exists the possibility of rotation of the device around the vertical axis by 180°, will allow the car both, to enter inside a parking place and out of it.

The ability to change the angle (α) of the plane of disks rotating will change the parameters of the lateral movement of the car according to the coefficient of friction between the tires and the road surface.

For heavy goods vehicles the inertial unit with four or more rotating inertia bodies can be used. The schematic drawings of such units are shown in **Figure 13** or in **Figures 6** and **9** in the patent 4050527 "Vibrodriving apparatus" [4].

3. Brief Discussion about Linear Displacement of the Vehicle

As it is obvious from the description of the operating principle of the proposed inertial drive, it can also be used for different other vehicles. A necessary condition is the creation as much as possible difference in the friction forces between the vehicle and the road, acting in the various phases of the rotation of the inertial drive. Rectilinear movement of the vehicle with the inertial drive on the normal wheels in the presence of rolling friction (which is much less than sliding friction) is not effective. To use the above described model with the inertial drive for the rectilinear motion of the vehicle, its design shall ensure the presence of the forces of sliding friction.

Experimental Model Examination

An example of this design is the model of a trolley with inertial drive, which is shown in the following video clip [5].

In this trolley, the back axle was stalled and was loaded with additional weight, for increasing the force of sliding friction. Also, the line of action of the inertial force passes through the rear axle of the trolley that increases the effect of the vertical component of the force of inertia on value of pressure on the rear axle (**Figure 14**).

The front swivel wheels provide a stable direction of the trolley motion because in the direction of rectilinear motion acts the force of rolling friction, but in the transverse direction acts the force of sliding friction between rubber tires and the floor, which is much larger than rolling friction.

Calculations using the algorithm described above for car parking (for the sliding motion) gave results similar to those observed in the video clip. The trolley has an average speed of 11.3 cm/second, slightly more than in the video clip (~10.8 cm. per second.). Measuring the distance traveled by the trolley is represented by the graph in **Figure 15**.

Approximation section $S(t)$ (interrupted line) obviously can be described by a simple formula of the following form:

$$S(t) = 10.8 \cdot t - 15 \tag{25}$$

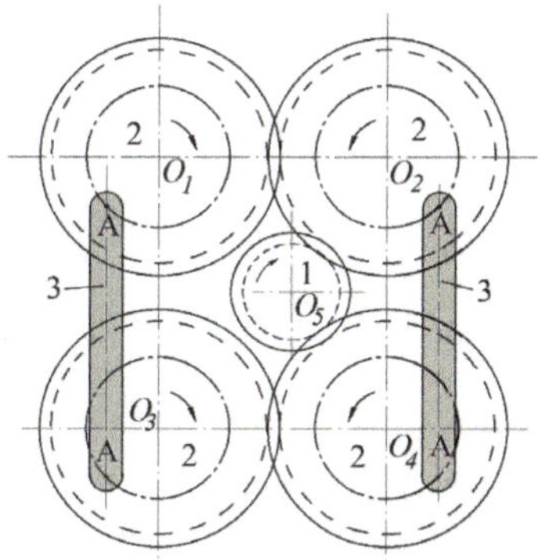

Figure 13. Inertial unit, consisting of four rotating discs (2) with inertial bodies (3), driven by gear (1).

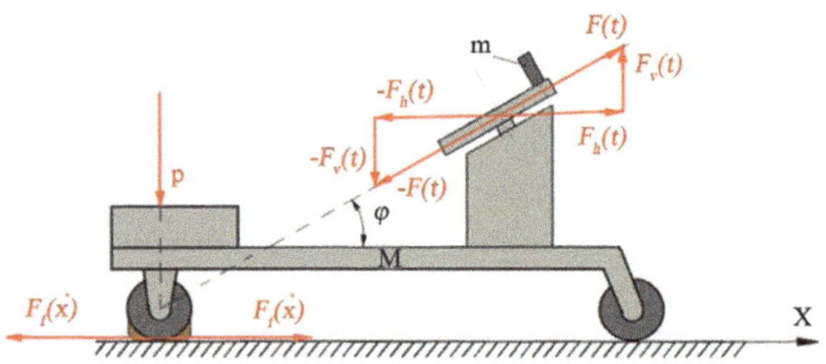

Figure 14. Lay-out of a platform moving by the inertial driving mechanism.

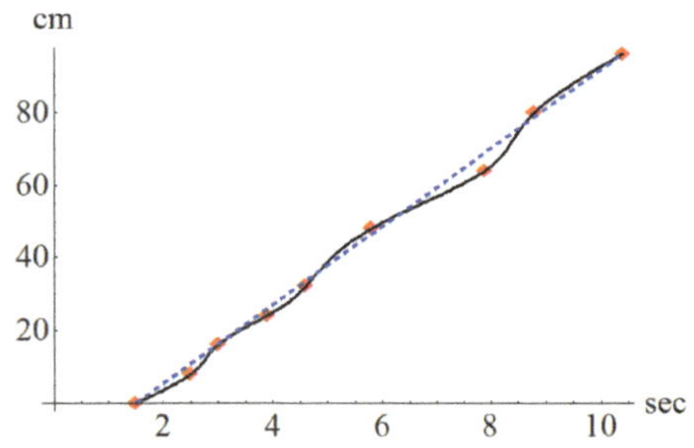

Figure 15. Displacements of the trolley versus time (according to test, shown in the video clip). A linear approximation of the "motion versus time" is also shown in this Figure.

(Time is taken from about ~1.5 seconds without movement, which is about 15 cm, while further it moves with the speed about 10.8 cm/sec for about $t = 10$ sec).

4. Conclusion, Some Possible Designs

We consider this paper as an example mainly describing the first steps in the design process. The step, where the main idea of the solution, answering the formulated problem the design task requires, must be found, at least one answer, or a number of them. At this stage, the main accent is put on the lay-out or model of the solution. It can be found in already existing designs, or inversed existing idea, or somehow changed, improved existing idea. A consideration of this approach and description of it are given, for instance in some author's books [1,2] and others.

Technical Proposals for Transportation

We are speaking about mechanical dynamic vibrations as a reason to create certain displacement of a car, improving the parking possibilities. This dynamic mechanical effect, as is here reminded, is widely used in industry.

We inversed this idea by deciding to vibrate the back of the car (not the surface the car is standing on) in such a way moving it to the needed place. Another change, which we propose in our design, is not to use electromagnets as a vibration source, which is often used in industrial machines, usually automatic, for transportation parts. We propose pure mechanical vibrator design (see **Figure 2**). The simplest consideration brings us to think about counterbalance fly-wheels used in rail-way engines or locomotives. The same solution is sometimes used in crank-shaft mechanisms. The cranks and wheels usually rotate in vertical plane, which makes the inertial forces appearing here not appropriated to create driving the mechanism to certain direction. To reach the effect of certain movement direction, the orientation of the vibrating device in space must be changed and the direction of the driving force must be strictly defined. This is achieved in the given here solution (**Figure 2**).

An idea to drive by the device, considered in **Figure 2**, placing it on elastic bases, say, on 4 rubber balls. This may be a specific device, on which the vibrator can be directed in the needed way. Such a device may be a simple toy or a special used carriage.

REFERENCES

[1] B. Sandler, "Creative Machine Design, Design Innovations and the Right Solutions," Paragon House Publishers, New York 1985.

[2] M. L. Dull, "Method and Device for Imparting Lateral Movement to an Automobile," US Patent No: 2639777, 1953.

[3] Video Clip in YouTube. http://www.youtube.com/watch?v=B3p4zH4992c&feature=relmfu

[4] J. L. Lebelle, "Vibrodriver Apparatus," US Patent: 4050527, 1977.

[5] Video Clip in YouTube. http://youtu.be/Owu2ANuwPLw

3

Application and Analysis of Force Control Strategies to Deburring and Grinding

Frank Domroes, Carsten Krewet, Bernd Kuhlenkoetter
Institute of Production Systems, TU Dortmund University, Dortmund, Germany

ABSTRACT

This paper aims to give an insight into the development and evaluation of force controlled machining processes with a commercially available setup. We will focus on a deburring and a grinding scenario, representing the major applications in today's robot machining. Whereas the deburring use-case implements a force dependent feed-rate control, the grinding use-case implements an orthogonal force (pressure) control. Both strategies will be evaluated and the influence of general machining and robot specific parameters will be discussed.

Keywords: Robot Machining; Deburring; Grinding; Force Control Strategies

1. Introduction

Industrial robots are more and more used for machining tasks. The main reason for this is that the investment costs that are required to purchase these robots are much smaller compared to the costs for purchasing a CNC machining center. Other advantages are the relatively small footprint that is needed to build up, in relation to the large work space of the robot and the fact that the robot can easily carry out preceding and subsequent handling processes. Thus, it is not necessary to implement additional devices to load and unload the machining centers.

Of course, the comparison is not complete without considering the robot's limitations, e.g. a significantly lower stiffness compared to a machining center. This lack of rigidity, among other, has negative effects on the accuracy of the robot during the machining process [1]. On the other hand, recently developed force control systems for industrial robots have proven very good performance, e.g. high sampling rate and low input-output latency, in a range of different machining applications [2].

2. Related Work

To face the occurring problems during machining robot manufacturers have optimized some of their robot kinematics for the use in machining processes. The manufacturer ABB Ltd. has developed a particularly stiffened robot that is explicitly marketed for "pre-machining": the IRB 6660-208/1.9 [3]. The manufacturer Stäubli is following a different approach with its RX170 hsm. This robot has a spindle directly connected to its fifth axis, in order to achieve an improved rigidity [4].

In addition to mechanical aspects, process control technology is an important factor. Each robot manufacturer uses a different programming language for the definition of motion commands. This means that the data generated by CAM systems can't be used directly with a robot, but must be translated into the native robot language. Some offline programming systems for robots are able to import NC programs and then create appropriate programs for the robot.

With the software option KUKA, CNC the manufacturer KUKA integrated a complete CNC control into the robot controller. Therefore the NC programs do not need to be converted into robot programs but run directly on the CNC control, which also results in a different motion behavior of the robot [5].

A different approach has been presented from ABB Robotics with the application package "Force Control for Machining" [6]: A 6D force torque sensor at the robots flange was integrated into the control loop. This makes it possible to apply a defined contact force between the robot and its environment during the movement. This functionality is particularly used in grinding and polishing processes [7].

Many university research efforts aim to increase the accuracy of robot machining processes. The Institute PTW at TU Darmstadt is analyzing the interactions between the robot structure and the executed milling process to develop a coupled simulation. Core component of the simulation is an offline correction in order to minimize the deviations of the trajectories that are caused by the process [8,9].

In the EU-funded project COMET (plug-and-produce components and methods for adaptive control of Industrial robots enabling cost effective, high precision manufacturing in factories of the future) an international consortium of industrial companies and research institutions are working on the development of components and methods for high-precision robotic machining. The approach of this project is to capture the inaccuracies of the industrial robot online and directly compensate these inaccuracies by applying an offset to the clamping system of the work piece using an additional high precision kinematic [10].

Additionally to the efforts on improved robot accuracy, force control strategies were developed for machining processes which are less geometrical constrained. In [11] a force dependent feed-rate control, among other, was presented, who shifts the robot speed in discrete steps depending on the measured force. If even at reduced speed the interaction forces exceed the defined range, a recover routine can avoid tool breakage. This function is very capable and we will later demonstrate how this enables online adaption of deburring paths.

Aside from the possibilities in process control, force control in robotics is used to reduce teaching effort. As teaching the paths on the geometry of e.g. casted work pieces in general can be very time consuming. Therefore for work pieces which vary in shape or size different approaches have been presented in literature to reduce the teaching effort.

Zhang *et al.* presented a force and vision sensor based deburring approach where the robot follows an afore marked path and thereby accomplished a time reduction compared to manual teaching from 8 to 10 weeks to 1 hour [12]. Solvang *et al.* took the expertise of the worker into account to define a path on a 2D camera image of the real work piece and map this input to a CAD- model in an offline programming environment to create the robot paths [13].

Although these methods allow a significant improvement to conventional teaching methods regarding both time savings and accuracy gain, they do fit exactly to the requirements of small and medium enterprises (SME's). As these companies mostly rely on small batch production, a large variety of work pieces has to be treated by the automated system with as little teaching effort and user interaction as possible. At the same time systems have to be smart and reliable. To keep the investment on an acceptable level, new approaches should focus on inherent flexibility before the integration of more sensor technology is considered.

3. Deburring of Water Pump Impellers

3.1. Background

In a typical deburring application the aim is to remove primary burr along the edges of the work piece, e.g. from casting or machining processes, with a rotary cutting tool mounted on a spindle-drive. Additionally, it is often desired to break the edge with a 45° angle, so that the worker will not be harmed during handling and/or assembly.

The application considered here is a deburring task of water pump impellers from various shapes and sizes. The impellers are created from high strength bronze in a foundry process with various subsequent stages of traditional CNC machining. Burr appears mostly along impeller blade edges and, considering the measurable influence of edge shape on pump efficiency, has to be removed with care. This job is traditionally done by hand labor aligned with high physical demands on the worker in a rough environment.

The experimental system is built around an industrial robot (ABB IRB 4400) with 60 kg maximum payload and equipped with a force control application package. Attached to the robot's flange are a force torque sensor (ATI Omega 160) and a 3-finger-gripper (SOMMER AUTOMATIC GD316) to unload the impellers directly from a CNC machine. An electric spindle drive mounted on a table can provide speeds between 5000 and 36,000 rpm at a maximum output power of 8 kW. The actual tool for deburring is a cone shaped hard-metal cutter.

3.2. Path Generation and Adaption

What makes the impeller deburring process especially suited for the development of a force controlled machining strategy is the fact, that precise 3D geometry information usually provided by CAD-files is not available for the majority of impeller types. Therefore the sole use of geometrical path planning is simple not an option. Instead a generalized task description was derived and deburring trajectories are corrected online based on contact force information. This should theoretically allow handling a high work piece variety without high offline programming and teaching effort as well as provide better and more constant process quality compared to manual deburring.

The edges to deburr on the impeller are marked in **Figure 1**. A comparison among the possible impellers showed that the path consists of 3 sections and can be described as a lower circular path between a start point (1) and the next impellers blade (2) and an upper circular path, from start point (3) with identical diameter but with

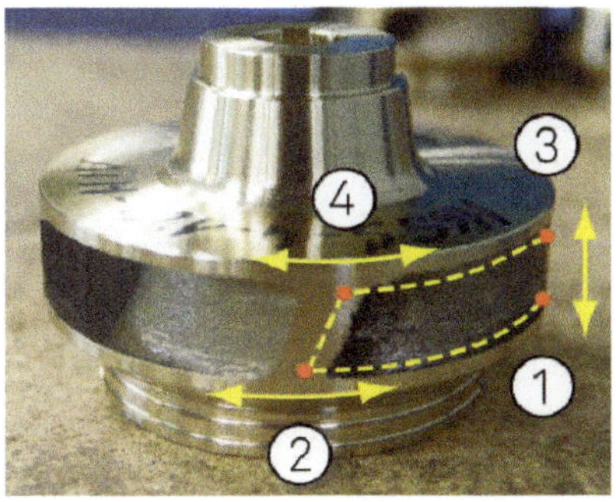

Figure 1. Impeller with marked deburring path and directions of path adaption.

a different endpoint (4) due to the inclination of the blade. The third section is defined by the linear connection of the two endpoints (2), (4). Among the parameters available for every impeller are diameter, number of blades and height. No a-priori information about edge positions, blade position and blade angle is available.

A robot equipped with force control can offer several possibilities for the detection of the missing information and the execution of the general task fitted to the specific impeller geometry: Two different searching methods were developed: one method ahead of deburring and one right in the course of a deburring path.

The first method is a linear sensing movement which represents a relatively simple way of finding a surface and updating a robot pose. Only the searching direction is force controlled in which the robot will move according to the defined reference force. It will stop and move back immediately when tool and work piece get into contact. The benefit from this method is that any error or change in tool or work object (tool wear or tolerances) will be included in the measurement and therefore compensated.

To ensure that neither tool nor work piece are damaged the robot control system has to be highly responsive. A low reference force of 15 N is used, but the main factor here is that the robot's force control system allows to define (position and/or force) conditions on axis controller level. During searching movement the force condition is checked in constant 4 ms cycles rather than 25 to 50 ms on high level control and therefore high force peaks are avoided. The accuracy of the searching movement depends on the work piece surface, especially with harsh casting parts. To perform a search the robot moves the impeller until in contact to the either rotating or static deburring tool mounted on the spindle drive. A comparison of 4 different searching configurations and the acquired accuracy is presented in the results section.

Based on the knowledge about the edge position in axial direction and the diameter, a circular path for deburring can be calculated with an estimated endpoint. However the true end point is unknown and an inaccurate path would cause either incomplete removal of burr or damage due to collision.

For this purpose the second searching method was developed, using the feed-rate control. The robot move towards the estimated endpoint at a relatively high speed as long as measured force stay below a user defined value (e.g. 40 N). When the amount of burr increases or the blade edge is reached, process force will suddenly rise and the feed rate will be automatically lowered. At reduced speed the robot will move on as long as the forces don't exceed the specified limit even at minimum feed rate. This allows control over the amount of material to be removed from the impeller edges (compare endpoints in **Figure 1**).

When the force limit is reached at minimum feed rate, a recover routine is triggered, which will stop the circular movement, store the current position and let the robot move on to the next instruction in program. This is performed twice for upper and lower edges and lastly between the former stored end positions.

With just these two methods, we can provide automated deburring for a wide range of different impeller models with a minimum of a-priori knowledge. Basically the only individual parameters used for the internal calculation of path and search starting point are diameter and height which the robotic system can access through a ".txt" file or similar. The generic path (**Figure 1**, dashed line) has to be defined just once and then it is automatically "shrinked" or "stretched" based on the sensed contact information before an in course of the deburring.

3.3. Results

The forces in movement direction of the tool as well as signals for speed change are shown in **Figure 2**. In the first section the robot is deburring the horizontal edge of the impeller at 10 mm/s with relatively low contact forces. After 4 seconds the vertical edge is reached and force rises due to the much higher amount of material to be removed. The feed-rate controls aimed to keep the material removal rate at high level as long as the measured force remains below a user defined level [11].

In the presented case, when the vertical edge was reached, the measured force exceeded the reference of 40 N and the feed-rate control kicked in. The robots speed was reduced stepwise (4 steps are used in this case) until lowest level (5%). At this reduced speed, the cutting continued about 4 seconds and the force rose slowly. At 130 % of reference the recover function was triggered (speed ratioshifts to zero in **Figure 2**) and the robot was stopped and moved out of the impeller.

From the graph it becomes also clear, that the force signals are increasingly affected by shatter when cutting

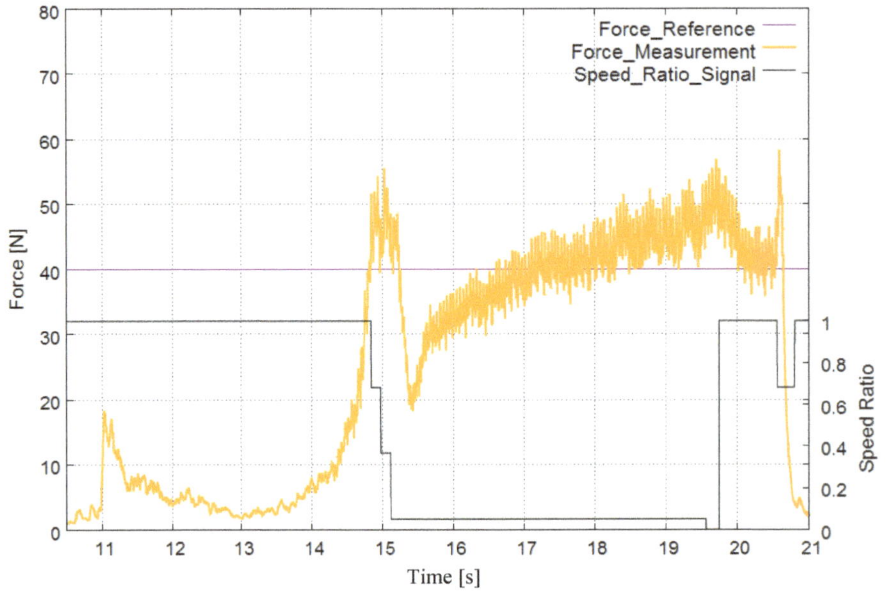

Figure 2. Forces and speed level during deburring.

deeper into material. Controlling this shatter is among the most challenging aspects in machining, as it affects surface quality and it can damage the robot. In our case this tendency was not yet correct active. To achieve higher stability of deburring results the force measurement are internally low pass filtered with 5 Hz.

Obviously, with the right combination of force control parameters the feed rate control allows the robot to keep forces under reference level even with sudden contacts and increasing depth of cut. The fast reaction eliminates high stress on the tool and the robot. From the force trend after initial contact it can also be concluded, that it's possible to control the amount of material to be removed by adjusting the reference force. This is important to the customer who naturally wishes not only to remove burr but also to control chamfer size.

As explained in the path generation section, the deburring paths rely on the sensed contact points during automated teaching. Therefore the repeatability of the searching procedure is of high interest and was examined during tests. We analyzed the results of a total amount of 80 edge touching actions in four different configurations: Touching the vertical impeller edge either from below or from above and with a rotating or a static tool. Tool rotation should reduce possible damage of the cutter but it is also of interest whether this can equalize the effect of burr along the touched edge on measurement quality.

The positions stored by the robot are shown in **Figure 3**. It is important to notice, that the positions are calculated from the robots internal resolver values for each axis, and therefore don't represent necessarily the absolute position of the mechanical end-effector. We will repeat testing with an external optical measurement system in future. So far, interpretation of captured data focuses on the repeatability of force triggered positions and disturbances from spindle rotation and gravity.

With the static tool the distance from starting point to the lower edge was measured after 20 repetitions as 4.5068 ± 0.0105 mm in z-axis (searching direction) and 0.0157 ± 0.0031 mm in x-axis. For the lower edge distances were measured as 4.4918 ± 0.0212 mm and 0.0140 ± 0.0028 mm for z and x-axis. Upper and lower search results show a similar perpendicular deviation to the searching direction.

The situation with a running spindle is more complex as with every touching the test constrains change. Material on the edge is removed and therefore the z-distance and standard deviation of contact points naturally increases. But still two effects can be derived from the data: Contact points show a significant drift in x-axis of 0.0153 ± 0.0054 mm and 0.0428 ± 0.0028 mm, measured in relation to starting position. In z-axis corresponding values are 4.9238 ± 0.1345 mm for lower and 5.5134 ± 0.2294 mm for upper edge respectively. This can be identified as influence of the spindle, as the clockwise rotating tool pushes the work pieces sideways when in contact.

The second effect is visible when comparing the depth of cut during touching. Mean Values for upper (1.022 mm) and lower edge (0.417 mm) indicate that the depth more than doubles when touching against gravitation compared to touch in direction of gravity force.

3.4. Discussion

The performance found in the touching experiments is very good, especially for a rather large and heavy robot.

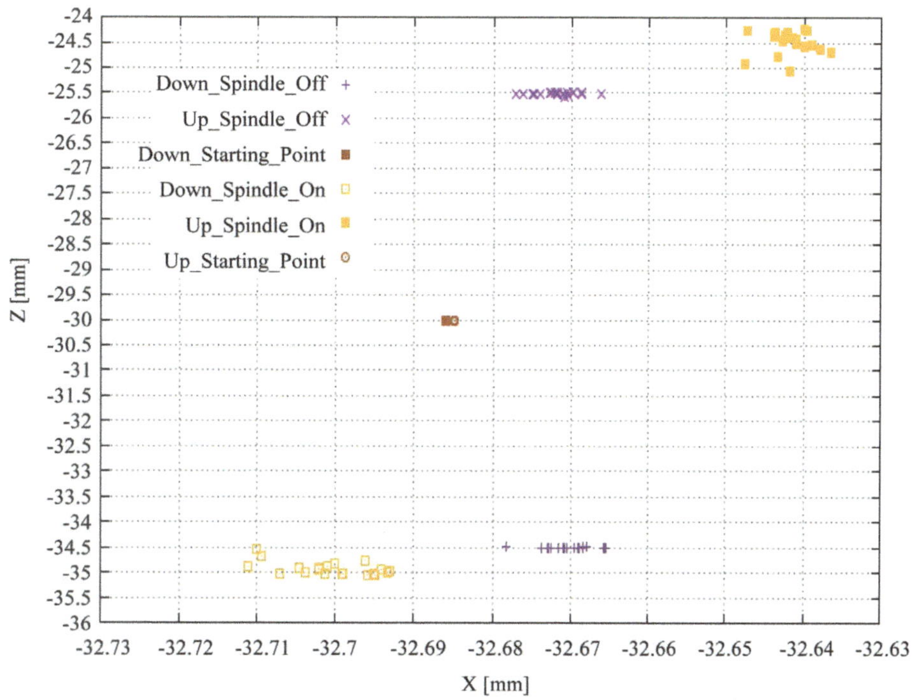

Figure 3. Contact points for different searching configurations.

It has to be considered though, that touching was realized always with equal robot arm configuration and with very limited movement of the primary robot axes. This is a common way of controlling position repeatability on a higher level than the manufacturer is stating in product data sheet.

The deviation perpendicular to the searching direction is marginally small compared to the standard repeatability values for a robot like this of about 0.1 mm.

Possible explanations for the drift in x-direction are errors during load identification and sensor calibration. The load on the sensor from gripper and work piece is around 20 kg and therefore rather large compared to the reference force during touching of just 15 N. But the influence of gravity might be just one factor among others here.

The test procedure shows that touching within one robot arm configuration and with limited searching range provides a reliable and accurate way of letting a robot learn its environment. The responsiveness of the control system meanwhile prevents that high forces exceed to the robot even than the upcoming contour in deburring is not known.

4. Grinding of Forming Tools

The goal of the subproject A6 in the Collaborative Research Centre 708 of the DFG (German Research Association) at the Institute of Production Systems (IPS) at TU Dortmund University is the development of a robot based grinding process to optimize the geometry and the surface properties of hard-coated surfaces.

The focus of this research project is the modeling of a robotic grinding process, a process-adapted simulation-based robot path planning and the use of a force controlled process execution. The aim of the project is to improve the robotic process execution through a holistic, systematic optimization of the process parameters of the grinding process, and the robot control algorithms, especially of those dealing with robotic force control, in order to meet the requirements according to dimensional accuracy and surface quality.

To verify the applicability of industrial robots in grinding processes with grinding tips, among other things, the influence of the robot system on the result of the machining process must be examined. One of the first steps is the derivation of a model for industrial robotic grinding with grinding tips, which describes the dependence of the removal rate of various process parameters (normal force, cutting speed, feed-rate, etc.). Such a model is intended to be used to predict the volume of the removed material and the shape of the product after machining.

A complete mathematical description of the grinding process is hardly possible, because all relevant factors, such as size and shape of each abrasive grain, can't be determined accurately. As grinding is defined as a machining process with geometrically undefined cutting edges, the factors of a process model can only be defined statistically. Therefore, in the context of this project only an empirical process model for robotic grinding with grinding pins is been developed, which is based on the results of the experiments carried out in a pilot work cell. A first process model has been published in [14].

Another main goal of the project is the optimization of the control of the machining process using a force control system. Therefore the controller action and the influence of this action on the machining process have to be looked at. The first results of this research work will be presented in this paper. Furthermore it is necessary to look at the advantages and disadvantages of a force controlled versus a position controlled grinding process.

4.1. Experimental Setup

To perform the necessary modeling experiments and to analyze the robot behavior during the force controlled machining process, a pilot robot cell was set up. **Figure 4** shows the setup and individual components.

As a part of the force control extension module a 6 axis force torque sensor is mounted between the robot flange and the electric spindle. The use of the force control option allows the execution of the machining process with a defined constant force. Thereby a uniform ablation along the machining path can be obtained.

The forcetorque sensor can capture and record the forces generated during the machining process. As tools, electroplated cBN grinding tips are used with a diameter of D = 12 mm with various grain sizes. The high-frequency electric spindle with a speed of up to 36,000 rpm provides high speeds (up to 20 m/s) during processing. Coated flat specimens with dimensions of 70 mm × 50 mm × 10 mm and layers applied with a minimum thickness of about 400 µm were used in the underlying test series.

4.2. Position-Controlled versus Force-Controlled Grinding Process

Since the robot system can be used with both control approaches, position-controlled and force-controlled, it should initially be considered which of these two approaches is more appropriate for the process of robotic based grinding with grinding tips.

Looking at the position controlled grinding process the infeed of the grinding paths is the primary control parameter of this process type (in contrast to the force controlled process type, where a defined contact force is the primary control variable). As the amount of the material that is removed in the present grinding application is very small (50 - 300 µm), the accuracy of the robot plays a crucial role. As a constant contact force is hard to be realized due to the lack of stiffness of the system, a deviation in the removal rate is occurring.

In order to compensate manufacturing inaccuracies that may occur, it may be necessary to scan the work piece surface before the machining process with a special

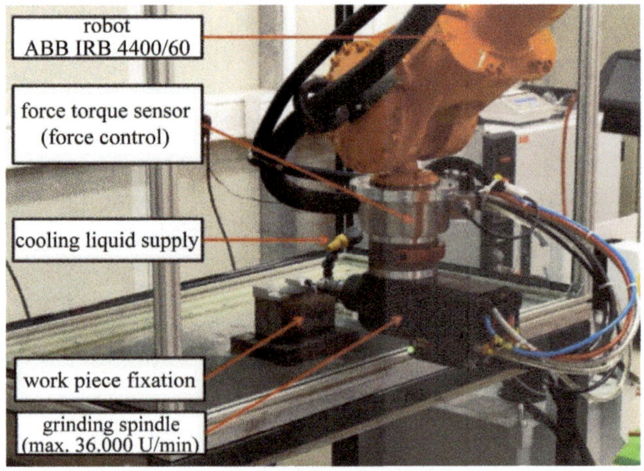

Figure 4. Experimental setup.

sensor or camera system and to adapt the robot paths appropriately, so that it is possible to obtain a material removal that is as constant as possible. Specially adapted strategies (gradual approximation) will also lead to useful machining results. In contrast to grinding in machining centers, where these strategies have been used long term, these strategies can also lead to problems in robotic machining due to the lower accuracy of the industrial robot. There is always the possibility of a too large increase of the infeed which leads to a large increase in the contact force, which can result in damages at the used grinding tool or the work piece that is machined.

In the force controlled grinding process it is theoretically possible to reach higher removal rates because the maximum removal is not limited by the infeed, and the contact pressure does not decrease during the machining process. A compensation of the position error in the direction of the contact force is also possible. In robotic based grinding the control of the contact force can be carried out by the force control of the robot. In this case, the goal is to establish a constant defined force between the tool and the surface of the work piece, independent of the feed movement.

The use of a modern force control system for industrial robots (e.g. the Force Control extension module from ABB Robotics [15,16]), which have a control cycle of 4 ms, allows the successful implementation of the force controlled process type in practice. Because of these advantages, the force controlled process type is preferred in the further work.

In order to get an impression of the correlation between the two process variants, in a first experiment the dependence of the normal contact force on the infeed between the work piece and the tool was determined. The grinding tips were moved over the surface of the work piece with different infeed values and different feed-rate using the robot. The contact force was measured by the forcetorque sensor. The experiments show a linear relationship between infeed and contact force [14].

4.3. Influence of the Force Deviation on the Deviation of the Removal Rate

In the following part of the paper we want to have a look on the influence of the force deviation on the deviation of the removal rate. Therefore several experiments have been executed in order to analyze if the contact force which is applied by the force-controlled robot system to the surface directly leads to a similar removal rate.

Using the same experimental setup as described in chapter 4.1 all process parameters have been fixed (constant infeed, vf = 15 mm/s, spindle speed: 16,000 rpm, width of the grinding tip: 10 mm, grain size: B126) with the exception of the contact force. Starting by a force value of 5 N, the contact force value has been increased up to 30 N. Furthermore a smooth increase of the force (22 N/s) along the grinding path has been implemented.

During the grinding process the contact force values have been recorded. Additional to the shape of the surface the work piece was measured before the grinding process and after the grinding process. For the measurements a coordinate measurement machine was used (Wenzel LH 54).

In **Figure 5**, the measured force progressions along the process time for some experiments of the experimental series are shown. All measured force progressions show the smooth increase of the forces. When the desired force value is reached, this force is maintained relatively constant until the end of the particular experiment.

The higher the desired force value is the later this constant phase is reached. It can also be seen that in the constant phase the values are also deviating around the adjusted force with amplitude of approx. ±1.5 N. For the lowest force (5 N) this value is higher (3 N).

A similar situation exists for the material removal along the robot path. In **Figure 6**, the force related removal progression of an experiment with a contact force of 22.5 N is shown. Clearly visible is the increase of the force at the beginning as well as the according progression in removal. Additionally, the removal at constant force is varying around a mean value. This deviation is not negligible and will be topic of further analysis.

5. Further Work

The results of our presented force controlled machining applications should allow the reader to compare the advantages and performance of the two presented force control strategies: force dependent feed-rate control and orthogonal force (pressure) control.

A force controlled robotic system of today's state of the art can be successfully adapted to robotic deburring and grinding, which are major applications in robotic machining. But the demonstrated performance cannot be delivered out of the box. With the capabilities of robot force control the demands in programming are rising and so does the number of parameters which determine the robots behavior and process quality. Many tests under well defined circumstances in a test bed have to be performed to verify correct program execution and the op-

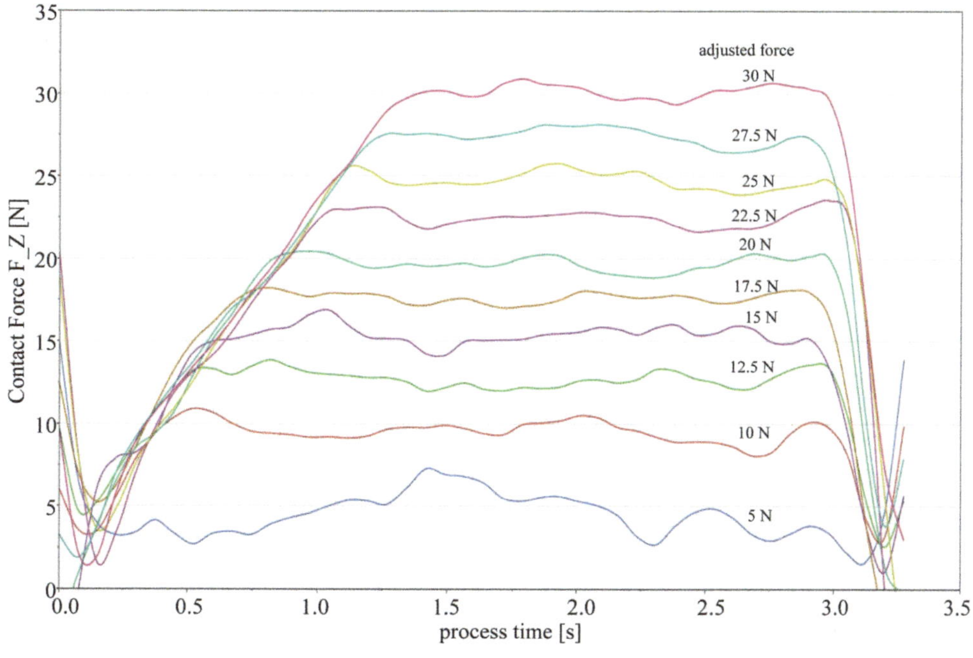

Figure 5. F_z progression of experimental series.

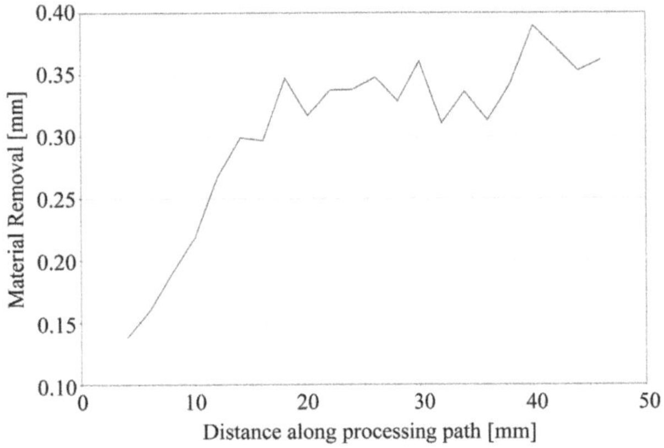

Figure 6. Material removal along the robot path at 22.5 N contact force.

timal parameters. The achieved results are limited to the inspected application or even one kind of work piece.

Based on our implemented use-cases we are able to investigate how the individual parameters, e.g. for the feed rate control, affect robot motion execution in contact with the environment as well as disturbances from the machining process itself. But instead of presenting another custom tailored solution, the aim is to present the acquired knowledge in an end-user friendly way, helping future operators on their way to optimal force control performance.

6. Acknowledgements

The authors gratefully acknowledge the financial support of the DFG (German Science Foundation) within the Collaborative Research Centre SFB 708, project A6.

Furthermore parts of the described work are funded in the project ProBeLa through the German Federal Ministry of Economics and Technology as part of the Central Innovation Programme SME (ZIM).

REFERENCES

[1] J. Pandremenos, C. Doukas, P. Stavropoulos and G. Chryssolouris, "Machining with Robots: A Critical Review," *Proceedings of DET*2011 *7th International Conference on Digital Enterprise Technology*, Greece, September 2011, pp. 614-621.

[2] A. Robertsson, T. Olsson, R. Johansson, A. Blomdell, K. Nilsson, M. Haage, B. Lauwers and H. De Baerdemaeker, "Implementation of Industrial Robot Force Control Case Study: High Power Stub Grinding and Deburring," 2006 *IEEE/RSJ International Conference on Intelligent Robots and Systems*, Beijing, 9-15 October 2006, pp. 2743-2748.

[3] "IRB 6660 Industrial Robot," ABB Robotics, Västerås, 2010.

[4] Stäubli International AG, "Robotertechnik für Maschinelle Bearbeitung," 2008. http://www.staubli.com/en/robotics/6-axis-scara-industrial-robot/specialized-robot/high-speed-machining-robots/

[5] "KUKA.CNC for Maximum Robot Performance in Machining Processes," KUKA Roboter GmbH, Gersthofen, 2011.

[6] "Application Manual Force Control for Machining, Revision: H," ABB Robotics, Västerås, 2012.

[7] A. J. Wertz, "Robotic Grinding Using ABB's Force Control Technology," Benton Foundry Inc, Gettysburg, 2011. http://www.ductile.org/magazine/2011_2/alanwertz1.pdf

[8] T. Hemker, M. Friedmann, O. V. Stryk, M. Pischan, J. Bauer and E. Abele, "Prediction of the Tool Displacement for Robot Milling Applications Using Coupled Models of an Industrial Robot and Removal Simulation," *Proceedings of the CIRP* 2*nd International Conference Process Machine Interactions*, Vancouver, 10-11 June 2010.

[9] J. Bauer, "Analyse der Wechselwirkungen Zwischen Industrieroboterstruktur und Fräsprozess," TU Darmstadt, Institut für Produktionsmanagement, Technologie und Werkzeugmaschinen, Darmstadt, 2012.

[10] B. Olofsson, O. Sornmo, U. Schneider, A. Robertsson, A. Puzik and R. Johansson, "Modeling and Control of a Piezo-Actuated High-Dynamic Compensation Mechanism for Industrial Robots," 2011 *IEEE/RSJ International Conference on Intelligent Robots and Systems* (*IROS*), San Francisco, 25-30 September 2011, pp. 4704-4709.

[11] Z. Pan, "Intelligent Robotic Machining with Force Control," Stevens Institute of Technology, 2005. http://books.google.de/books?id=FU5SNwAACAAJ

[12] H. Zhang, H. Chen, N. Xi, G. Zhang and J. He, "On-Line Path Generation for Robotic Deburring of Cast Aluminum Wheels," *Proceedings of the* 2006 *IEEE/RSJ International Conference on Intelligent Robots and Systems*, Beijing, 9-15 October 2006, pp. 2400-2405.

[13] B. Solvang, P. Korondi, G. Sziebig and N. Ando, "SAPIR: Supervised and Adaptive Programming of Industrial Robots," *The* 11*th International Conference on Intelligent Engineering Systems*, Budapest, 29 June 2007-2 July 2007, pp. 281-286.

[14] C. Krewet, M. Cabaravdic, A. Hypki and B. Kuhlenkötter, "Empirische Abtragsmodellierung für das Robotergestützte Schleifen Thermisch Gespritzter Schichten,"2012.

[15] J. He, Z. Pan and H. Zhang, "Adaptive Force Control for Robotic Machining Process," *Proceedings of the* 2007 *American Control Conference*, New York, 9-13 July 2007, pp. 1-6.

[16] J. Ge, F. Tan and H. Zhang, "Robot Machining: A Force-Control-Based Fast Programming Method," 2008 *IEEE Conference on Robotics, Automation and Mechatronics*, Chengdu, 21-24 September 2008, pp. 730-735.

4

Intelligent Fault Detection of Retainer Clutch Mechanism of Tractor by ANFIS and Vibration Analysis

Ebrahim Ebrahimi[1], Payam Javadikia[2], Mohammad Hadi Jalili[1], Nasrolah Astan[1], Majid Haidari[3], Mojtaba Bavandpour[1]

[1]Department of Mechanical Engineering of Agricultural Machinery, Islamic Azad University, Kermanshah Branch, Kermanshah, Iran

[2]Department of Mechanical Engineering of Agricultural Machinery, Razi University, Kermanshah, Iran

[3]Department of Mechanical Engineering of Agricultural Machinery, Islamic Azad University, Takestan Branch, Takestan, Iran

ABSTRACT

In this study, ANFIS, as decision support system, is applied to detect the faults of MF 285 mechanism tractor clutch. Maintenance mechanisms include normal mode, rolling element failure, seal failure and attrition-based. Experiments were carried out in three speeds: 1000, 15,000, 2000 RPM and two conditions. The sensor was mounted vertically and horizontally. Vibrating spectrum of the time domain and the frequency of vibration data were obtained. Thirty-three statistical parameters of vibration signals in frequency domain and time were chosen as the sources attribute to detect errors. Finally, the top three features as input vectors to the ANFIS were evaluated. Using statistical parameters the performance of the system was calculated with the experimental data and training of ANFIS model. The system could not provide a seal to identify the fault. Regardless of the vibration data obtained from the classification of the seal, the overall classification accuracy of the ANFIS was 99.14% in the amount of 100% of the sensor installed vertically and horizontally. The results showed that this system can be used as an intelligent diagnosis system.

Keywords: Fault Detection; Maintenance Clutch Mechanism; Vibration Analysis; Neuro-Fuzzy Inference Systems

1. Introduction

Rotating machines are the mainstay of several industrial applications worldwide. They can be found automobile, water industry, wind turbines, agricultural machinery, etc. Failures associated with gears represent the cause of extended outages and are typically caused by gradual deterioration and wear [1].

An important objective of condition-based maintenance (CBM) is to determine the optimal time for replacement or overhaul of a machine. The ability to accurately predict the remaining useful life of a machine system is crucial for its operation and can also be used to improve productivity [2].

Vibration analysis is a powerful tool in machine fault diagnosis. It is a way to check the status of a machine using a special tool. Troubleshooting machines can be more identified in the future and before the crash.

Fault diagnosis of mechanical or electronic components or systems is a subject of expert system applications. Today, neuro-fuzzy devices are emerged by combining network based on adaptive neural networks and fuzzy models. These adaptive networks, based on fuzzy inference system, are called ANFIS. In such systems the desired output will be achieved by means of human knowledge, in the form of if-then rules for the desired input. Dealing with systems that are not well-defined it is not true model the system based on conventional mathematical tools and component identification and therefore the neuro-fuzzy systems could be applied.

Many researchers [3-12] have employed various methuds to detect weak periodic signals using the sensitivity of system parameters. Some of these methods are shock pulse monitoring, crest factor analysis, kurtosis, spectrum analysis, envelope spectrum analysis, wavelet analysis, filtration of signal, etc.

Among the researches carried on machine condition monitoring with vibration analysis using fuzzy logic and neuro-fuzzy networks there are rotary machines fault diagnosis, troubleshooting automotive active suspension system and taking into account the non-linear behavior of hydraulic actuator and significantly the impaired controller, an engine control ventilation, fault diagnosis of bearing rotary, fault diagnosis of rolling bearing [4,7,12], tapered gearbox fault diagnosis [11].

In agricultural tractors, fractures in different parts of the power transmission due to the lack of early detection and location of damage and repair or replace the damaged part will be caused waste of time, cost more to repair and loose time for farming operations which often are duly carried. MF285 Tractor, is equipped with a two stage clutch. The two main types of clutches are necessary. The first resembles a driven plate clutch to power takeoff but the latter is based on the sheath power takeoff. The shaft makes the gear box to move while the pad moves through two gears toward the empowerment shaft. **Figure 1** illustrates the position of each component.

Failure on this part will lead to some fraction in other parts specially in gearbox of tractor, on condition of late diagnosis. The overall objective of this research is to present the intelligent detection of some parts of MF285 tractor clutch holder mechanism with high accuracy.

2. Materials and Methods

Classical methods for fault diagnosis include measurable change of system, values of tolerance condition monitoring of machinery health state. This method requires the presence of a person qualified to comment on the health of the defective machine. This study intends to provide an intelligent fault diagnosis system, so that it could help to give a quick decision on machine safety to an expert. In this study the experiments on trial setup of MF285 tractor clutch holder mechanism in a single laboratory unit (**Figure 2**).

Intelligent system for Condition monitoring is generally, divided to three parts: data acquisition, signal processing and feature extraction and classification of defects. The proposed system for failure detection includes vibration signals collection in data acquisition, data transfer from time zone to frequency zone in signal processing, calculating the most important features in features extraction and finally data entrance to ANFIS and calculating the total classification accuracy.

2.1. Data Acquisition

Data was acquired in two accelerometer sensor direction

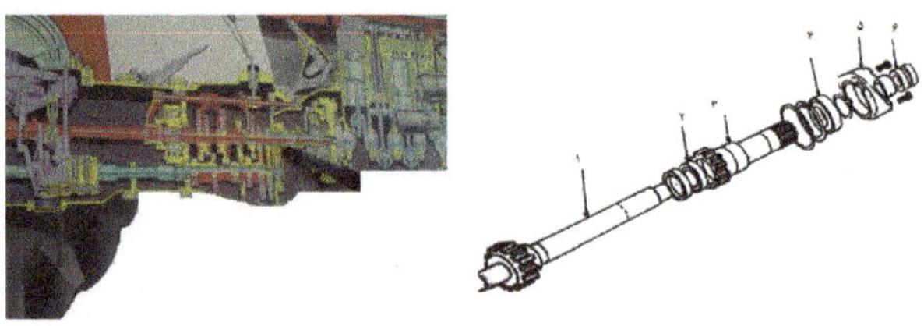

Figure 1. The position of clutch holder mechanism (right) and internal components of clutch holder mechanism (left).

and three rotation speeds include 1000, 1500 and 2000 RPM in safe mode on laboratory setup. The accelerometer was installed in two X and Y directions on the main holder. ARMA Software was used for this purpose and 10,240 voltage data were recorded and saved in an Excel file. Data collected for each sample took 10 s. Total 24 states and 130 vibration data for each state (780 for safe mode and 2340 for down mode and totally 3210 samples) were extracted.

Different failures that may be happen on mechanism were occurred on different parts of clutch holder mechanism (**Figure 3**) and the vibration data acquired separately. **Table 1** shows the specifications of each failure.

Due to the breakdown in rotating machines, time domain signals vary in each fault conditions. Vibrational spectra taken during the time of each of the modes are obtained. A typical vibration spectrum obtained at a moment when the photograph is taken. **Figure 4** shows a sample of vibration spectrum obtained during data mapping taken momentarily.

Figure 2. Setup tools and parts used in the laboratory.

Figure 3. Clutch holder mechanisms failures: seal failure (top-right), roller failure (top-left), PTO shaft friction (bottom).

Table 1. Specifications of each failure.

Failure type	Failure profile
Seal failure	The inner seal is torn and the coil is damage
PTO shaft friction	Shaft diameter is reduced upto 10 percent
Roller failure	Three roller balls and the inner ring are destroyed

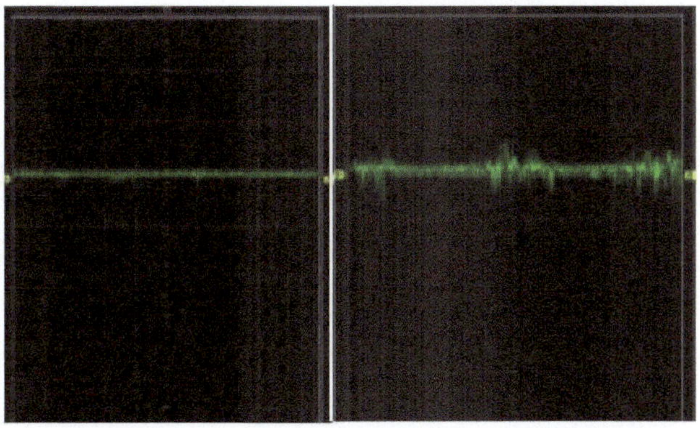
(a) (b)

Figure 4. Vibrationsignal from health condition (a) andvibrationsignal from fault condition (b) horizontal sensor and 1000 rmp.

2.2. Signal Processing

Signal processing and feature extraction are the most important parts of the monitoring process by which the system feature including safety or failure and the system type could be specified. One common way to provide useful features is to transfer signal from time domain to frequency domain. The analyzer, using the Fourier series, changes the analogous signals of time zone into digital data in frequency zone. For this purpose, at first program the necessary codes in the form of MATLAB. This Excel program contains vibration data and using the filter applied, the data was filtered the noises were eliminated and finally the time and frequency statistical parameters were calculated. So, some conversion feature of Fourier series (FFT) with 10240 points was calculated for each time signal. Also, the power spectrum density (PSD) and phase angle (FFT) of vibration signals were calculated using MATLAB.

Using MATLAB, the diagrams related to vibration spectrum in frequency domain in safe and failure modes were acquired to be analyzed and compared with each other (**Figures 5** and **6**).

2.3. Feature Extraction

The main purpose of feature extraction techniques is to

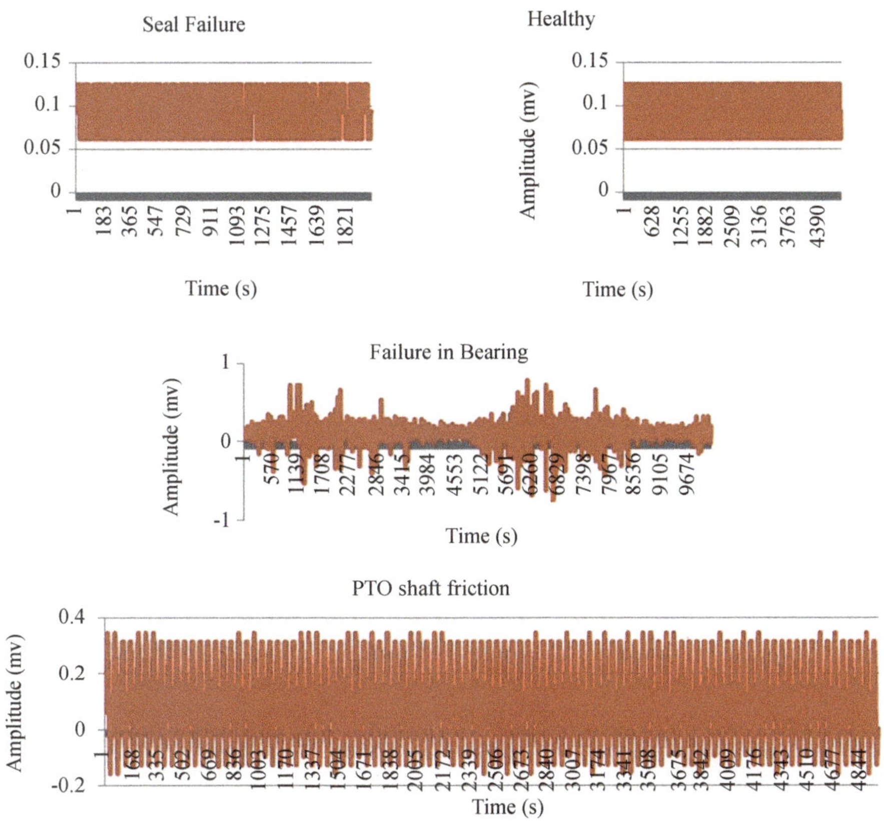

Figure 5. Typical vibration spectra obtained in time domain.

present failure features in signals. In this research, the time domain data with 10240 formed the sample. Also, the frequency domain data points, which is PSD data and FFT phase angle were each 10,240. These signals were used to extract thirty-three features of safe and failure modes. Eleven features (T1 - T11) of time domain statistical parameters and rest of parameters (P1 - P11) and (A1 - A11) form the frequency domain statistical parameters.

2.4. Fault Classification with ANFIS

For better definition of output result each of safe and fault modes have a code to classify the failures (**Table 2**).

Then in order to data model in ANFIS, 8 models (A - H) were considered for modeling in ANFIS and experimental data were placed in classification (**Table 3**).

During the initial data mapping and using seal fault in

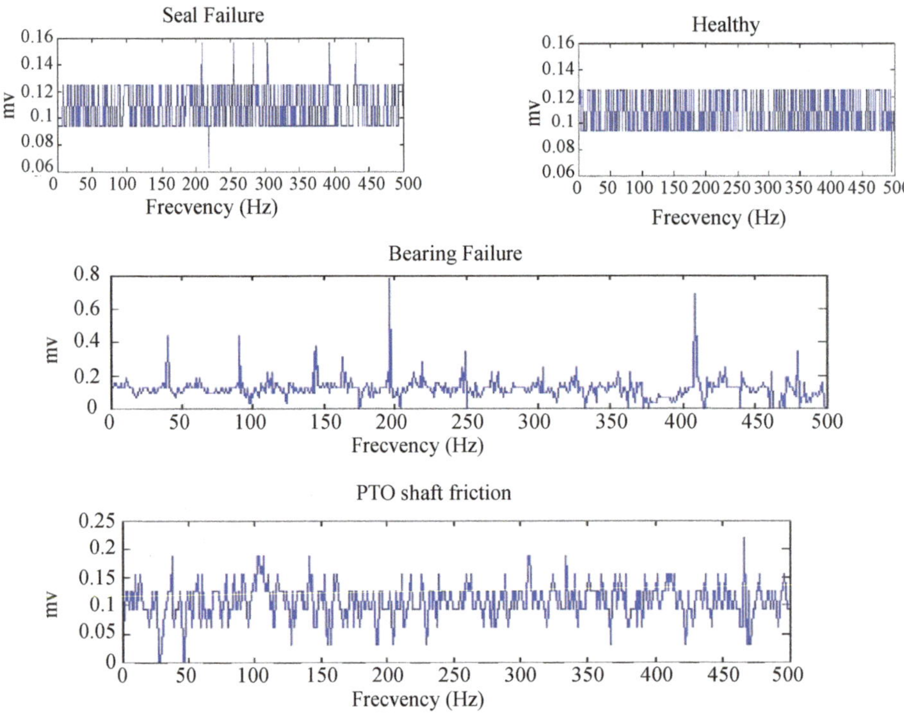

Figure 6. Frequency spectra obtained in frequency domain.

Table 2. Code definition for each safe and failure mode.

Mode	Safe	Roller fault	Seal fault	Axis friction
Defined code in classification	0	1	2	3

Table 3. Modes classification and input data to ANFIS.

Experiment model	Sensor status	(rpm)	Classification code	Number of experimental data	No. of training data	Input to ANFIS
A	Vertical	1000	3210	468	1092	A_9 P_{11} A_8
B	Vertical	1000	310	351	819	A_9 P_{11} A_8
C	Vertical	1500	3210	468	1092	A_9 P_{11} A_8
D	Vertical	2000	3210	468	1092	A_8 P_9 A_7
E	Horizontal	1000	3210	468	1092	T_5 A_9 A_8
F	Horizontal	1500	3210	468	1092	A_9 P_{11} A_8
G	Horizontal	2000	3210	468	1092	T_5 A_9 A_8
H	Horizontal	2000	310	351	819	T_5 A_9 A_8

laboratory model, the vibration resulted from seal fault were felt less than safe mode and no significant difference was observed from initial comparison of data obtained from seal failure and safe mode and their vibration spectra. On the other hand, after investigating data related to seal fault, in order to select super features, less significant features from data on seal fault were observed. So, it was assumed that the present system could not have a good show of defects resulted from seal fault. Therefore, the data related to seal fault were eliminated in B and H models and other data were entered into the network.

3. Results

The FFT diagram is the most applicative device for the vibration analysis. But much information on machine behavior may be eliminated in FFT diagram for various results like: some frequencies filtering, the adjacent frequencies overlap regarding the selective clarity. The neuro-adaptive techniques provide some method to make a phase modeling trend along with information learning from a data collection. In order to evaluate the proposed fault detection system the vibration data were placed in classification and 8 models were considered for modeling and entering the ANFIS network. The ANFIS feature vector was extracted for each model in different sensor modes and rotations per each three features. For the experiment models, in order to train and test the data, better significant features which made better results using the failure and try method were applied to enter the network. 70% data in each model were assigned to train and 30% were loaded to test in the network. To obtain better results the input and output membership features and different training periods were considered. Also, the initial value of 1% as the step size was formed for the parameters adaptation. After loading for all models, the failure diagram was drawn based on didactic periods. Then, the rules monitor form, fuzzy rules, fault detection topology and levels diagram were extracted for each separate model. Meanwhile, from the results obtained of ANFIS, the mean square error (MSE), sum of square error (SSE), correlation coefficient (R), mean absolute error (MAE), the degree of significance (P) and the prediction were obtained (**Table 4**).

By observing the result of ANFIS and counting correct answers, confusion matrix for the entire model was established. In **Table 5** the confusion matrix is illustrated for model B.

Diagonal elements in matrix represent the number of correct answers.

Overall classification accuracy, sensitivity and statistical are two statistical criteria to elevate the performance of the system in classification. These criteria are defined as follows:

Table 4. The best results of the ANFIS models in different experiment.

Model Testing	The results of the ANFIS			
	R	SSE	MAE	MSE
A	0.80	77.9	0.44	0.49
B	0.98	1.02	0.05	0.06
C	0.82	69.00	0.39	0.44
D	0.66	124.6	0.60	0.79
E	0.78	254.4	0.53	0.48
F	0.89	135.9	0.32	0.26
G	0.73	112.5	0.49	0.72
H	0.99	0.002	0.0015	2.44

Table 5. The confusion matrix for flaw detection in model B.

Output/Desired	Safe	Roller failure	Axis friction
Safe	39	0	0
Roller failure	1	38	0
Axis friction	0	0	39

Sensitivity equals to the number of correct positive decisions divided by total possible decisions. Total classification accuracy equals to the number of correct decisions divided by total possible decisions. Finally, the statistical criteria were obtained to detect system performance in classifying each model (**Table 6**).

In order to compare the success of failure detection of the models together, a description of overall classification accuracy in each case is shown as diagram (**Figure 7**). In model B the total classification accuracy were 99.14% and the highest amount of accuracy 100% was obtained in model D, which is desired. Also, the total classification mean accuracy of fault detection system with vertical sensor 76.41 and horizontal sensor 71.47 was obtained (**Figure 8**).

4. Conclusion

In this study, vibration data resulted from experimental setup were obtained using the data acquisition system. Statistical features of time and frequency domain were obtained for different failures detection and among them three super features were selected to enter the ANFIS network. 8 models were places in classification to detect the failure. In two models (models B and H), the seal fault was not placed in classification. After calculating statistical parameters, the total classification accuracy in

Table 6. Statistical criteria and classification accuracy obtained in all test models.

Statistical criteria		Status	Model
Total percent	Sensitivity percent		
	54.8	0	
	100	1	
67.9	17.5	2	A
	100	3	
	100	0	
	97.43	1	
99.14	100	3	B
	43.6	0	
	100	1	
	43.6	2	
71.79	100	3	C
	30.76	0	
	100	1	
	64.1	2	
66.66	71.8	3	D
	15.38	0	
	82	1	
	0	2	
46.15	87.17	3	E
	46.15	0	
	100	1	
	33.33	2	
69.78	100	3	F
	51.28	0	
	100	1	
	30.76	2	
69.78	97.43	3	G
	100	0	
100	100	1	H
	100	3	

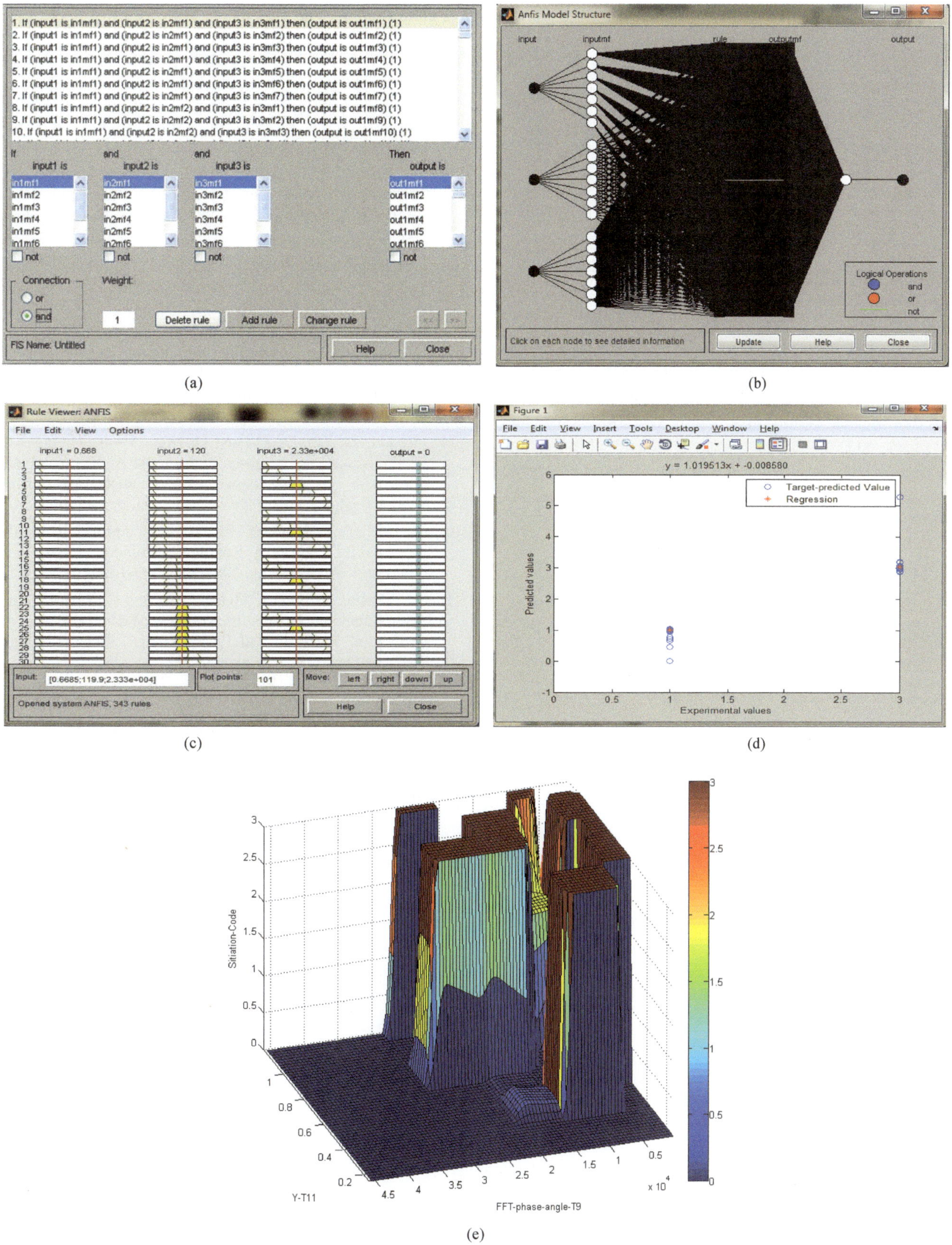

Figure 7. (a) The extraction of fuzzy rules; (b) Designed ANFIS topology; (c) Rules monitor, regression diagram on predictions; (d) The extraction of levels monitor in model B; (e) The topography of model.

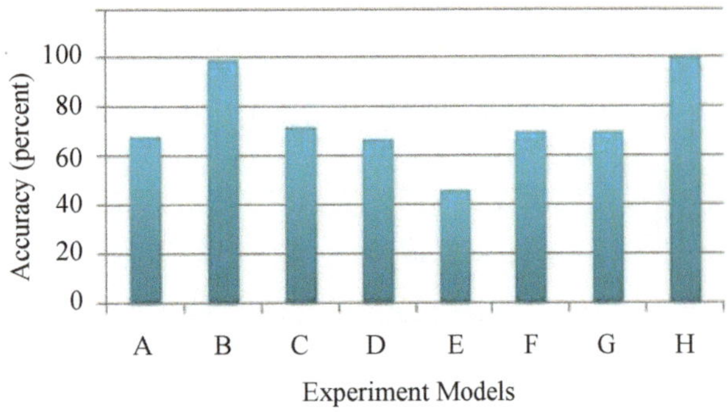

Figure 8. Diagram of total classification accuracy based on experiment models.

model D and H were, respectively, 99.14 and100 percents. Comparing the results of total classification accuracy obtained from ANFIS models showed the assumption that the seal failure makes little vibration is true and the proposed fault detection system was not able to show the seal failure. As a whole, the results showed that using intelligent neuro-fuzzy logic could be an effective method to detect the failures in clutch holder mechanism.

5. Acknowledgements

The authors would like to thank Islamic Azad University, Kermanshah Branch for its support to this research.

REFERENCES

[1] M. Elforjani, D. Mba, A. Muhammad and A. Sire, "Condition Monitoring of Worm Gears," *Applied Acoustics*, Vol. 73, No. 8, 2012, pp. 859-863.

[2] H.-E. Kim, A. C. C. Tan, J. Mathew and B.-K. Choi, "Bearing Fault Prognosis Based on Health State Probability Estimation," *Expert Systems with Applications*, Vol. 39, No. 5, 2012, pp. 5200-5213.

[3] D. G. Lewicki and H. J. Decker, "Transmission Bearing Damage Detection Using Decision Fusion Analysis," Glenn Research Center, Cleveland, Army Research Laboratory, NASA/TM, 2004, pp. 1-20.

[4] G. Goddu, B. Li, M. Chow and J. Hung, "Motor Bearing Fault Diagnosis by Fundamental Frequency Amplitude Based Fuzzy Decision System," Development of Electrical and Computer Engineering, North Carolina University, 1998, pp. 1961-1965.

[5] N. Hotwai, "Vibration Analysis of Faulty Beam Using Fuzzy Logic Techique," Department of Mechanical Engineering National Institute of Technology, Rourkela, 2009, pp. 23-27.

[6] Ch. Kong, J. Ki, S. Oh and J. Kim, "Trend Monitoring of a Turbofan Engine for Long Endurance UAV Using Fuzzy Logic," *KSAS International Journal*, Vol. 9, 2008, pp. 64-70.

[7] Z. Kiral and H. Karagülle, "Simulation and Analysis of Vibration Signals Generated by Rolling Element Bearing with Defects," Department of Mechanical Engineering, Dokuz Eylul University, Turkey, 2003, pp. 667-678.

[8] Y. Lei, Z. Hea, Y. Zia and Q. Hua, "Fault Diagnosis of Rotating Machinery Based on Multiple ANFIS Combination with Gas," Department of Mechanical Engineering, Xi'an Jiaotong University, Xi'an, 2003, pp. 667-678.

[9] Y. A. Lei, Z. He and Y. Zi, "A New Approach to Intelligent Fault Diagnosis of Rotating Machinery," Expert Systems with Applications, 2008, pp. 1593-1600.

[10] R. Martins Marcal, K. Hatakeyama and A. Susin, "Managing Incipient Faults in Rotating Machines Based on Vibration Analysis and Fuzzy Logic," Electrical Engineering Department, 2006, pp. 1-6.

[11] N. Saravanan and S. Cholairajan, "Ramachandran Vibration-Based Fault Diagnosis of Spur Bevel Gear Box Using Fuzzy Technique," Department of Mechanical Engineering, Amrita Vishwa Vidyapeetham, India, 2009, pp. 3119-3135.

[12] J. H. Singonahalli and N. R. Iyer, "Etection of Roller Bearing Defect Using Expert System and Fuzzy Logic," Department of Mechanical Engineering, California University Sacramento, Sacramento, 1996, pp. 596-613.

5

Developing an Intelligent Fault Diagnosis of MF285 Tractor Gearbox Using Genetic Algorithm and Vibration Signals

Ebrahim Ebrahimi[1], Payam Javadikia[2], Nasrolah Astan[1], Majid Heydari[3], Mojtaba Bavandpour[1], Mohammad Hadi Jalili[1], Ali Zarei[4]

[1]Department of Mechanical Engineering of Agricultural Machinery, Kermanshah Branch, Islamic Azad University, Kermanshah, Iran

[2]Department of Mechanical Engineering of Agricultural Machinery, Razi University, Kermanshah, Iran

[3]Department of Mechanical Engineering of Agricultural Machinery, Takestan Branch, Islamic Azad University, Takestan, Iran

[4]Department of Mechanical Engineering of Agricultural Machinery, East Azarbaijan Sience and Research Branch, Islamic Azad University, Tabriz, Iran

ABSTRACT

This article investigates a fault detection system of MF285 Tractor gearbox empirically. After designing and constructing the laboratory set up, the vibration signals obtained using a Piezoelectric accelerometer which has been installed on the Bearing housings are related to rotary gear number 1 in two directions perpendicular to the shaft and in line with the shaft. The vector data were conducted in three different speeds of shaft 1500, 1000 and 2000 rpm and 130 repetitions were performed for each data vector state to increase the precision of neural network by using more data. Data captured were transformed to frequency domain for analyzing and input to the neural network by Fourier transform. To do neural network analysis, significant features were selected using a genetic algorithm and compatible neural network was designed with data captured. According to the results of the best output mode for each position of the sensor network in 1000, 1500 and 2000 rpm, totally for the six output models, all function parameters for MATLAB Software quality content calculated to evaluate network performance. These experiments showed that the overall mean correlation coefficient of the network to adapt to the mechanism of defect detection and classification system is equal to 99.9%.

Keywords: Fault Detection; Gearbox; Vibration Analysis; Genetic Algorithm

1. Introduction

Rotating machines are the mainstay of several industrial applications worldwide. They can be found in automobile, water industry, wind turbines, agricultural machinery, etc. Failures associated with gears represent the cause of extended outages and are typically caused by gradual deterioration and wear [1]. An important objective of condition-based maintenance (CBM) is to determine the optimal time for replacement or overhaul of a machine. The ability to accurately predict the remaining useful life of a machine system is crucial for its operation and can also be used to improve productivity [2]. Vibration analysis is a powerful tool in machine fault diagnosis. It is a way to check the status of a machine using a special tool. Troubleshooting machines can be more identified in the future. Fault diagnosis of mechanical or electronic components or systems is a subject of expert system applications. Today, neuro-fuzzy devices are emerged by combining network based on adaptive neural networks and fuzzy models. These adaptive networks, based on fuzzy inference system, are called ANFIS. In such systems the desired output will be achieved by means of human knowledge, in the form of if-then rules for the desired input [2]. Neuro Fuzzy system can be applied as a true model based on conventional mathematical tools and component identification. Many researchers [3-12] have employed various methods to detect weak periodic signals using the sensitivity of system parameters. Some of these methods are shock pulse monitoring, crest factor analysis, kurtosis, spectrum analysis, envelope spectrum analysis, wavelet analysis, filtration of signal, etc. Among the researches carried on machine condition monitoring with vibration analysis using fuzzy logic and neuro-fuzzy networks there are rotary machines fault diagnosis, troubleshooting automotive active suspension system and taking into account the non-linear behavior of hydraulic actuator and significantly the impaired controller, an engine control ventilation, fault diagnosis of bearing rotary, fault diagnosis of rolling bearing [13], tapered gearbox fault diagnosis [14].

In recent years, the development of mechanical equipments and maintenance has always been regarded as an essential aspect and different ways to improve safety and reduce vehicle maintenance have always been so important. On the other hand, technological advances in complicated, powerful and expensive machines caused the maintenance programs and systems to be improved [15-17]. The aim of maintenance based on the status is to gain the symbols of the industrial machinery at a time of being in progress, to provide economic and safety maintenance or to make it stay active [18-20]. Condition monitoring is performed using different techniques and methods. Each technique is particularly focused on some characteristic symptoms of the status and function of the machine and measures certain parameters that indicate the health status of the machine. Various condition monitoring techniques include vibration analysis, noise analysis, oil analysis and thermal analysis. At the technique of vibration analysis, vibration levels are measured using specific sensors [21,22]. To this end, one of the parameters of displacement velocity or acceleration would be measured. The vibration amplitude of each machine monitors the health status of this machine. Furthermore, vibration has high stability and low influence of environmental factors and since it can be an appropriate parameter for condition monitoring methods. In addition, vibration analysis can be used to troubleshoot the machine. Most problems in rotary machines are displayed as vibrating symptoms or changes in the vibration characteristics of the system. Technical advances in mechanical equipment and strong machines, machines and industrial equipment complexity and the high cost of these equipments make it necessary to develop and consider new methods of maintenance including condition monitoring. What is certain is that developing and implementing the methods for machine condition monitoring have had slower rates as a comprehensive way to maintain and troubleshoot the complex mechanical systems during the past years in developing countries which mostly rely on the imports of industrial machinery, construction, and transportation. Due to increase of the accuracy and decrease of the mistakes resulted from human judgment, today, the modeling methods are used for defects classification and detection. Regarding the considerable and increasing development in science, the older modeling and relations, more or less, have lost their influence. Besides, considering the dynamical modeling which works in various parts, caused the use of typical simple models to be complicated. The un-linearity of the problems is also an important and unignorable factor. In this regard, several modeling methods have been proposed, including ant colony, fuzzy logic theory and effective methods like artificial neural networks and genetic algorithm. Each method has its own advantages and disadvantages [23-25]. In developed agriculture, it is so important to do agricultural works just on time. Time is a vital factor in agricultural systems, which could affect total function [26,27]. Therefore, it is necessary for the machines to be available at the time required. Whenever in Tractors of a certain farm a defect is occurred in the gear box, including the important issues that we will be encountered with, too much time and fee would be spent to transmit, open and close and repair the gear box, and since it is a vital part of the tractor and there is no more work without the gear box, so we will have to replace or repair it. However, sometimes a certain human error would cause the maintenance time and cost to be doubled. Since, for years, the experts have followed the discovery of defects in various parts of the tractor in order to reduce the time and cost of economy and raise the work efficiency. This study has been also conducted to make it easier to troubleshoot in gearbox of MF285 Tractor.

2. Materials and Methods

In this study a smart diagnostic system to make decisions faster than normal due to impair function machines and to increase efficiency and overall lifetime of the device is presented. After adjusting the environment settings, the test was performed in a healthy condition of gears and two defective conditions of the gear worn and the gear broken, and in three iterations of input to the axle along with installing the accelerometer sensor in both directions of Perpendicular to the axis and aligned with the axis on the bearing. For each case, the sample test was repeated 130 times and the data related to each sample were recorded and registered. In total, 2340 experimental data set were recorded as the files on Excel software and it took 20 days to collect data, as a whole. After designing and building profiles needed to position the pieces, all the necessary settings on the device and on the desktop were conducted to start the data analyzing process (**Figure 1**).

Once the engine rotation been measured using the optical rpm gauge, we provide different rotations in 3 various conditions and in two sensor directions in motor, implementing the dimmer, and record the relevant data.

The gear vibrating signals are measured by the accelerometer installed on bearings on the shaft. The engine has one central axis which is coupled with the gear input and the resulted rotation speed of motor is transformed according to the proportion of the base gear. In order to make failures on gear, we used the gear with 44 gears in condition of gear 1 on the input axis (**Figure 2**).

In order to collect the vibrating signals, containing the healthy condition and fraction and breakage defects in both perpendicular sensor and axial sensor directions in three motor inputs 1500, 1000 and 2000 and under the loading for 4 seconds, the data were collected in all conditions. In all conditions, every 4 seconds 10,240 samples of voltage were measured in time range of 0.0004. The software used, ARMA, recorded 10,240 voltage data in time range of 4 seconds based on the adjustments in an Excel file. After gaining the data related to healthy condition, the defects and failures considered were applied and the vibrating data were gained, separately.

3. Signal Processing

Signal processing and the feature extracting part are the most important parts in the condition monitoring procedure, based on which the system feature including being healthy or fractured and type of the system are determined. The current method is to provide the sufficient features, signal transferring from time to frequency domain. For this purpose necessary codes have been transformed in the form of MATLAB software. This program selects Excel files containing the vibration data (based on time/voltage) and filters the data, eliminates the noises and finally calculated the statistical parameters of time and frequency extent, implementing the filter used. Therefore, a fast Fourier function (FFF) of 10,240 points was calculated for each time signal, separately. Also, the power spectral density (PSD) and phase angle (FFT) of vibration signals were calculated using the software MATLAB.

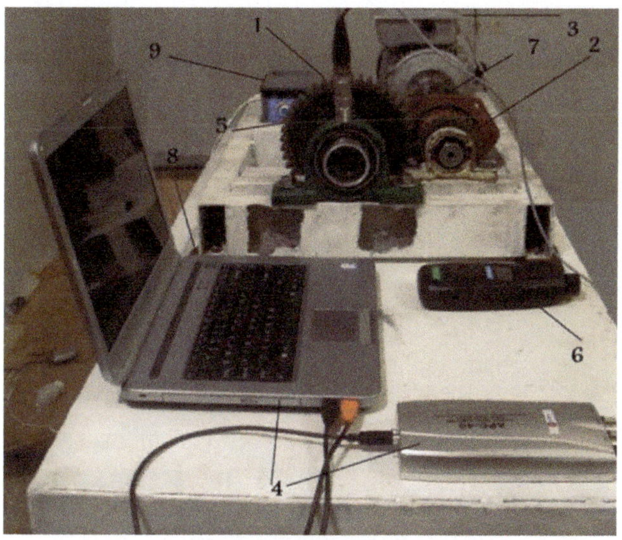

Figure 1. Gearbox set up.

Figure 2. Broken conditions of gears.

4. Feature Extraction

The main aim in feature extraction methods is to provide a way or arrangement that could monitor the components hidden in a signal. In this study, 10,240 data sample from time extent formed the data collection. Also, the frequency extent data points, let's say PSD and phase angel FFT, were each 10,240. These signals were used to extract 33 features of healthy and failure conditions. 11 features (T1 - T11) were from statistical parameters in time extent and the remaining (P1 - P11) and (A1 - A11) were from statistical parameters in frequency extent (**Table 1**).

Then, using MATLAB software, the diagrams related to vibration spectrum in frequency extent in healthy and failure conditions were gained due to get analyzed and compared to each other. In order to make the differences in the frequency of normal and defective devices more clear, either axial rotation 1000 rpm and 2000 rpm on the two healthy gear broken at the same diagrams were classified (**Figure 3**).

Figure 3 shows that in safety condition in both 1000 and 2000 rpm, the amplitude for each diagram remains unchanged during the work and only one or two cases have a maximum that could be justifiable due to the gear involvement. But in unhealthy conditions while the gear with broken gear is replaced with the safe one, the frequency-voltage diagram will immediately have many maximum points which are occurred almost in a regular order. It could be observed that in increasing the rotary axis rotation there are some fixed amplitude increase, while in the broken gear the increase in amplitude is iterated in an unusual way in certain coefficients. The difference resulted from the broken gear has been illustrated in frequency-voltage diagrams against the time-voltage diagram.

5. Classification of Defects

In MATLAB software using the neural network, the chromosomal genetic with 20 gens was designed and the

Table 1. Features of time and frequency domain.

Frequency extent features		Time extent features
phase angel FFT	PSD	
$A_1=\frac{\sum_{j-1}^{J} i(j)}{J}$	$P_1=\frac{\sum_{K-1}^{K} s(k)}{K}$	$T_1=\frac{\sum_{n-1}^{N} x(n)}{N}$
$A_2=\sqrt{\frac{\sum_{j-1}^{J}(i(j)-A_1)^2}{J-1}}$	$P_2=\sqrt{\frac{\sum_{K-1}^{K}(s(k)-P_1)^2}{k-1}}$	$T_2=\sqrt{\frac{\sum_{n-1}^{N}(x(n)-T1)^2}{N-1}}$
$A_3=\sqrt{\frac{\sum_{j-1}^{J}(i(j))^2}{J}}$	$P_3=\sqrt{\frac{\sum_{K-1}^{K}(s(k))^2}{K}}$	$T_3=\sqrt{\frac{\sum_{n-1}^{N}(x(n))^2}{N}}$
$P_4=\max\lvert i(j)\rvert$	$P_4=\max\lvert s(k)\rvert$	$T_4=\max\lvert x(n)\rvert$
$A_5=\frac{A_3}{\frac{1}{J}\sum_{j-1}^{J}\lvert s(j)\rvert}$	$P_5=\frac{P_3}{\frac{1}{K}\sum_{k-1}^{K}\lvert s(k)\rvert}$	$T_5=\frac{T_3}{\frac{1}{N}\sum_{n-1}^{N}\lvert x(n)\rvert}$
$A_6=\frac{A_4}{\frac{1}{J}\sum_{j-1}^{J}\lvert s(j)\rvert}$	$P_6=\frac{P_4}{\frac{1}{K}\sum_{k-1}^{K}\lvert s(k)\rvert}$	$T_6=\frac{T_4}{\frac{1}{N}\sum_{n-1}^{N}\lvert x(n)\rvert}$
$A_7=\frac{A_4}{A_3}$	$P_7=\frac{P_4}{P_3}$	$T_7=\frac{T_4}{T_3}$
$A_8=\sqrt{\frac{j\sum_{j-1}^{J} i^2(j)-\sum_{j-1}^{J}(i(j))^2}{j(j-1)}}$	$P_8=\sqrt{\frac{K\sum_{k-1}^{K} s^2(k)-\sum_{k-1}^{K}(s(k))^2}{k(k-1)}}$	$T_8=\sqrt{\frac{n\sum_{n-1}^{N} x^2(n)-\sum_{n-1}^{N}(x(n))^2}{n(n-1)}}$
$A_9=(A_8)^2$	$P_9=(P_8)^2$	$T_9=(T_8)^2$
$A_{10}=\frac{\sum_{j-1}^{J}(i(j)-A_1)^4}{J\times(A_9)^2}$	$P_{10}=\frac{\sum_{k-1}^{K}(s(k)-P_1)^4}{K\times(P_9)^2}$	$T_{10}=\frac{\sum_{n-1}^{N}(x(n)-T_1)^4}{N\times(T_9)^2}$
$A_{11}=\frac{j}{(j-1)(j-2)}\sum_{j-1}^{J}\left(\frac{i(j)-A_1}{A_8}\right)^3$	$P_{11}=\frac{k}{(k-1)(k-2)}\sum_{k-1}^{K}\left(\frac{s(k)-P_1}{P_8}\right)^3$	$T_{11}=\frac{n}{(n-1)(n-2)}\sum_{n-1}^{N}\left(\frac{x(n)-T_1}{T_8}\right)^3$
In which i(j) is a signal for j = 1, 2, …, J. J indicates the number of data points.	in which s(k) is a signal for k = 1, 2, …, K. K indicates the number of data points.	in which x(n) is a signal for n = 1, 2, …, N. N indicates the number of data points.

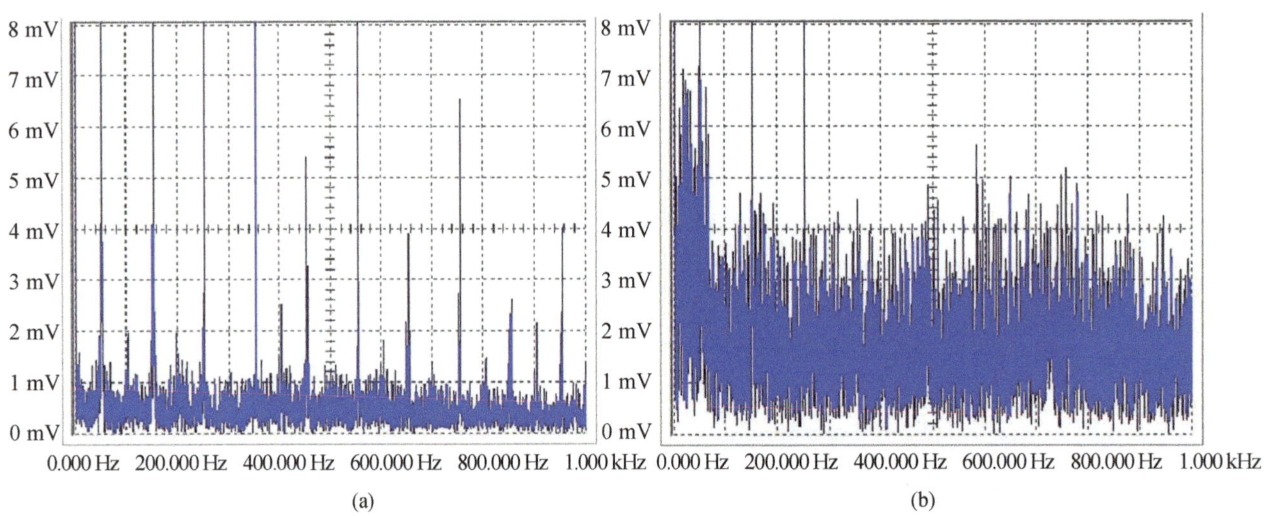

Figure 3. FFT diagram of (a) healthy condition (b) fault condition.

proper neural network structure was created by determining the number of layers, neurons, epoxies, biases and gens implementing the numbered gens capacity, which studied using various statistical parameters of efficiency and error percent of neural network structure. Applying the genetic-neural network hybrid, the necessary classification and calculations were performed to create a network compatible with the test. In this study, one chromosome with 20 gens was designed. The first gen determines the number of inputs. Because, it was tried to investigate each input parameter on the model output and if any of these inputs have trivial impact on the model output it would be deleted from the model and in other words the unvalued inputs genetic algorithm will be eliminated (**Figure 4**).

Gens 2 and 9 illustrate two considerable issues from the neural network structure, one shows the number of layers hidden in network and the other illustrates the number of neurons in each hidden layer, and the amount could be variable from 0 to 12. The transfer functions include: competitive functions, two-valued stimulation threshold, two-valued symmetrical stimulation threshold, sigmoid, converse, positive linear, linear, radial, saturated linear, symmetrical saturated linear, triangular and sigmoid hyperbolic tangent. The 18th gen determined the type of training function, and the amount of this gen could be variable from 0 to 18. The training functions include: the combined gradient algorithm Powell Beale, the combined gradient algorithm Polak-Ribirere, Resilient Back propagation, BFGS algorithm, One Step Secant

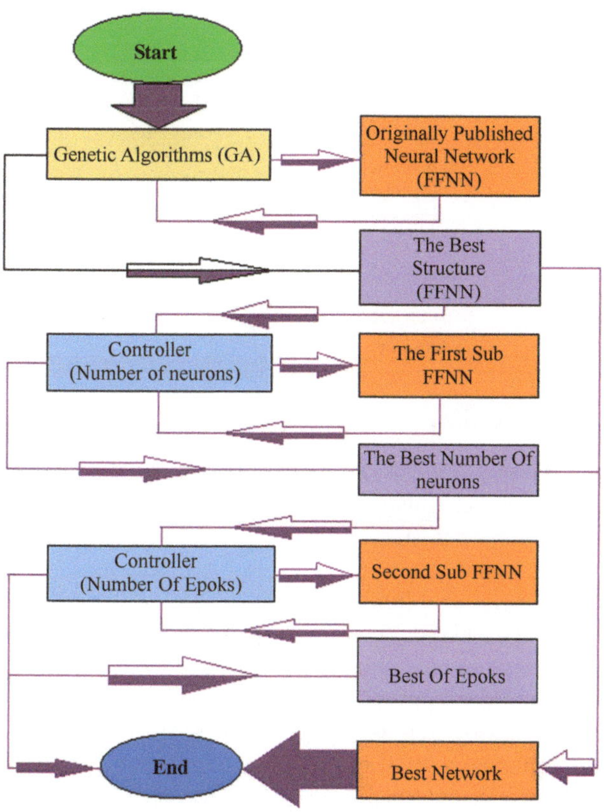

Figure 4. A sample of genetic algorithm structure and procedure to form the best neural network.

algorithm, Levenberg-Marquardt algorithm, the scaled combined gradient algorithm, quasi-Newton back propagation algorithm, monomentum algorithm, combined gradient algorithm Flathcher-Reeves and etc. the 19th gen determines the learning function that have the range of 0 to 14. The learning functions include: bias reduction, weigh or bias of descent gradient, the descent gradient with Monomentum weigh or bias, weigh given, the rule of learning hob with descent weigh, the Instar weigh, the Kohnen weigh, the vector quantization learning weigh LVQ1, LVQ2, outstar weigh, percepteron weigh and bias, percepteron normalized weigh and bias, self-organizer mapping weigh, collective self-organizer mapping weigh and the rule of Vidro-Hauf weigh and bias learning. The 20st gen determines the epoxies to train each network and the range is variable from 10 to 500. Regarding the genetic algorithm adjustments, the best population number for each generation was selected as 50. So, the scale of terminating the genetic algorithm was selected as 100 and the competitive method and two-pointed integration were implemented. All data were included 30 percent of test and 70 percent of training and all the calculations were performed by hybrid genetic-neural network. The optimizing program was used in software MATLAB version Release 2010a, as a whole. In order to classify and analyze the failures and relevant parameters, using the neural network and gaining effective and necessary inputs to be applied in neural-genetic network, the nerve solution was implemented. This software is a sort of developed environment for neural networks, which includes the models based on network design, implementation using advanced learning methods like conjugate gradient, Lonberg-Marquadet and the error propagation during the time. This designing, training and establishing the neural networks (learning under supervision and training without supervision), is used to perform an extent range of tasks like: data analysis, classification, multi-variable regression and time series prediction. Implementing the nerve solution software sensitivity analysis the considered inputs from 55 inputs we had calculated, were gained and the analysis diagram of sensitivity to the average for all data related to the condition of perpendicular to shaft at 1000 rpm was drawn .According to previous analysis using MATLAB for which the results were recorded as the same names in different columns in Excel files, the name of each column is significant for the next calculation. Implementing the name of these variables, the relevant addresses were used as the neural network inputs (**Table 2**), and then using genetic algorithm in software MATLAB, the best designing structure and it's features were registered (**Table 3**).

6. Result

To evaluate the accuracy of the neural network model,

Table 2. Features of selected inputs for neural network models.

inputs	models	Number
psd-value-y-FFT phase angle-y	Perpendicular-1000 rpm	**1**
fft-values-FFT phase angle-psd-value	Perpendicular-1500 rpm	**2**
psd-value-fft-values-fft-values-y-psd-value	Perpendicular to the axis at 2000 rpm	**3**
fft-values-psd-value-FFT phase angle-y	Aligned with the axis at 1000 rpm	**4**
psd-value-psd-value-y-FFT phase angle-y	Aligned with the axis at 1500 rpm	**5**
FFT phase phase-FFT phase angle-fft-values	Aligned with the axis at 2000 rpm	**6**

Table 3. Neural network, designed and optimized using genetic algorithm.

Number of epoks	Quality function	bias/weigh training function	network training function	output layers		hidden layers		No. of inputs	Neural network, designed and optimized using genetic algorithm
				transfer function	No. of neurons	transfer function	No. of neurons		
287	the mean square error	descent gradient with monomentum weigh or bias	Lonberg-Marquidet back propagation	Hyperbolic tangent sigmoid	1	Hyperbolic tangent sigmoid	5	4	1
143	the mean square error	descent gradient with monomentum weigh or bias	Lonberg-Marquidet back propagation	Hyperbolic tangent sigmoid	1	Hyperbolic tangent sigmoid	12	3	2
82	the mean square error	descent gradient with monomentum weigh or bias	Lonberg-Marquidet back propagation	Hyperbolic tangent sigmoid	1	Hyperbolic tangent sigmoid	9	5	3
272	the mean square error	descent gradient with monomentum weigh or bias	Lonberg-Marquidet back propagation	Hyperbolic tangent sigmoid	1	Hyperbolic tangent sigmoid	7	4	4
251	the mean square error	descent gradient with monomentum weigh or bias	Lonberg-Marquidet back propagation	Hyperbolic tangent sigmoid	1	Hyperbolic tangent sigmoid	8	5	5
146	the mean square error	descent gradient with monomentum weigh or bias	newton back propagation network	Hyperbolic tangent sigmoid	1	Hyperbolic tangent sigmoid	14^*1^*1	3	6

the data collection in training and test sets were analyzed statistically based on features extracted from the network. MATLAB software was used to perform classification. Statistical parameters such as mean-square error (MSE), Mean-square normalized error (MSNE) and multi-layer network error were compared (**Figures 5** and **6**), (**Table 4**). The best network outputs for each sensor condition are divided into three 1000, 1500 and 2000 rpm. There are, totally, 6 output models for which all the system function statistical parameters have been calculated (**Table 4**). Due to compare two diagnostic positions of each model, general description of the correlation coefficient in each case is given in diagram (**Figure 7**). The highest correlation coefficients in Model 4 and 5 are about 99% and the lowest one in model (6) is about 97%.

7. Conclusion

In this study a network-compatible device has been designed using artificial neural networks, to detect and classify defects in a Massey Ferguson 285 tractor gear. According to the results, the best network outputs for each sensor mode are divided into three categories in 1000, 1500 and 2000 rpm. Totally, outputs for six cases and the function of all the quality parameters have been computed in order to troubleshoot the gear 1 for gear box in tractor Massey Ferguson 285 in the worst case (aligned with the axis at 2000 rpm) with a correlation coefficient of 97% and the mean square error of 0.000047%. The numbers indicate the high accuracy and low error in trial, respectively. These experiments showed that the overall

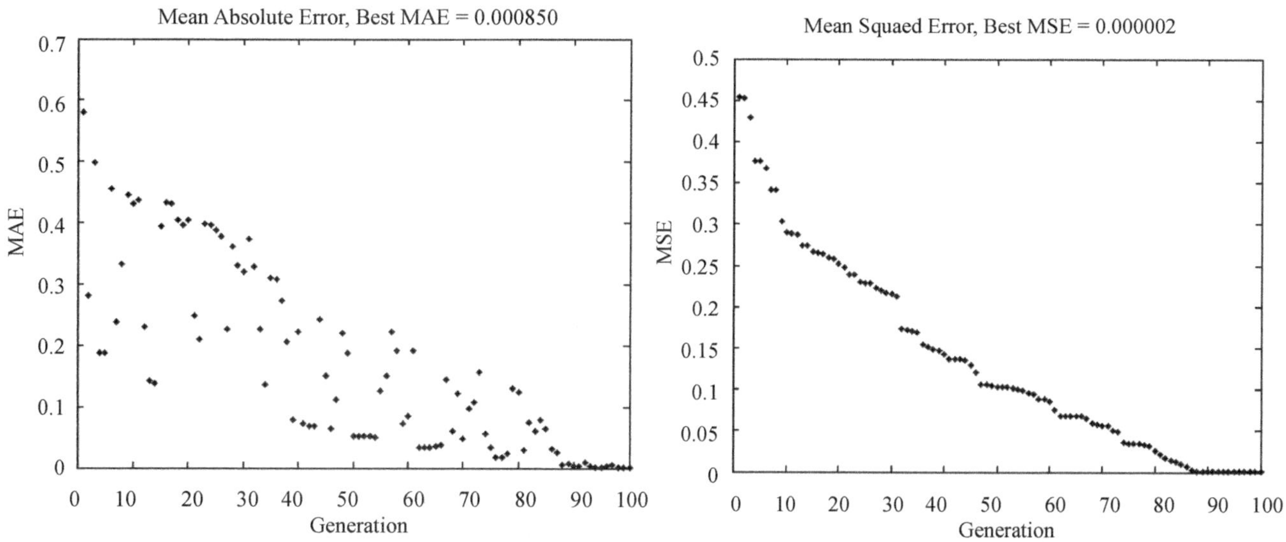

Figure 5. Mean-square error network output results at 1000 rpm in the status of perpendicular on the axis.

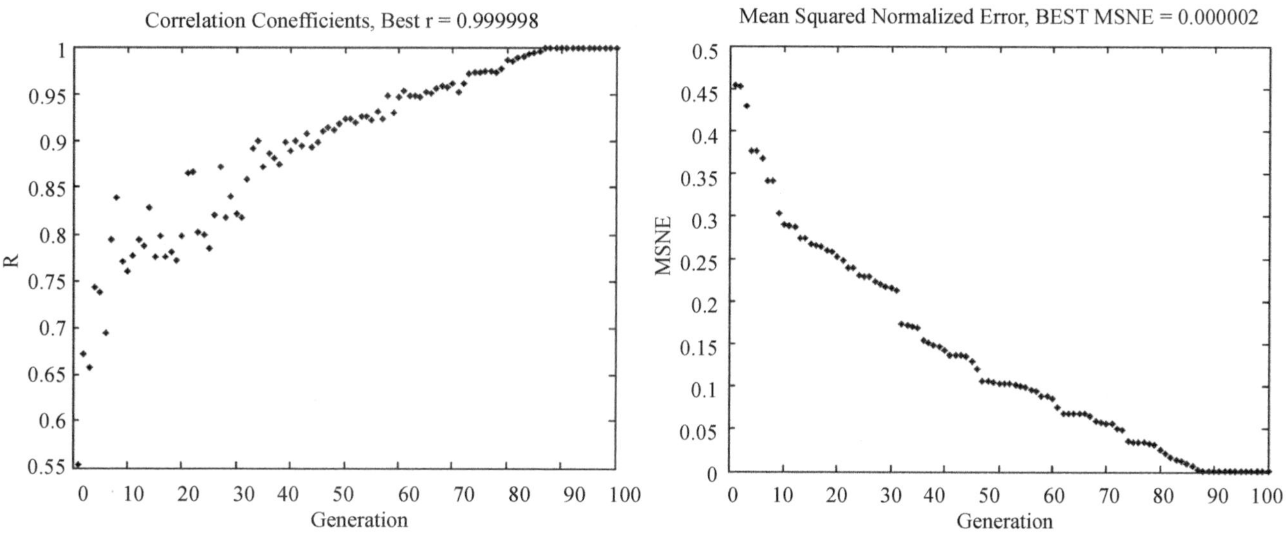

Figure 6. Mean-square normalized error and the correlation coefficient in network output results at 1000 rpm in the status of perpendicular on the axis.

Table 4. The best network results gained in graphs.

correlation coefficient (r)	mean absolute error (MAE)	Mean-square normalized error (MSNE)	mean-square error (MSE)	The best network design using genetic algorithm	NO
0.999998	0.000850	0.000002	0.000002	perpendicular on the axis at 1000 rpm	1
0.999998	0.001442	0.000004	0.000004	perpendicular on the axis at 1500 rpm	2
0.999998	0.001442	0.000006	0.000006	perpendicular on the axis at 2000 rpm	3
0.999999	0.000147	0.000001	0.000001	aligned with the axis at 1000 rpm axis at	4
0.999999	0.000207	0.000001	0.000001	aligned with the axis at 1500 rpm	5
0.999979	0.004012	0.000047	0.000047	aligned with the axis at 2000 rpm	6

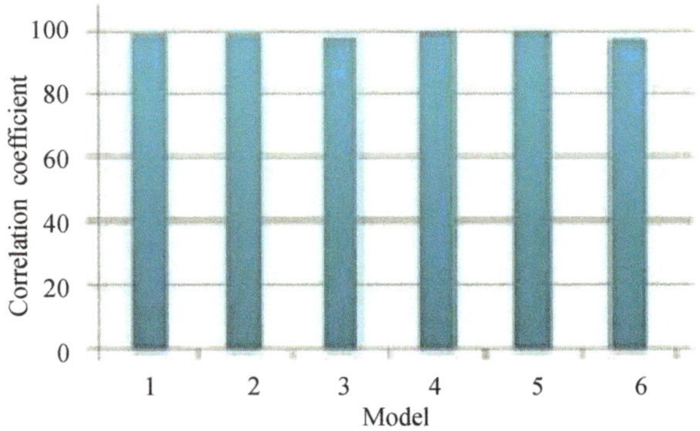

Figure 7. The overall correlation diagram based on tested models.

mean correlation coefficient of the network is equal to 99%, to adapt to the mechanism of defect detection and classification system, and according to the definition of correlation coefficient, this number is near 1 that is the desired mode and the mean error occurred in defect detection is nearly 0. These statistical results show the efficiency and applicability of this method to detect the defects in gearbox.

8. Acknowledgements

The authors would like to thank Islamic Azad University, Kermanshah Branch for its support to this research.

REFERENCES

[1] M. Elforjani, D. Mba, A. Muhammad and A. Sire "Condition Monitoring of Worm Gears," *Applied Acoustics*, Vol. 73 No. 8, 2012, pp. 859-863.

[2] H.-E. Kim, A. C. C. Tan, J. Mathew and B.-K. Choi, "Bearing Fault Prognosis Based on Health State Probability Estimation," *Expert Systems with Applications*, Vol. 39, No. 5, 2012, pp. 5200-5213.

[3] P. J. Dempsey, D. G. Lewicki, and H. J. Decker, "Decker Transmission Bearing Damage Detection Using Decision Fusion Analysis," Glenn Research Center, Cleveland, Army Research Laboratory, NASA/TM, 2004, pp. 1-20.

[4] G. Goddu, B. Li, M. Chow and J. Hung, "Motor Bearing Fault Diagnosis by Fundamental Frequency Amplitude Based Fuzzy Decision System," North Carolina University. 1998, pp. 1961-1965.

[5] N. Hotwai, "Vibration Analysis of Faulty Beam usig Fuzzy Logic Techique," BTech Thesis, National Institute of Technology, Rourkela, 2009, pp. 23-27.

[6] Ch. Kong, J. Ki, S. Oh and J. Kim, "Trend Monitoring of a Turbofan Engine for Long Endurance UAV Using Fuzzy Logic," *KSAS Enternational Gournal*, Vol. 9. 2008, pp. 64-70.

[7] Z. Kiral and H. Karagulle, "Simulation and Analysis of Vibration Signals Generated by Rolling Element Bearing with Defects," *Tribology International*, Vol. 36, No. 9, 2003, pp. 667-678.

[8] Y. G. Lei, Z. J. He and Y. Y. Zi and X. Hu, "Fault Diagnosis of Rotating Machinery Based on Multiple ANFIS Combination with Gas," *Mechanical Systems and Signal Processing*, Vol. 21, No. 5, 2007, pp. 2280-2294.

[9] Y. G. Lei, Z. J. He and Y. Y. Zi, "A New Approach to Intelligent Fault Diagnosis of Rotating Machinery," *Expert Systems with Applications*, Vol. 35, No. 4, 2008, pp. 1593-1600.

[10] R. Martins Marcal, K. Hatakeyama and A. Susin, "Managing Incipient Faults in Rotating Machines Based on Vibration Analysis and Fuzzy Logic," Electrical Engineering Department, 2006, pp. 1-6.

[11] N. Saravanan, S. Cholairajan and K. I. Ramachandran "Vibration-based Fault Diagnosis of Spur Bevel Gear box Using Fuzzy Technique," *Expert Systems with Applications*, Vol. 36, No. 2, 2009, pp. 3119-3135.

[12] T. I. Liua, J. H. Singonahallib and N. R. Iyerb, "Detection of Roller Bearing Defect Using Expert System and Fuzzy Logic," *Mechanical Systems and Signal Processing*, Vol. 10, No. 5, 1996, pp. 595-614.

[13] E. Ebrahimi and K. Mollazade, "Intelligent Fault Classification of a Tractor Starter Motor Using Vibration Monitoring and Adaptive Neuro-Fuzzy Inference System," *Insight—Non-Destructive Testing and Condition Monitoring*, Vol. 52, No. 10, 2010, pp. 561-566.

[14] G. H. Payganeh, M. N. Khajavi, R. Ebrahimpour, and E. Babaei, "Troubleshooting of Rotating Machines Using Neural Networks," *6th National Conference on Maintenance and Repair*, Tehran, Iran.

[15] A. Bavi and M. Salehi, "Genetic Algorithms and Optimization of Composite Structures," Tehran, Abed.

[16] D. Boostan and N. Pariz, "Analysis of Vibration Signals Using Short Time Fourier Transform, Neural Networks and Fuzzy Inference to Identify the Type and Severity of Damage in the Bearing," *16th Annual International Conference on Mechanical Engineering*, Tehran, Shahid Bahonar University.

[17] M. Behzad, K. Sepanloo, M. Asayesh and E. Rohani, "Fundamentals of Vibrations in the Maintenance, Repair and Troubleshooting of Rotating Machinery," NPC Publication.

[18] H. Beigy, "MF 285 Vibrating Gearbox Condition Monitoring and Fault Features and Its Classification Using Artificial Neural Networks," M.Sc. Thesis, Tehran University.

[19] E. Rohani, M. Alikhani, E. Mostafavi Manesh, M. R. Darbandi and M. E. A. Behzad, "Troubleshooting by the Help of Turbo Compressors Gas Pipeline Vibration Analysis, Condition Monitoring and Diagnostics of Machines," *4th Conference of Iran*, Sharif University of Technology, Tehran.

[20] M. Ryahi and A. Tarbareh, "Designing and Manufacturing the System of Information Recording and Preserving on Heavy Machinery to Monitor and Analyze the Status, Condition Monitoring and Diagnostics of Machines," *4th Conference of Iran*, Sharif University of

Technology, Tehran.

[21] B. V. Zahraei and M. Hosseyni, "Genetic Algorithms and Engineering Optimization," Tehran, Gothenbur.

[22] M. Shakeri and M. Nozad, "Remote Diagnostics of Rotating Machineries Based on Vibration Condition Monitoring and Troubleshooting of Machines," *4th Technical Conference and Equipment of Iran*, Sharif University of Technology, Tehran.

[23] E. Ghafar Nejad Mehraban and M. Homayoon Sadeghi, "Smart Troubleshooting for Car Gearbox Vibration Using Analysis and Artificial Neural Networks by Analyzing Violet," *16th Annual Conference on Mechanical Engineering*.

[24] A. Farshidian Far, S. Maleki and E. Andakhshideh, "Vortex-Induced Vibration Modeling and Optimization of Cylinders by a Neural Network and Genetic Algorithm," *7th Annual Conference of the International Association of Aerospace Iran*, Sharif University of Technology, Tehran.

[25] M. Kia, "Neural networks in Matlab," Kian Publication of Rayaneh Sabz, Tehran.

[26] H. Mohammadi Monavar, "Vibration Condition Monitoring in Agriculture Electro-Motors with Fans," Master's Thesis, Tehran University, Tehran.

[27] K. MollaZadeh, "Vibration Condition Monitoring of Hydraulic Steering Pump MF 285 and the Defects Classification Using Fuzzy Logic," Master's Thesis, Tehran University, Tehran.

6

Optical Microstructure Design Optimization for Display Backlighting

Chi-Chang Hsieh[1], Yan-Huei Li[2]
[1]Department of Mechanical and Automation Engineering, National Kaohsiung First University of Science and Technology, Taiwan
[2]Instiute of Engineering Science and Technology, National Kaohsiung First University of Science and Technology, Taiwan

ABSTRACT

This study proposes an innovative design method for functional optical film microstructures used in displays and applies this design to LCD backlighting to replace multi-layer optical film functional integration with composite optical film. We design a novel optical film microstructure based on light uniformity and wide-angle lumination distribution and determine the optimal optical microstructure parameters by combining the global optimization of a genetic algorithm with ray tracing. The purpose of this study is to develop substitutes for traditional multi-layer prism brightness enhancers and light-diffusing film stacks and to examine the structural changes during calculations and summarize the characteristics. In this study, we focus on determining the optimal light uniformity of new optical film microstructures. The seven-inch LED backlight module of the method proposed in this study achieved 94.59% uniformity and 168° lumination distribution while reducing thickness by 66% to 82% compared to a traditional multi-layer optical film stack.

Keywords: Optical Film; Microstructure; Display; Backlight

1. Introduction

In recent years, green energy-saving and ultra-thin displays have become the demand for the mainstream market. Therefore, backlight modules with extremely high cost and assembling ratios have become fiercely contested items among major manufacturers because of the technological innovation requirements. Multiple optical films with various effects are stacked in layers in backlight modules to focus the light energy and equalize the light distribution. However, this increases the cumulative thickness and overall cost of backlight modules. Currently, backlight modules for displays typically comprise brightness enhancing films with prismatic structures to enhance luminance in the normal direction. However, this approach is likely to cause significant declines in luminance at minor angles away from the normal direction. Therefore, diffusion films are commonly added to reduce the significant decline in luminance and improve uniformity. However, this reduces the axial luminance. During this back-and-forth tuning process, the backlight module tends to become excessively thick because of the addition of extra optical films, which also dissipate the light energy. Thus, the goal of this study was to design a compound optical film with an innovative structure emphasizing light uniformity. The aim was to widen the luminance distribution viewing angle and resolve the uneven luminance in the central viewing angles of older structures to achieve one-to-two or one-to-many performances, replace the traditional process of stacking multiple optical films together, reduce the thickness of backlight modules, design an easy-to-manufacture structure, and effectively reduce costs.

This study employed a genetic algorithm (GA) as the model computational framework. GAs were first proposed in 1975 by Holland. Using the GA method enables a variety of problems to be converted into genotypes [1]. In 2009, Li *et al.* combined a neural network and real-valued GA to optimize the design of the bottom structure of a light guide plate. In their study, the luminance uniformity reached 92.09% and showed greater accuracy than that of results provided by the modeling software ASAP [2]. In 2010, Yamada *et al.* employed an evolutionary algorithm (EA) to automatically generate a set of

Fresnel lenses of varying geometries and densities. This method achieved the highly uniform flux concentration required for the design of photovoltaic cells [3].

Genetic algorithm has less been applied in backlight module with a light guide structure design. In 2008, Fan *et al.* applied genetic algorithms to design LED lens in direct type LED Backlight Unit to achieve high uniformity [4]; In 2009, Li *et al.* applied genetic algorithm to optimize the distribution density of the prism pattern to improve the uniformity of the light guide plate [2]. In 2013, Kim *et al.* applied genetic algorithm to numerical model of a planar prism LED luminaire, and to the analysis of the best angle prism [5]. This study proposes an innovative design and optimize structure applied Genetic algorithm, which can achieve high uniformity of the illuminance distribution with a large angle.

2. Design Analysis of optical Film Structure

2.1. Configuration of Backlight Modules

This study used sidelight-type backlight modules with 7-in no-wedge angle light guide plates with the following dimensions: 92 mm in width, 158 mm in length, and 2.5 mm in thickness. The material used was PMMA with a refractory index of 1.49. The reflectivity was approximately 95%. The distance between the light exit face of the light guide plate and the receiver board was 0.45 mm. Only the no-wedge angle light guide plates contained concave hemispheric structures, the rest were all unstructured optical surfaces. The diameter of the structure was 0.1 mm. The distance between the structures was 1 mm in both the X and Y directions. The light source was an LED, specifically item number NESW020A from NICHA, which emits 3.7 lm with a 20 mA electrical supply at a light-emitting angle of 120°. The light pattern was close to a Lambertian distribution. Additionally, 7 LEDs were arranged equidistantly along the short edge of the light guide plate.

2.2. Design Method for the optical Structure Geometry

The structural geometry concept of this study was based on achievements in surface microstructure design for LED packaging completed by our laboratory [6,7]. Results from the research have indicated that stepped structures have light focusing effects. Applying this characteristic to the design of optical films has a potential to provide excellent performances. The original stepped structure (shortened to original structure hereafter) in the patent awarded to our laboratory in 2008 for composite optical film can achieve a wider luminance distribution throughout the viewing angle [8]. However, the axial luminance remains inferior to that of a prismatic sheet combined with a diffusion film. Because our laboratory had already generated a conceptual structure, which had room for improvement, we focused on introducing light-focusing novel structures to enhance the brightness and performance. Therefore, this study used the concept of geometric shape combinations to combine the prismatic structure providing brightness-enhancing effects with stepped structures providing broader luminance distributions throughout the viewing cone, in the hopes that multiple performance improvements could be achieved. **Figure 1** shows the design concept diagram for combining geometric shapes. Square structures were introduced into the original prismatic structure, forming stepped geometric structures similar to oblique prisms. Novel structures were created through equidistant division, changing angles, or changing the structural width and height. This enabled the structural parameters to be defined, which facilitated the experimental process to identify optimal solutions.

The shape and dimension of the optical film structure refer to the structural dimensions of the prismatic brightness-enhancing film 3M BEFII 90/50. The structural shape of 3M BEF 90/50 comprises equilateral triangles 23 μm in height, with a tip angle of 90°, and arranged in parallel rows. The novel geometrical structure in this study was based on this geometry and further divided equidistantly into 16 sections in the horizontal direction and 8 sections in the vertical direction. The equidistant division lines along the edges of the prismatic surfaces were joined to form the stepped structure proposed in this study. **Figure 2** shows a conceptual diagram of the design of this novel structure.

2.3. Defining the structural Parameters of the optical Geometry

Figure 3 shows a diagram that defines the structural shape of the stepped structure. In this figure, W represents the base width of a single structure. The width of each exposed step was W/16. H represents the overall height of the structure, and the height for each step was H/8. θ represents the angle from the edge of the stepped structure to the horizontal plane. When θ was 45°, the

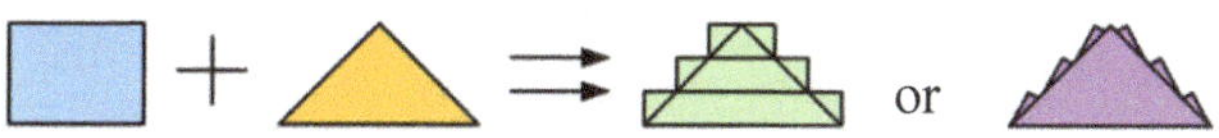

Figure 1. Design concept for the combination of geometries.

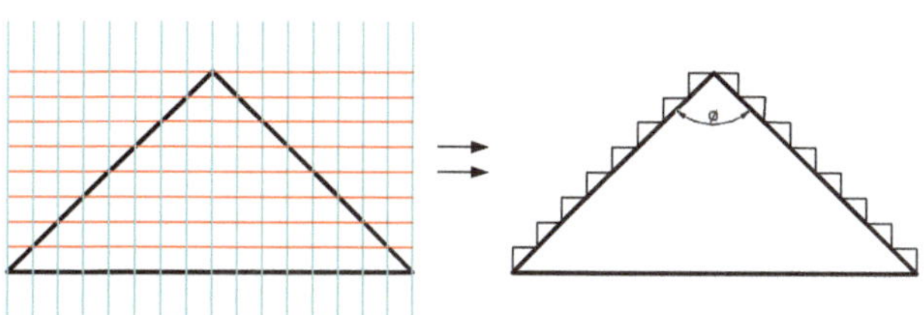

Figure 2. Design concept for the new type of structure.

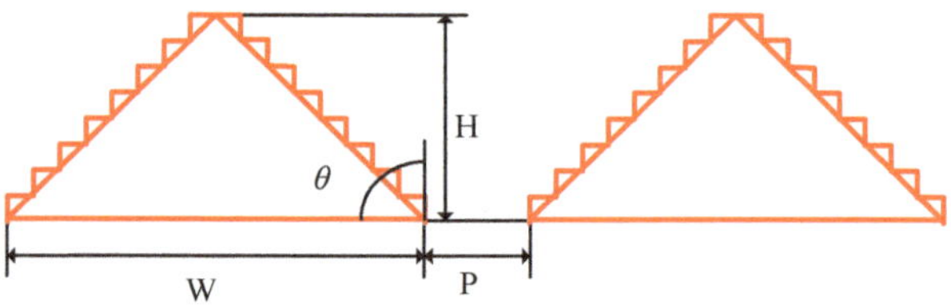

Figure 3. The external dimensions of the stepped structure.

formed structure was a triangular prism. P represents the distance between each structure. The 8-stepped structural arrays could be defined completely using these 4 parameters. Once the parameters were defined, the machining limit or the size of commercially available optical film sheets was used as the boundary conditions for the reasonable ranges of structural parameters. This mitigates the manufacturing difficulties caused by excessively small structures, or an inability to effectively downscale the thickness of backlight modules because of excessively large structures.

2.4. Methods for Measuring Uniformity

This study used the 9-point VESA measurement method as a reference for uniformity measurements of the 7-inch backlight module. The luminance value of each point was first measured, then the 9-point VESA measurement method commonly employed throughout the industry was applied [9]. The minimum and maximum luminance values of backlight modules measured using the 9-points were shown as percentages, which were the level of uniformity of the panel. The uniformity formula is shown below:

$$\text{Uniformity} = 100\% \times \frac{L_{min}}{L_{max}} \quad (1)$$

2.5. Full Width at Half Maximum

In 2009, Sang-HoonBaik used the full width at half maximum (FWHM) to define the angular luminance distribution of optical films to determine the size of the view angle [10], which was then examined against an orthogonally arranged dual prismatic sheet for performance comparison. This study referenced this definition of view angle to define the view angle of the angular luminance distribution curve. This luminance angle differs from the commercially available viewable angle determined by the level of contrast. For TFT-LCDs, a higher luminance angle is less likely to cause uneven brightness and luminance declines when viewed from multiple angles.

3. Results and Discussion

3.1. Results from the genetic Algorithm

Figure 4 is a fitness function curve for the GA. The figure shows that by generation 33, the difference between chromosomes was sufficiently minor that neither mating nor mutation could achieve superior solutions. The iterating computation using the optimal value obtained in this GA was stopped at generation 100 because the termination criteria were satisfied. **Figure 5** shows the optimized exterior structure through a global search method using the GA.

3.2. Analysis of structural form and Performance

3.2.1. Discussion on the Relationship between structural form and View Angle Luminance Distribution

Table 1 shows the interval combinations and uniformity values for intervals of 0.03 mm, with uniformity values above 92%. Because of the many combinations, only combinations with uniformity values above 92% were included in the examination. Based on the 26 combinations of structural shapes and light trace charts, we found that the structural shapes could be roughly classified into 3 types. The various shape types are shown in **Figure 6**. TYPE 1 comprises prismatic structures, TYPE 2 comprises step-shaped structures with larger side angles, and TYPE 3 comprises step-shaped structures with smaller side angles.

Figure 6 shows that the prismatic structure of TYPE 1 shapes characteristically corrects light rays at greater angles of deviation from the normal line to within plus or minus 30° from the center. Luminance at the central view angles is higher and more concentrated than the lumi-

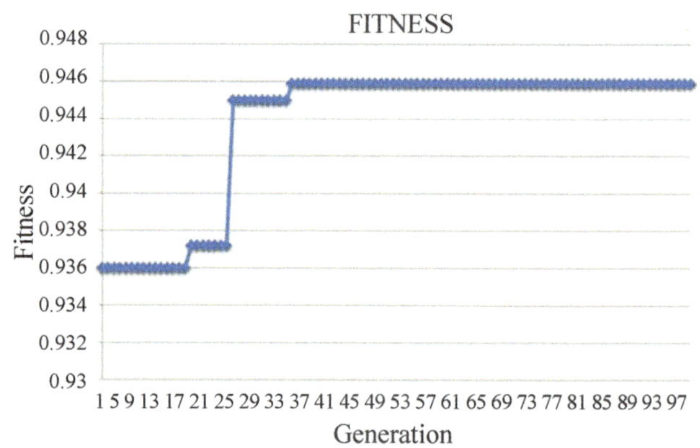

Figure 4. Fitness function curve of the genetic algorithm.

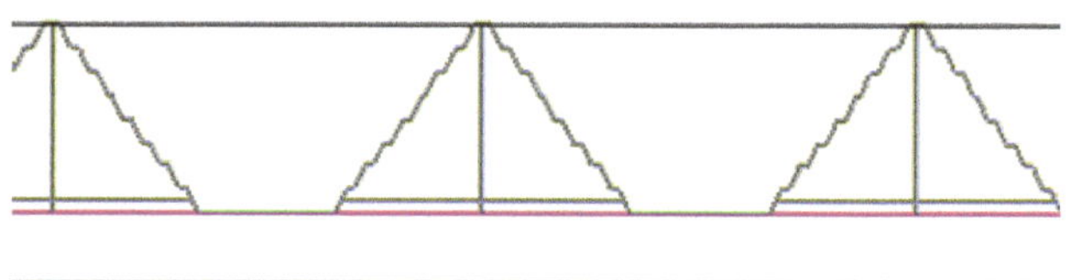

Figure 5. The optimal structural shape obtained using the GA.

Table 1. Parameter combinations with intervals of 0.03 mm and respective uniformity values.

Width	Height	Angle	Interval	Uniformity (%)	Generation
0.858	0.265	79.131	0.030	92.01	G34
0.061	0.038	80.520	0.030	92.02	G50
0.061	0.038	85.367	0.030	92.03	G59
0.061	0.038	85.168	0.030	92.04	G74
0.061	0.038	66.625	0.030	92.07	G69
0.061	0.019	65.000	0.030	92.09	G42
0.122	0.038	65.000	0.030	92.09	G43
0.061	0.038	72.855	0.030	92.09	G69
0.061	0.031	65.000	0.030	92.12	G65
0.061	0.038	87.724	0.030	92.12	G71
0.137	**0.038**	**65.000**	**0.030**	**92.13**	**G62**
0.061	0.021	65.000	0.030	92.35	G55
0.061	0.038	74.936	0.030	92.35	G62
0.106	0.038	65.000	0.030	92.35	G62
0.061	**0.038**	**84.619**	**0.030**	**92.38**	**G63**
0.636	0.300	65.000	0.030	92.76	G39
0.061	0.021	65.000	0.030	92.81	G72
0.083	0.038	65.000	0.030	92.84	G58
0.061	0.026	65.000	0.030	92.85	G41
0.064	0.038	65.000	0.030	92.86	G99
0.161	**0.096**	**65.000**	**0.030**	**92.92**	**G45**
0.090	0.038	65.000	0.030	93.27	G53
0.061	0.064	65.000	0.030	93.51	G51
0.136	0.038	65.000	0.030	93.63	G67
0.061	0.035	65.000	0.030	93.88	G42
0.061	0.040	65.000	0.030	94.59	G33

nance at larger view angles, ranging between 180 nit and 190 nit. The luminance view angle was measured as 80° using the FWHM measurement method.

TYPE 2 comprised stepped structures with larger side angles. The light trace chart of TYPE 2 shapes in **Figure 6** shows that the overall luminance distribution of these structures was generally wider, with weaker luminance in the central view angles, ranging between approximately 150 nit and 160 nit. However, the luminance view angle was wider at approximately 178°. This indicates that the characteristic of TYPE 2 structures is a wider viewing angle from the light source.

TYPE 3 also comprises stepped structures, but these structures have smaller side angles that are generally below 80°. The light trace chart of TYPE 3 structures in **Figure 6** shows that, similar to that of TYPE 2 structures, the light travel angle is slightly enlarged, causing the light to proceed in directions with greater viewing angles. When light travels through the side inclination angles of stepped structures, the law of refraction guides the light toward more central viewing angles, achieving the same light-focusing effect as that of prismatic structures. This causes the view angle luminance distribution chart for TYPE 3 structures to show higher central luminance at between approximately 180 nit and 190 nit. The luminance values for larger viewing angles are significantly higher than the values for prismatic structures with viewing angles of approximately 168°. These comparison results indicate that stepped TYPE 3 structures have the light-focusing characteristic of prismatic structures with the wide-angle characteristic of TYPE 2 stepped structures. This improved the low luminance value in central viewing angles. With the structural parameter combinations derived from the GA, an excellent uniformity of 94.59% was obtained.

3.2.2. Comprehensive Comparisons

Figure 7 is a comprehensive comparison chart of various optical film combinations and viewing angle luminance distributions. The results were used to evaluate the luminance distribution of various types of optical films in various angles. Compared to optical films with the original stepped structure, regardless of whether the original stepped structure had a viewing angle approximately 10° wider than that of the GA structure, the axial luminance of the GA structure was 17.2% higher than that of the original step structure, with a 1.99% increase in uniformity.

Backlight modules generally use light-focusing prismatic structures in brightness-enhancing films. However, uneven brightness occurs at greater viewing angles with extremely steep declines. Therefore, another diffusion film is typically used to equalize the light distribution and reduce the significant decline in luminance. Light smoothing also improves the uniformity of backlight modules. As shown in **Figure 6**, the viewing angle for the GA structure was 67° wider than that of a single brightness enhanced film (BEF) with increased membrane, and shows a significant difference in luminance gap beyond the plus or minus 50° viewing angels.

Regarding uniformity, comparing the uniformity results obtained using the GA and the addition of diffusion film to prismatic sheets showed that the 94.59% of the GA structure was the optimum value. This uniformity value was also 2.52% higher than that of adding a diffusion film to a single BEF, and 3.48% higher than that of adding a diffusion film to dual BEFs.

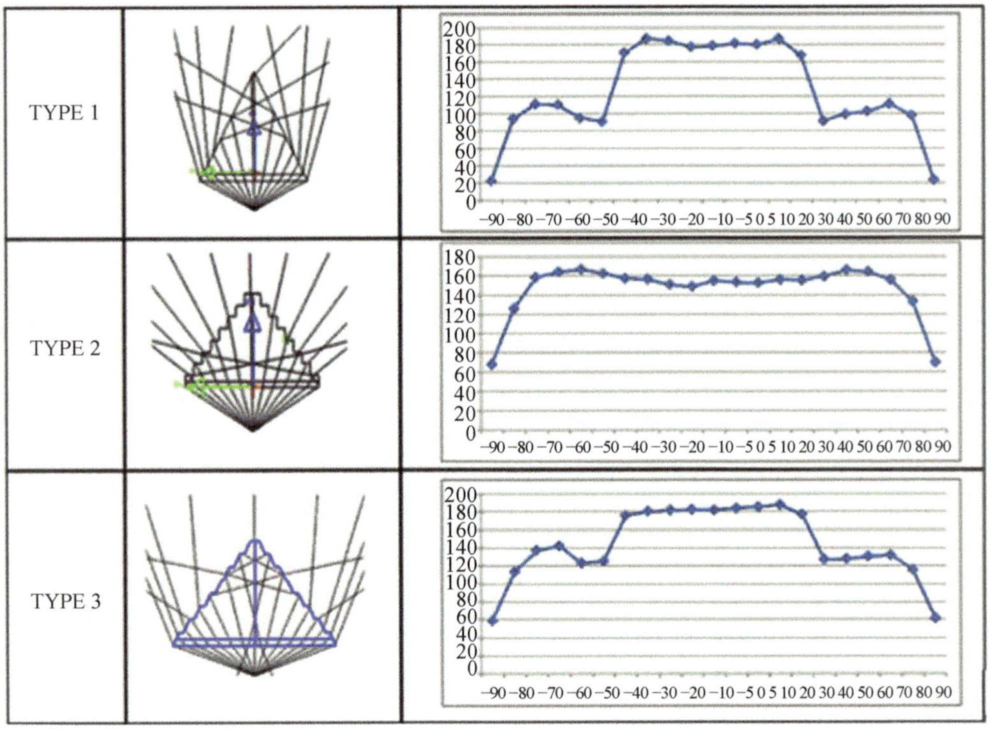

Figure 6. Comparison of the light trace charts and viewing angle luminance distribution curves for each type of structural form.

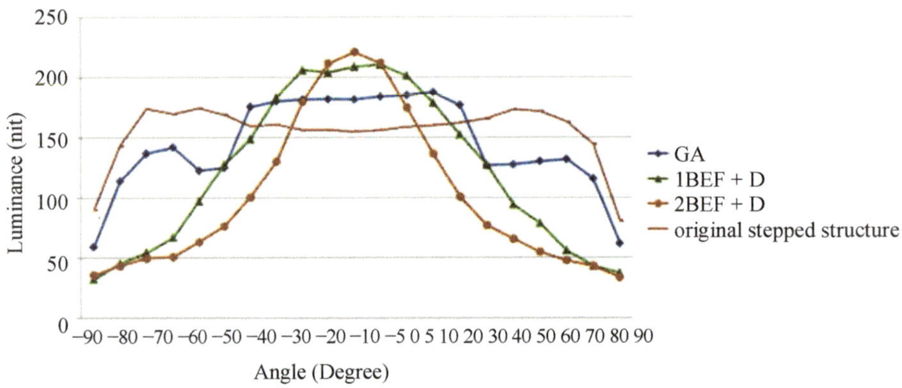

Figure 7. The viewing angle luminance distribution curve for various groups.

These comparisons show that the GA structures provide performances that are comparable or even superior to those of commercially available optical films. Regarding the number of optical films used, the GA structure optical films achieved the optimal performance with a single layer. Compared to the 100 μm to 200 μm thickness and high cost of diffusion films commonly used in the market, the height of the GA structure was merely 41 μm. The prismatic sheet plus diffusion film measured 123 μm to 223 μm in thickness. The GA structure reduced the thickness by 66% to 82%. Thus, the film thickness and costs were substantially reduced, and various issues associated with stacking multiple optical films together were also avoided.

4. Conclusion

In this study, we proposed the idea of combining geometric shapes to design new compound function optical microstructures. Real-valued coding was used to introduce the GA, which then identified a set of optimized optical microstructures with superior uniformity at 94.59%. Conclusions based on analysis of the modeling results and consolidated data are provided below:

1) The height of the designed structure was 40 μm, which is only 17 μm more than that of the prismatic M BEFII 90/50 structure. Because the addition of diffusion films with a thickness between 100 μm and 200 μm is not required to achieve good uniformity, a good lumi-

nance view angle, and a good axial luminance, the overall thickness is reduced by approximately 66% to 82%. By only requiring the use of a single optical film, the structural alignment and abrasion damage problems involved in stacking multiple optical films are avoided, and the light source use rate is improved through the reduced dissipation of light energy. Thus, the goal of reducing the thickness of the backlight module through the design of a compound optical film is achieved.

2) The influence of the shapes on the angular luminance distribution can be observed by classifying the structural forms. The results indicate that the prismatic structure is extremely effective for focusing light, but a steep angular luminance decline occurs at greater viewing angles. Additionally, the width of the viewing angle is insufficient. Stepped structures with larger side angles are more effective for dispersing light, showing greater viewing angles. However, the brightness at central viewing angles tends to decline. The GA structure was a stepped structure with smaller side angles. When light passes through the higher part of the stepped structure, the prismatic collimation effect was exhibited. The light was dispersed when passed through the wide part. Therefore, this structure can combine these two features to achieve the optimal design of an optical microstructure.

3) After optimization, the GA structure proposed in this study had a luminance viewing angle of approximately 168°. Although it was approximately 10° smaller than that of the original structure, the axial luminance was increased by 17.2%, which resolved the problem of low central luminance in the original structure. The uniformity was also approximately 1.99% higher than that of the original structure.

Comparing the GA structure optimized in this study with the traditional pairing of prismatic sheets and diffusion films, the viewing angle of the GA structure was 67° better than that of the 1BEF + diffuser, and 118° greater than that of the 2BEF + diffuser configuration. Regarding uniformity, the GA structure was superior to both film combinations, increasing uniformity by 2.52% and 3.45%, respectively. The results show that using a single sheet of GA-structured optical film effectively improves the uniformity to provide good luminance viewing angles without requiring diffusion films.

REFERENCES

[1] J. H. Holland, "Adaptive in Natural and Artificial Systems," University of Michigan Press, Ann Arbor, 1975.

[2] C. J. Li, Y. C. Fang and M. C. Cheng, "Study of Optimization of an LCD Light Guide Plate with Neural Network and Genetic Algorithm," *Optics Express*, Vol. 17, No. 12, 2009, pp. 10177-10188.

[3] N. Yamada and T. Nishikawa, "Evolutionary Algorithm for Optimization of Non-Imaging Fresnel Lens Geometry," *Optics Express*, Vol. 18, No. S2, 2010, pp. A126-A132.

[4] Y. H. Fan, Y. L. Liao, R. J. Chen, S. K. Chang and J. C. T. Hsieh, "Optimal Design of LED Backlight Unit with Double-Triangular Structure Using a Genetic Algorithm," *Fourth International Conference on Natural Computation*, Jinan, 25-27 August 2008, pp. 625-629.

[5] Y. S. Kim, A. S. Choi and J. W. Jeong, "Applying Micro Genetic Algorithm to Numerical Model for Luminous Intensity Distribution of Planar Prism LED Luminaire," *Optics Communications*, Vol. 293, 2013, pp. 22-30.

[6] P. Y. Tsai, C. C. Hsieh and Y. H. Li, "Application of a Novel Focused Microlens to Light-Emitting Diode Packaging," *Journal of Micro/Nanolithography, MEMS, and MOEMS*, Vol. 9, No. 3, 2010, Article ID: 033012.

[7] C. C. Hsieh, P. Y. Tsai, Y. H. Li and C. W. Fan, "Investigation of the Influence of Structural Parameters of a Fresnel-Type Lens on the Narrowing of Light-Emitting Diode Light Beams," *Journal of Microlithography, Microfabrication, and Microsystem*, Vol. 10, No. 3, 2011, pp. 1-9.

[8] C. C. Hsieh, C. K. Yen and C. Y. Wang, "Composite Optical Film," Patent M330483, 2008.

[9] P. Downen, "A Closer Look at Flat-Panel-Display Measurement Standards and Trends," *Information Display*, Vol. 22, 2006, pp. 16-21.

[10] S. H. Baik, S. K. Hwang, Y. G. Kim, G. Park, J. H. Kwon, W. T. Moon, S. H. Kim, B. K. Kim and S. H. Kang, "Simulation and Fabrication of the Cone Sheet for LCD Backlight Application," *Journal of the Optical Society of Korea*, Vol. 13, No. 4, 2009, pp. 478-483.

7

Study on Interface Friction Model for Engineering Materials Testing on Split Hopkinson Pressure Bar Tests

Yubin Lu[1*], Songyan Zhang[2]
[1]Key Laboratory of Testing Technology for Manufacturing Process (Ministry of Education), Southwest University of Science and Technology, Mianyang, China
[2]Student Office of School of Arts and Literature, Southwest University of Science and Technology, Mianyang, China

ABSTRACT

Split Hopkinson pressure bar (SHPB) has become a frequently used technique to measure the uniaxial compressive stress-strain relation of various engineering materials at high strain-rates. The accuracy of an SHPB test is based on the assumption of uniaxial and uniform stress distribution within the specimen, which, however, is not always satisfied in an actual SHPB test due to the existence of some unavoidable negative factors, e.g., interface friction constrains. Kinetic interface friction tests based on a simple device for engineering materials testing on SHPB tests are performed. A kinetic interface friction model is proposed and validated by implementing it into a numerical model. It shows that the proposed simple device is sufficient to obtain kinetic interface friction results for common SHPB tests. The kinetic friction model should be used instead of the frequently used constant friction model for more accurate numerical simulation of SHPB tests.

Keywords: Kinetic Interface Friction Model; Sliding Velocity; Contact; Split Hopkinson Pressure Bar (SHPB); Engineering Materials

1. Introduction

Split Hopkinson pressure bar (SHPB) technique, as shown in **Figure 1**, is commonly used to determine the strain-rate dependence of the dynamic compressive properties for engineering materials such as metals, concrete-like materials and polymers. Due to the significant importance and great success of SHPB in the dynamic testing of mechanical properties of engineering materials, SHPB has been named by American Society of Mechanical Engineers (ASME) as a historical engineering landmark [1]. It was estimated from the most comprehensive engineering science database, Engineering Village, that more than 2500 papers have been published on SHPB related researches since 1949. It was shown by both Engineering Village and Science Citation Index that the number of publications on SHPB has increased dramatically since 1980s, as shown in **Figure 2**. But it is expected that a valid SHPB test should give the uniaxial stress-strain curve at the planned strain-rate, which imposes requirements of uniaxial and uniform stress distribution in the SHPB specimen while the strain-rate variations in the specimen during the effective range of an SHPB test are small. However, these requirements are not satisfied in some SHPB tests, and therefore, the strain-rate effects due to the non-strain-rate factors cannot be considered as genuine strain-rate effects in an SHPB test.

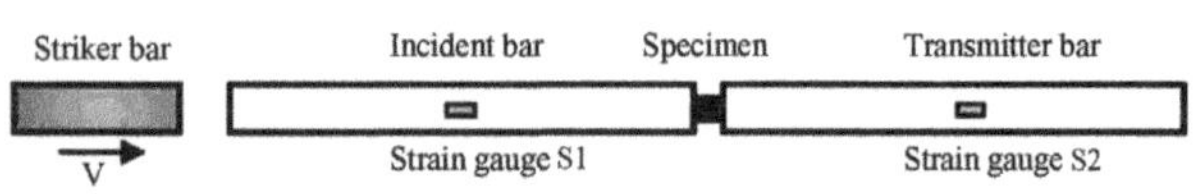

Figure 1. The schematic of SHPB technique.

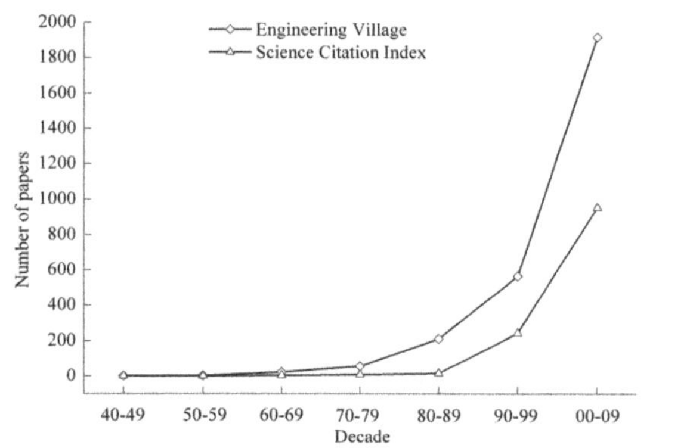

Figure 2. Estimation of the total number of papers published in each decade between 1940 and 2009 where an SHPB test was used to obtain the high strain-rate mechanical properties of materials.

Friction between specimens and anvils has long been

*Corresponding author.

recognized as a problem in compression testing, both at low [2] and high strain-rates of deformation [3,4]. It has three major effects, *i.e.*, 1) it generates a shear stress at the specimen/anvil interface and so changes the stress state in the specimen from uniaxial to triaxial [5] which leads to the measured stress being higher than the true strength (yield stress) of the material [6]; 2) the specimen does not preserve its original geometry but 'barrels' [5,7-10]; 3) localization of deformation within the specimen is promoted due to frictional locking of the surface resulting in truncated cones of non-deforming material which slide over the unconstrained material at the sides [5,11,12]. These shear bands are often precursors of fracture [13]. Accordingly, interfacial friction constrains between SHPB specimen and pressure bars are one of the non-strain-rate factors that may significantly affect the SHPB results by violating the uniaxial stress state through boundary constrains [3,14-19].

Thus, constructing a proper interface friction model is vital for the accurate determination of material properties. Over the years two main methods have been developed for the study of friction relevant to compression testing, *i.e.*, 1) the ring test, e.g. [2,3] and 2) the aspect ratio test where cylinders with a variety of different height/diameter ratios are used. Of the two, probably the ring test is the most used with at least 160 published studies. The Avitzur analysis [20-26] has been checked using high-speed photography [27-29]. In this study, kinetic friction coefficients for various engineering materials sliding against stainless steel measured from a simple slip-way set-up are presented, and a kinetic interface friction model for engineering materials testing on SHPB tests is proposed in Section 1. The kinetic friction model is used to interpret and correct the compression stress-strain curves obtained from SHPB tests in Section 2 followed by discussion in Section 3. Conclusions are drawn in Section 4.

2. A Kinetic Interface Friction Model in SHPB Tests

The Coulomb friction model is widely used in numerical simulations of contact problems

$$\sigma_t = \mu \sigma_n \tag{1}$$

in which σ_t is the tangential friction force per unit area, σ_n is the pressure in normal direction of the contact area, and μ is the coefficient of friction, which has been taken as a constant in many previous numerical simulations of SHPB tests (e.g. [14,19]). Researchers like Siebel [30], Briscoe and Nosker [31], Malinowski and Klepaczko [15] and Wang *et al.* [32] have also proposed theoretic expressions for the influence of the interface friction on the dynamic compressive behavior of engineering materials in SHPB tests. However, they also used a constant friction coefficient.

In fact, in an SHPB test, the relative velocity (sliding velocity) between the end surfaces of the SHPB specimen and the pressure bars varies continuously at different time and different position of the surface, which causes the variation of the friction coefficient since it is dependent of the relative velocity between two contacting surfaces. The kinetic friction coefficient is normally smaller than the static friction coefficient (for some macromolecular materials, e.g. Polyurethane, their kinetic friction coefficient is larger than the static one [33], which will not be studied in this paper). In an SHPB test, when assuming the specimen to be incompressible, the maximum relative radial velocity on the interfaces between the SHPB specimen and the end surfaces of the pressure bars can be estimated by [34]

$$V_r = \frac{d_0 \dot{\varepsilon}_z}{4(1-\varepsilon_z)} \tag{2}$$

in which d_0 is the original outer diameter of the specimen, $\dot{\varepsilon}_z$ is the engineering axial strain-rate and ε_z is the engineering axial strain. Thus, for a mortar specimen of $d_0 = 74$ mm, **Figure 3** shows the variation of the maximum relative radial velocity (V_r, at the edge of specimen) versus strain at different strain-rates in SHPB tests. It can be seen that the maximum relative radial velocity between the specimen and the pressure bars varies in the range of 0.62 to 12 m/s for a common SHPB test when measured strain is less than 3.5%.

In this velocity range, the so-called Stribeck curve is suitable to describe the variation of the friction coefficient with the relative velocity [35]. The friction coefficient has its maximum value at $V_r = 0$ (*i.e.* the static friction coefficient, μ_s) and then decreases considerably to its minimum value with the increase of the relative velocity in a small range and followed by a slow increase with the relative velocity. However, details of the relationship between the kinetic friction coefficient and the relative velocity depend upon the degree of surface roughness

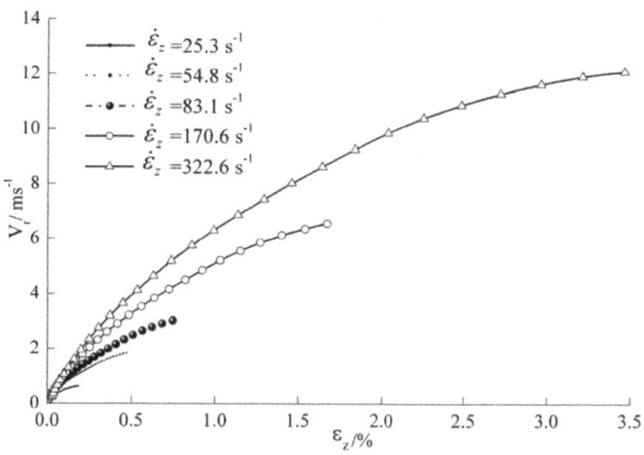

Figure 3. Variation of V_r with engineering axial strain ε_z with different values which are attained under various strain-rates in an SHPB test on mortar specimens.

and lubrication. For example, as the longitudinal pressure is applied, the lubricant will be pressurized, which could squeeze the lubricant out from the interface and/or increase the pore pressure of a thin layer lubricant fluid on the specimen. Such effects could influence the local stress state, and therefore, affect the SHPB measurements. The effects of the hydrodynamics of lubricants in SHPB tests were partly addressed in [36-39], which, however, demand further investigations.

In the 1960's, the American Society for Lubrication Engineers (ASLE) listed more than 200 types of friction tests and equipment in use [40], and the list has since grown. However, within the test configurations described by Benzing [40], the relative velocity can only vary from 2.5 mm/s to 2.54 m/s depending on the apparatus [34]. SHPB techniques were also used to test the friction in high velocity (e.g. [32,41-46]). However, it is difficult to get a single friction-velocity equation over a wide range of relative velocities. Friction measurements under changed relative velocities will be conducted in order to relate the kinetic friction coefficient to the relative velocities for different engineering materials. The experiments here were designed to find the dependence of the kinetic friction coefficient on the relative velocity. Meng [34] used a simple apparatus which consists of a slip-way, a specimen and some measuring devices, as shown in **Figure 4**, for the measurement of the kinetic friction coefficient at a velocity of several meters per second. A highly sensitive capacitive accelerometer on a ceramic base with built-in electronics was used in the test, which can measure acceleration down to 0 Hz. The contact surface was lubricated by Shell Helix Super oil to simulate a similar situation in an SHPB test. The specimen slips from the top of the slip-way. Using the data of the acceleration history, $a(t)$, the kinetic friction coefficient, $\mu_d(t)$, between the specimen and the stain-less steel slip-way and the corresponding $V_r(t)$ history can be obtained using the following formula [34],

$$\mu_d(t) = \tan\theta^* - \frac{a(t)}{g\cos\theta^*}, V_r(t) = \int_0^t a(t) \quad (3)$$

where θ^* is the slope angle of the slip-way, and g is the acceleration of gravity.

A series of tests were performed based on the equipment shown in **Figure 4** to construct the kinetic friction model for engineering materials in common SHPB tests, and the results are shown in **Figure 5**, where the vertical coordinate is the ratio of the kinetic friction coefficient μ_d to the static friction coefficient μ_s, μ_d/μ_s, and the horizontal coordinate is V_r. In a practical numerical analysis, it is necessary to propose a simplified friction model to describe experimental results. Comparing with the Stribeck curve [34], it is reasonable to use an exponential-linear function to fit the testing results, *i.e.*

$$\mu_d/\mu_s = C_1\exp(-V_r/C_2) + C_3V_r + C_4 \quad (4)$$

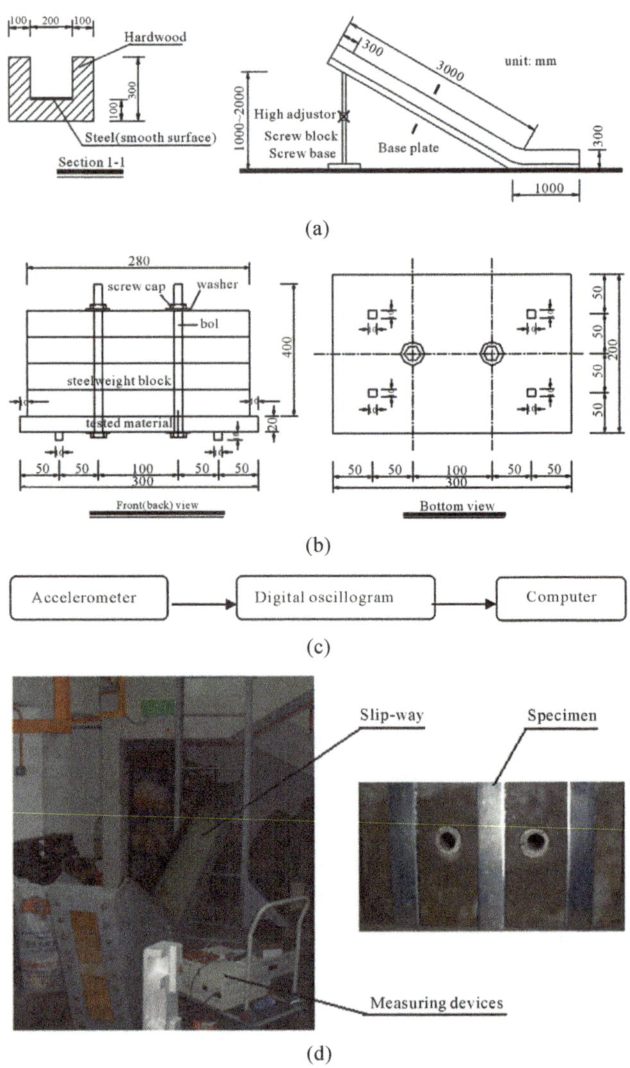

Figure 4. Test apparatus: (a) Slip-way (side view); (b) Specimen; (c) Measuring devices; (d) Photo of the test apparatus [40].

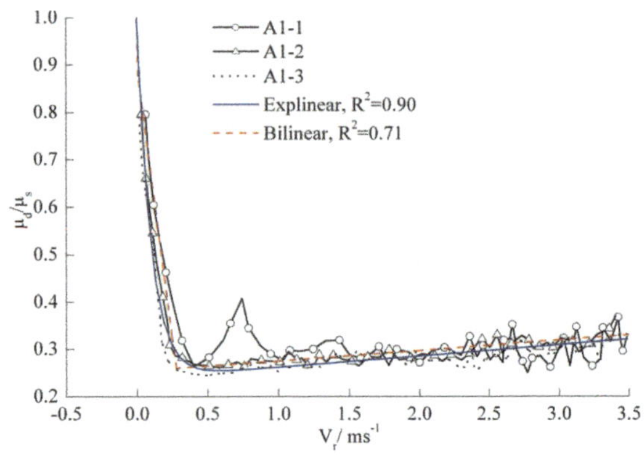

(a) $\sigma_n = 17.4$ **kPa,** $\theta^* = 23.7°$

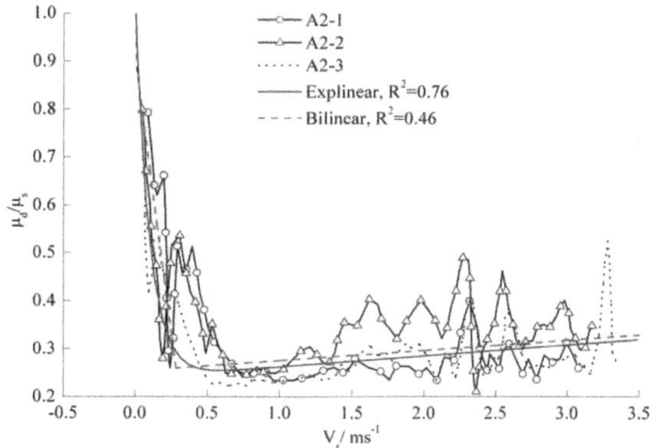

(b) $\sigma_n = 31.1$ kPa, $\theta^* = 23.7°$

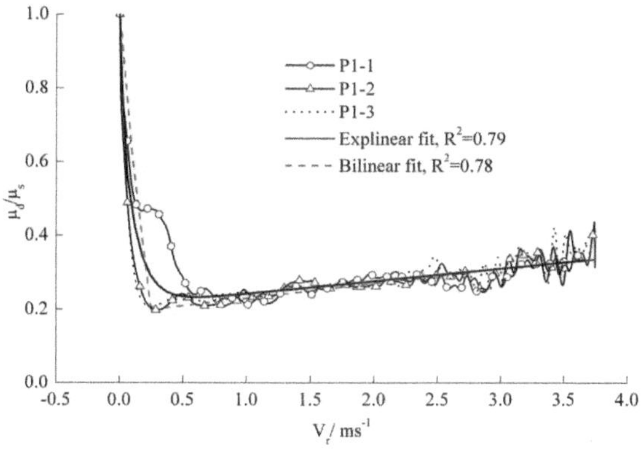

(c) $\sigma_n = 15.4$ kPa, $\theta^* = 23.7°$

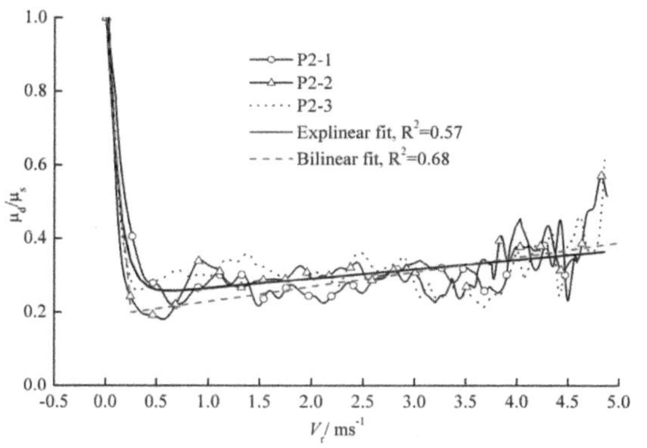

(d) $\sigma_n = 13.2$ kPa, $\theta^* = 38.3°$

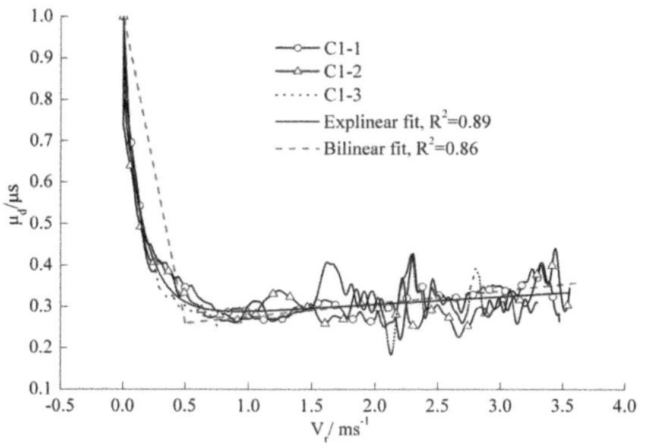

(e) $\sigma_n = 17.7$ kPa, $\theta^* = 23.7°$

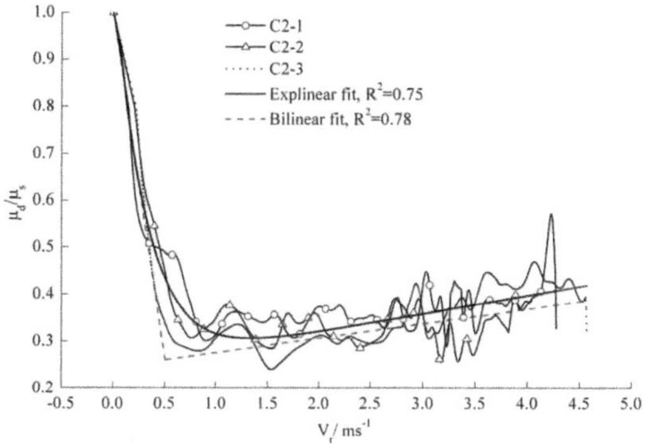

(f) $\sigma_n = 15.2$ kPa, $\theta^* = 38.3°$

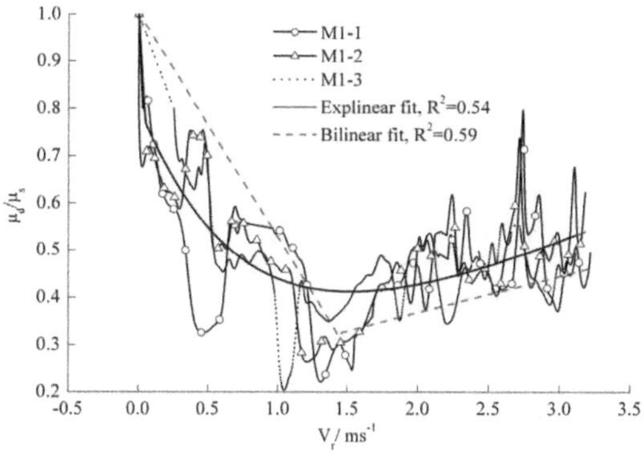

(g) $\sigma_n = 15.5$ kPa, $\theta^* = 23.7°$

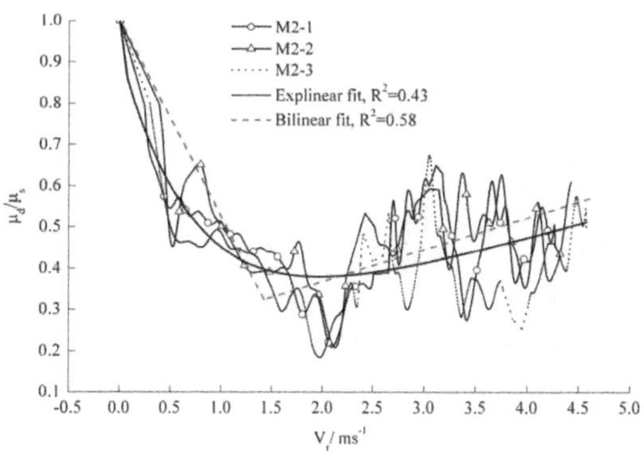

(h) $\sigma_n = 13.3$ kPa, $\theta^* = 38.3°$

Figure 5. Variations of μ_d/μ_s with V_r based on repeat tests for (a) and (b) aluminum-alloy specimens (*i.e.* A1-# and A2-# where #∈1, 2, 3), for (c) and (d) polycarbonate (PC) specimens (*i.e.* P1-# and P2-# where #∈1, 2, 3), for (e) and (f) concrete specimens (*i.e.* C1-# and C2-# where #∈1, 2, 3), and for (g) and (h) mortar specimens (*i.e.* M1-# and M2-# where #∈1, 2, 3), under different normal stresses and slope angles, and fitted curves of these variations.

in which $C_1 - C_4$ are constants obtained from the test results between μ_d/μ_s and V_r, as shown in **Table 1**, and $\mu_d/\mu_s = 1$ when $V_r = 0$.

The coefficient of determination (Adj. R-Square, R^2) between the experimental data and fitted curves obtained from Origin 8.0 is also shown in **Figure 5**. It is observed

Table 1. Constants in Equation (4).

Material	μ_s	C_1	C_2	C_3	C_4
Aluminum-alloy	0.337 [40]	0.761	0.101	0.0233	0.239
PC	0.40 [40]	0.776	0.114	0.0290	0.224
Concrete	0.434 [40]	0.697	0.246	0.255	0.303
Mortar	0.430 [40]	0.720	0.780	0.145	0.280

that under the same slope angle, the value of R^2 will be reduced with the increase of normal pressure; while under the nearly same normal pressure, the value of R^2 will also be reduced with increasing the slope angle. The exponential-linear function is better than the bilinear function used in Ref. [34] to fit the kinetic friction coefficient, as the former can be easily input into numerical simulations. The maximum relative velocity in the present tests (up to 4.6 m/s) covers the relative velocity range in typical SHPB tests of engineering materials studied in this paper because Equation (2) is based on the incompressibility of the specimen and the neglect of Poisson's effect in the pressure bar material. When the elastic deformation of an SHPB specimen and the lateral movement of the pressure bars are considered, the relative velocity between the specimen faces and the pressure bars is smaller than that estimated by Equation (2).

3. Discussion

To study the friction effect on the SHPB results using the constant friction model and the proposed kinetic friction model, taking mortar as an example, four cases were simulated with different values of μ based on the constitutive equation and numerical SHPB model of mortar as described in Ref. [18], *i.e.* 1) $\mu = 0$, which corresponds to the frictionless case; 2) $\mu = 0.163$, corresponding to $\mu_d/\mu_s = 0.38$, which is the minimum value of Equation (4) for mortar; 3) $\mu = \mu_d$ where μ_d is given by Equation (4), which is implemented into ABAQUS using a user subroutine VFRIC; 4) $\mu = \mu_s = 0.43$ for mortar, which is the static friction coefficient. The simulation results under the strain-rate of 170 s^{-1} corresponding to the peak strength of mortar specimens with the ratio of length to diameter $\lambda_s = 0.5$ and $d_0 = 74$ mm, are shown in **Figure 6**. It is observed that the maximum difference among cases 1-3 is about 15%. However, a large difference between case 4 and three other cases is observed. A constant friction coefficient of zero and the kinetic friction model are applied in the finite element analysis of SHPB tests for mortar specimens of $d_0 = 74$ mm and $\lambda_s = 0.5$. Comparison results of friction contribution to the dynamic compressive strength enhancement are shown in **Figure 7**. In **Figure 7**, the vertical coordinate is the dynamic increase factor (DIF), defined as the ratio of dynamic compressive strength under the

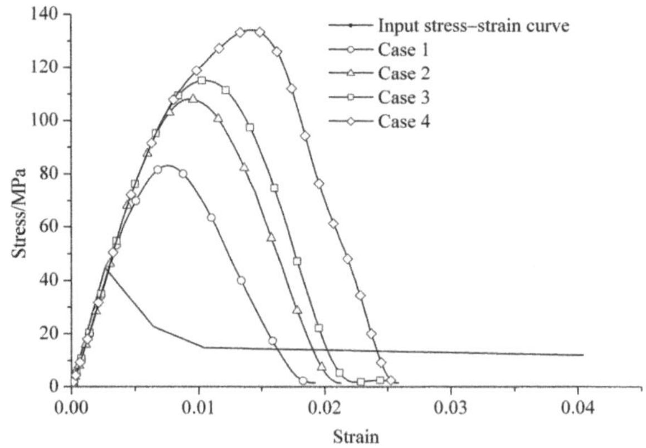

Figure 6. Comparison of calculated results for different friction models at the strain-rate of 170 s^{-1} corresponding to the peak strength of mortar specimens with $d_0 = 74$ mm and $\lambda_s = 0.5$.

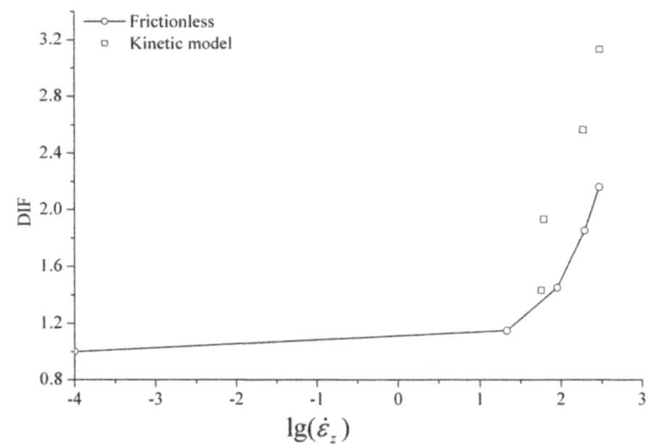

Figure 7. Contribution of friction to the dependence of the dynamic increase factor (DIF) on strain-rate in the numeric SHPB tests when using different friction models for mortar specimens with $d_0 = 74$ mm and $\lambda_s = 0.5$.

testing strain-rate to quasi-static compressive strength, which represents the dynamic compressive strength enhancement of mortar specimens testing on SHPB set-up. It is observed from **Figure** 7 that DIFs obtained from numerical SHPB simulations when using the kinetic friction model are greater than those predicted based on the frictionless assumption in the strain-rate range of 10^1 - 10^3 s^{-1}, especially when the strain-rate is greater than the transition strain-rate (the critical strain-rate where strain-rate dependence of DIF starts to increase apparently). There seems to be no remarkable difference in the transition strain-rate between the frictionless model and kinetic model. The end friction effect described by the kinetic friction model induces higher hydrostatic pressure on the mortar specimens comparing with not considering the end friction effect (frictionless), and higher dynamic compressive strength of mortar specimens will be obtained from numerical SHPB simulations as the mechanical behavior of mortar is hydrostatic-pressure-dependent.

Therefore, the DIF predicted from the kinetic friction model is higher than that from the frictionless model at the same strain-rate. Thus, a more reliable assessment and correction procedure for SHPB tests should be based on the kinetic friction model whenever the results of kinetic friction coefficient versus the relative velocity are available.

4. Conclusion

Friction on the bar/specimen interfaces is one of the possible error sources of the SHPB tests and should be examined in the assessment and correction procedure for SHPB results through numerical analyses. The simple device is capable of providing the kinetic interface friction test results for common SHPB tests. Based on the numerical analyses which implement the proposed kinetic interface friction model, it is found that the difference between the constant friction model and kinetic interface friction model may cause different results. Thus, whenever it is possible, a more reliable assessment and correction procedure for the SHPB test should be based on a kinetic friction model.

5. Acknowledgements

We acknowledge the support by Scientific Research Fund of Sichuan Provincial Education Department (10zd1012), and the Doctorate Foundation of Southwest University of Science and Technology (10zx7141).

REFERENCES

[1] ASME, "#242 Split-Hopkinson Pressure Bar Apparatus," 2010. http://www.asme.org/Communities/History/Landmarks/241_SwRI_SplitHopkinson.cfm

[2] P. Hartley, C. E. N. Sturgess and G. W. Rowe, "Influence of Friction on the Prediction of Forces, Pressure Distributions and Properties in Upset Forging," *International Journal of Mechanical Sciences*, Vol. 22, No. 12, 1980, pp. 743-753.

[3] J. R. Klepaczko and Z. Malinowski, "Dynamic Frictional Effects as Measured from the Split Hopkinson Pressure Bar," In: K. Kawata and J. Shioiri, Eds., *High Velocity Deformation of Solids*, Springer-Verlag, Berlin, 1978, pp. 403-416.

[4] D. A. Gorham, P. H. Pope and O. Cox, "Sources of Error in Very High Strain Rate Compression Tests," In: J. Harding, Ed., *Mechanical Properties at High Rates of Strain*, Institution of Physics Conference Series, Oxford, 1984, No. 70, pp. 151-158.

[5] J. Mescall, R. Papirno and J. Mclaughlin, "Stress and Deformation States Associated with Upset Tests in Metals," In: R. Chait and R. Papirno, Eds., *Compression Testing of Homogeneous Materials and Composite*, American Society for Testing and Materials, Philadelphia, 1983, pp. 7-23.

[6] J. A. Schey, T. R. Venner and S. L. Takomana, "The Effect of Friction on Pressure in Upsetting at Low Diameter-to-Height Ratios," *Journal of Mechanical Working Technology*, Vol. 6, No. 1, 1982, pp. 23-33.

[7] J. K. Bannerjee, "Barreling of Solid Cylinders under Axial Compression," *Journal of Engineering Materials and Technology*, Vol. 107, No. 2, 1985, pp. 138-144.

[8] J. K. Bannerjee and G. Cárdenas, "Numerical Analysis on the Barreling of Solid Cylinders under Axisymmetric Compression," *Journal of Engineering Materials and Technology*, Vol. 107, No. 2, 1985, pp. 145-147.

[9] R. Narayanasamy and K. S. Pandey, "Phenomenon of Barrelling in Aluminium Solid Cylinders during Cold Upset-Forming," *Journal of Materials Processing Technology*, Vol. 70, No. 1-3, 1997, pp. 17-21.

[10] E. Parteder and R. Bünten, "Determination of Flow Curves by Means of a Compression Test under Sticking Friction Conditions Using an Iterative Finite Element Procedure," *Journal of Materials Processing Technology*, Vol. 74, No. 1-3, 1998, pp. 227-233.

[11] W. Ohnson, G. L. Baraya and R. A. C. Slater, "On Heat Lines or Lines of Thermal Discontinuity," *International Journal of Mechanical Sciences*, Vol. 6, No. 6, 1964, pp. 409-414.

[12] Y. L. Bai and B. Dodd, "Adiabatic Shear Localization: Occurrence, Theories and Applications," 1st Edition, Pergamon, Oxford, 1992.

[13] A. Jenner, Y. L. Bai and B. Dodd, "A Thermo-Plastic Shear Instability Criterion Applied to Surface Cracking in Upsetting and Related Processes," *The Journal of Strain Analysis for Engineering Design*, Vol. 16, No. 3, 1981, pp. 159-164.

[14] L. D. Bertholf and C. H. Karnes, "Two Dimensional Analysis of the Split Hopkinson Pressure Bar System," *Journal of the Mechanics and Physics of Solids*, Vol. 23, No. 1, 1975, pp. 1-19.

[15] J. Z. Malinowski and J. R. Klepaczko, "A Unified Analytic and Numerical Approach to Specimen Behaviour in the Split-Hopkinson Pressure Bar," *International Journal of Mechanical Sciences*, Vol. 28, No. 6, 1986, pp. 381-391.

[16] D. A. Gorham, "The Effect of Specimen Dimensions on High Strain Rate Compression Measurements of Copper," *Journal of Physics D: Applied Physics*, Vol. 24, No. 8, 1991, pp. 1489-1492.

[17] Q. M. Li and H. Meng, "About the Dynamic Strength Enhancement of Concrete-Like Materials in a Split Hopkinson Pressure Bar Test," *International Journal of Solids and Structures*, Vol. 40, No. 2, 2003, pp. 343-360.

[18] Q. M. Li, Y. B. Lu and H. Meng, "Further Investigation on the Dynamic Compressive Strength Enhancement of Concrete-Like Materials Based on Split Hopkinson Pres-

sure Bar Tests, Part II: Numerical Simulations," *International Journal of Impact Engineering*, Vol. 36, No. 12, 2009, pp. 1335-1345.

[19] H. Meng and Q. M. Li, "Correction between the Accuracy of a SHPB Test and the Stress Uniformity Based on Numerical Experiments," *International Journal of Impact Engineering*, Vol. 28, No. 5, 2003, pp. 537-555.

[20] B. Avitzur, "Forging of Hollow Discs," *Israel Journal of Technology*, Vol. 2, No. 3, 1964, pp. 295-304.

[21] B. Avitzur, "Metal Forming: Processes and Analysis," McGraw-Hill, New York, 1968.

[22] B. Avitzur and F. Sauerwine, "Limit Analysis of Hollow Disk Forging. Part 1: Upper Bound," *Journal of Engineering for Industry*, Vol. 100, No. 3, 1978, pp. 340-344.

[23] B. Avitzur and C. J. Van Tyne, "Ring Forming: An Upper Bound Approach. Part 2: Process Analysis and Characteristics," *Journal of Engineering for Industry*, Vol. 104, No. 3, 1982, pp. 238-247.

[24] B. Avitzur and C. J. Van Tyne, "Ring Forming: An Upper Bound Approach. Part 3: Constrained Forging and Deep Drawing Applications," *Journal of Engineering for Industry*, Vol. 104, No. 3, 1982, pp. 248-252.

[25] B. Avitzur and R. A. Kohser, "Disk and Strip Forging for the Determination of Friction and Flow Strength Values," *Tribology Transactions*, Vol. 21, No. 2, 1978, pp. 143-151.

[26] B. Avitzur and C. J. Van Tyne, "Ring Forming: An Upper Bound Approach. Part 1: Flow Pattern and Calculation of Power," *Journal of Engineering for Industry*, Vol. 104, No. 3, 1982, pp. 231-237.

[27] S. M. Walley, J. E. Field, P. H. Pope and N. A. Safford, "A Study of the Rapid Deformation Behaviour of a Range of Polymers," *Philosophical Transactions of Royal Society A*, Vol. 328, No. 1597, 1989, pp. 1-33.

[28] S. M. Walley, J. E. Field, P. H. Pope and N. A. Safford, "The Rapid Deformation Behaviour of Various Polymers," *Journal de Physique III France*, Vol. 1, No. 12, 1991, pp. 1889-1925.

[29] S. M. Walley, P. D. Church, M. Furth and J. E. Field, "A High-Speed Photographic Study of the Rapid Deformation of Metal Annuli: A Comparison of Theory with Experiment," *Journal de Physique IV France*, Vol. 7, No. C3, 1997, pp. 317-322.

[30] E. Siebel, "Fundamentals for the Calculation of Power and Working Requirements in Forging and Rolling-Mill Practice (Grundlagen zur Berechnung des Kraft- und Arbeitsbedarf beim Schmieden und Walzen)," *Stahl und Eisen*, Vol. 43, 1923, pp. 1295-1298. (in German)

[31] B. J. Briscoe and R. W. Nosker, "The Influence of Interfacial Friction on the Deformation of High Density Polyethylene in a Split Hopkinson Pressure Bar," *Wear*, Vol. 95, No. 3, 1984, pp. 241-262.

[32] X. Y. Wang, F. Y. Lu and Y. L. Lin, "Study on Interfacial Friction Effect in the SHPB Tests," *Explosion and Shock Waves*, Vol. 26, No. 2, 2006, pp. 134-139. (in Chinese)

[33] F. Y. Lin, D. S. Zhang and X. D. Ma, "Study on the Friction Characteristics of Polyurethane for the Friction Linear of Hoisting," *Lubrication Engineering*, Vol. 2, No. 33-34, 2000, p. 60. (in Chinese)

[34] H. Meng, "Numerical Split Hopkinson Pressure Bar (NSHPB) Test and Its Applications in the Assessment and Improvement of SHPB Test Results," Ph.D. Thesis, Nanyang Technological University, Singapore, 2002.

[35] D. Pavelescu and A. Tudor, "The Sliding Friction Coefficient—Its Evolution and Usefulness," *Wear*, Vol. 120, No. 3, 1987, pp. 321-336.

[36] G. I. Taylor, "Plastic Strain in Metals," *Journal of the Institute of Metals*, Vol. 62, 1938, pp. 307-324.

[37] J. A. C. Martins, J. T. Oden and F. M. F. Simoes, "A Study of Static and Kinetic Friction," *International Journal of Engineering Sciences*, Vol. 28, No. 1, 1990, pp. 29-92.

[38] B. A. Hélouvry, P. Dupont and C. C. D. Wit, "A Survey of Models, Analysis Tools and Compensation Methods for the Control of Machines with Friction," *Automatica*, Vol. 30, No. 7, 1994, pp. 1083-1138.

[39] D. P. Hess and A. Soom, "Friction at a Lubricated Line Contact Operating at Oscillating Sliding Velocities," *Journal of Tribology*, Vol. 112, No. 1, 1990, pp. 147-152.

[40] R. Benzing, I. Goldblatt and V. Hopkins, *et al.*, "Friction and Wear Devices," 2nd Edition, ASLE, IL Park Ridge, 1976.

[41] K. Ogawa, "Impact Friction Test Method by Applying Stress Wave," *Experimental Mechanics*, Vol. 37, No. 4, 1997, pp. 398-402.

[42] V. Prakash, "Time Resolved Friction with Applications to High Speed Machining," *Tribology Transactions*, Vol. 41, No. 2, 1998, pp. 189-198.

[43] S. Rajagopalan and V. Prakash, "A Modified Torsional Kolsky Bar for Investigating Dynamic Friction," *Experimental Mechanics*, Vol. 39, No. 4, 1999, pp. 295-303.

[44] H. Zhang, A. Patanella and H. D. Espinosa, *et al.*, "Dynamic Friction of Nano-Materials," *AIP Conference Proceedings*, Vol. 505, No. 1, 2000, pp. 1225-1228.

[45] H. D. Espinosa, A. Patanella and M. Fischer, "A Novel Dynamic Friction Experiment Using a Modified Kolsky Bar Apparatus," *Experimental Mechanics*, Vol. 40, No. 2, 2000, pp. 138-153.

[46] Y. L. Lin, F. Y. Lu and Y. X. Cui, "Testing of Friction Coefficients of Material Loaded by Shock Wave," *Tribology*, Vol. 27, No. 1, 2007, pp. 64-67. (in Chinese)

8

The Study of Contact Pressure Analyses and Prediction of Dynamic Fatigue Life for Linear Guideways System

Thin-Lin Horng
Department of Mechanical Engineering, Kun-Shan University, Tainan, Chinese Taipei

ABSTRACT

The application of the linear guideways is very extensive, such as automation equipment, heavy-duty carry equipment, heavy-cut machining tool, CNC grinding machine, large-scale planning machine and machining center with the demand of high rigidity and heavy load. By means of the study of contact behavior between the roller/guideway and roller/slider, roller type linear guideways can improve the machining accuracy. The goal of this paper is to construct the fatigue life model of the linear guideway, with the help of the contact mechanics of rollers. In beginning, the analyses of the rigidity of a single roller compressed between guideway and slider was conducted. Then, the normal contact pressure of linear guideways was obtained by using the superposition method, and verified by the FEM software (ANSYS workbench). Finally, the bearing life theory proposed by Lundberg and Palmgren was used to describe the contact fatigue life.

Keywords: Linear Roller Guideways; System Analysis; System Stress Analyses; Prediction of Fatigue Life

1. Introduction

The linear guideway type (LGT) recirculating rollers shown in **Figure 1** has many advantages compared to ball guideways [1] and conventional sliding guides, such as flat ways and V-ways [2]. For instance, the ultimate loading of the roller guideway can be larger than that of a ball guideway, making the lubrication more efficient so that the abrasion of linear motion roller bearings is less than that of sliding guides. Linear motion roller bearings also have no stick slip. Linear motion roller guideways are widely used instead of ball guideways and sliding guides in heavy CNC machining centers, grinding machines, precision heavy-duty X-Y tables, and TFT-LCD transport systems [3]. In recent years, the speed of machines using linear motion rolling bearings has increased. Kasai *et al.* [4,5] found that carriages of preloaded recirculating linear roller bearings were moved periodically with the roller passing frequency due to roller circulation. Ye *et al.* [6] carried out a modal analysis for carriages of linear guideway type (LGT) recirculating linear roller bearings under stationary conditions and pointed out the existence of the rigid-body natural vibrations of the carriage. Schneider [7] developed a theory for the natural vibrations of an LGT recirculating linear ball bearing with a 45 degree contact angle. The study of prediction of dynamic fatigue life for linear guideways system, is not developed in the mentioned studies.

Horng [8-10] demonstrated stiffness equation and stress components for the circularly crowned roller compressed between two plates. Moreover, based on the previous results, Horng [11,12] developed stiffness expression, system stress distribution and vibration behavior for the linear guideway type recirculating rollers. In practice, the fatigue life of rollers composed in guideways plays a key role in the study of working life of linear guideways. Basically, two kinds of contact fatigue models have been developed over the past half century, *i.e.*, crack initiation model and crack propagation model. Before 1980, almost all contact fatigue models had been developed by assuming the fatigue life was dominated by crack initiation.

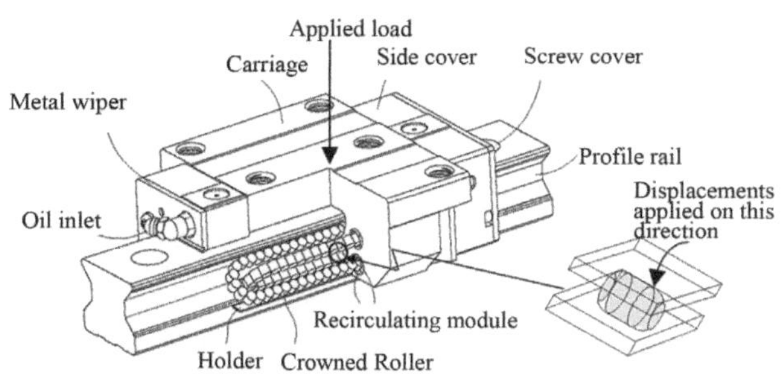

Figure 1. Configuration of a linear guideway type (LGT) recirculating roller bearing [1].

The most famous model was the Lundberg-Palmgren model [13]. The second kind of models is developed by assuming the fatigue life was dominated by crack propagation. Most crack propagation models follow a Paris law [14], which was developed from tests of structural materials. The represented models are the Keer-Bryant model [15] and the Tallian model [16]. In contact fatigue life calculations, different critical stresses have been used. For example, Lundberg and Palmgren used maximum orthogonal shear stress in their model. Some researchers found cracks initiated at the depth where maximum shear stress occurs, which prompted them to use maximum shear stress to calculate the fatigue life [17]. Popinceanu [18] recommended the use of von Mises equivalent stress as the critical stress. Zhou [19] used octahedral shear stress to develop his new contact fatigue model. Cheng [20] found that the fatigue life of roller bearings under the pure rolling condition can be predicted by simply knowing the Hertzian contact pressure and the contact width, which avoids complicated calculation of the subsurface stresses. Previous models derived from Hertizan contact mechanics theory ignored the role played by the reliability of precious calculation in contact deformation and stresses. However, Palmgren did not have confidence in the ability of the Hertzian equations to accurately predict rolling bearing stresses. Palmgren states, "The calculation of deformation and stresses upon contact between the curved surfaces is based on a number of simplifying stipulations which will not yield very accurate approximation values. For line contact the limit of validity of the theory is exceeded whenever edge pressure occurs". In other words, to calculate deformation and stresses accurately, especially edge stress occurred, it is essential to the prediction of fatigue.

In practice, the real experimental tests for the guideway were conducted to validate its quality. However, the real experimental tests require more cost and time-consuming. An alternative is to develop an analytical method that evaluates the dynamic fatigue life efficiently and easily. Moreover, the study of system stress analyses and prediction of dynamic fatigue life for linear guideway type (LGT), is not developed in the mentioned studies. In the present study, the analyses of deformation curve and the rigidity of guideway/slider for a single roller compressed between them were conducted. Then, the normal contact pressure of linear guideways was obtained by using the superposition method, and verified by the FEM software (ANSYS workbench). Finally, the bearing life theory proposed by Lundberg and Palmgren [21] is adopted to describe the contact fatigue life.

2. Determination of Normal Contact Pressure

Figure 1 shows the configuration of an LGT recirculating rollers. In order to analyze the normal contact pressure applied on rollers, the stiffness formula for a crowned roller compressed by a carriage and a profile rail is investigated. The model and dimensions are shown in **Figure 2** and the profiles of a crowned roller are shown in **Figure 3**. The crowned roller is divided into three parts, two crowned parts and one cylindrical part, in which the primary axes of the roller system, denoted by x and y, are shown in **Figure 3**. The assumptions of this stiffness formula are: 1) friction is neglected because of the lubrication of the roller; 2) the depth effect of the carriage and rail can be simulated as equivalent depth; 3) the solution is obtained for the small-strain and linear elastic conditions; 4) rollers compressed between a carriage and profile rail are simulated as springs with parallel connection; 5) each discrete normal spring was modeled as a crowned roller compressed between two plates; 6) the normal force is applied to each contact point of the carriage and profile rail; 7) stick and slip effects of the contact zone between rollers and profile are neglected; 8) the preload effect of the linear guideway type (LGT) recirculating rollers is neglected; and 9) the stiffness of recirculating rollers in motion is similar to that of recirculating rollers in static.

According to the results by Horng [10], the normal contact pressure of the half-space within a small piece of

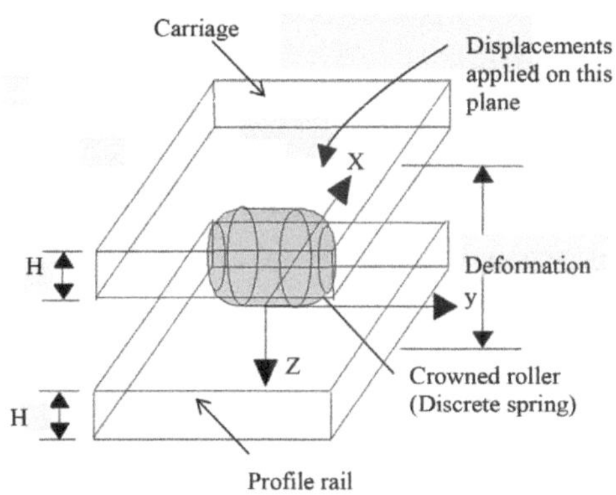

Figure 2. Model and dimensions of a crowned roller compressed between a carriage and profile rail, or one discrete spring.

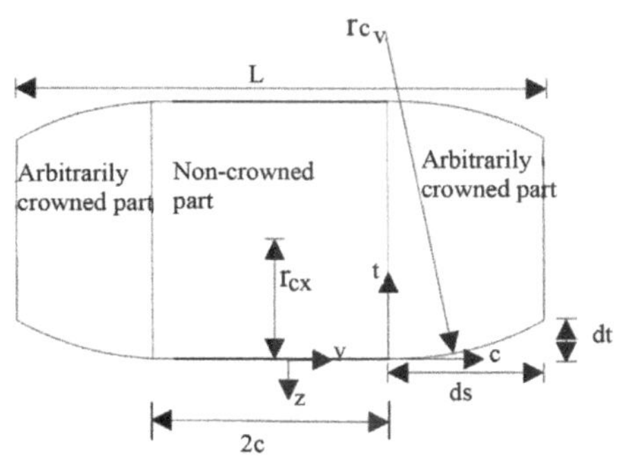

Figure 3. Geometry of arbitrarily crowned profiles.

distance interval along y coordinate demonstrated in **Figure 2**, is given by

$$\bar{p}(y) = -\frac{E(n+1)Bc^{n+1}}{2(1-\upsilon^2)\pi}\frac{I_n}{(c^2-y^2)^{1/2}} + \frac{P/b}{\pi(c^2-y^2)^{1/2}} \quad (1)$$

where E is Young's modulus, v is Poisson's ratio and c is the half length of the non-crowned part. B and n are coefficients of the polynomial. I_n is the principal value, P is the total load and b is the half width of the elliptical contact load.

In order to overcome a defect of the half space theorem in which infinite plate thickness is set, a weighting method is introduced to find the normal pressure under a finite plate thickness. From the comparison between the displacement of finite plate thickness $\bar{u}_z(y)$ and the displacement of the infinite plate thickness $\tilde{u}_z(y)$, it seems that the final normal pressure is equal to multiplying $\bar{p}(y)$, which is calculated in half-space, by a weighting factor $w(y)$. $w(y)$ can be written in empirical form [10]

$$W(y) = e^{\left[12\left(\frac{\tilde{u}(y)-\bar{u}(y)}{\tilde{u}(0)}\right)-9.4\right]} \quad (2)$$

Therefore, the final $p(y)$ measured from the center of contact to the edge can be expressed as:

$$p(y) = w(y) \times \bar{p}(y) \quad (3)$$

The distance $0.97c$ from the center of contact is set for the position which the edge stress occurred [10].

3. Lundberg-Palmgren Theory

The genesis of the Lundberg-Palmgren theory for bearing life prediction traces back to the 1924 Palmgren paper [22]. However, because the 1924 paper was missing two elements, it did not allow for a comprehensive bearing life theory. The first missing element was the ability to calculate *i* he subsurface principal stresses and, hence, the shear stresses below the Hertzian contact of either a ball on a nonconforming race or a cylindrical roller on a race. The second missing element was a comprehensive life theory that would fit the observations of Palmgren. Palmgren, as will be discussed, distrusted Hertz theory and depended on a load-life reation for ball and roller bearings. In 1930, V. Thomasand and H. Hoersch [23] at the University of Illinois, Urbana, developed an analysis for determining subsurface principal stresses under Hertzian contact. In 1939, W. Weibull [24] published his theory of failure. Weibull was a contemporary of Palmgren and shared the results of his work with him. Palmgren, in concert with Lundberg, incorporated his previous work along with that of Weibull and what appears to be that of Thomas and Hoersch into what has become known as the Lundberg-Palmgren theory. Please note that Lundberg and Palmgren do not reference the work of Thomas and Hoersch in their paper. Lundberg and Palmgren, using the theoretical work of Weibull assumed that the logarithm of the reciprocal of the probability of survival S could be expressed as: a power function of maximum orthogonal shear stress τ_0 (Gpa), fatigue life (Number of stress circle) N, depth to the maximum orthogonal shear stress z_0, and stressed volume V. That is.

$$\ln\frac{1}{S} = \frac{\tau_0^c N^e}{z_0^h} V \quad (4)$$

where

$$V = 2blz_0 \quad (5)$$

c is stress criteria exponent, e is Weibull slope, h is depth exponent, z_0 is depth of critical shear stress (mm), b is the half Hertzian contact width (mm) and l is the possible contact length (mm). Then,

$$\ln\frac{1}{S} = \frac{2\tau_0^c N^e bl}{z_0^{h-1}} \quad (6)$$

It is known from Lundberg and Palmgren theory that for the improved line contact condition three parameters e, c, and h are related by the following equation [24]:

$$(c-h+1)/2e = 10/3 \quad (7)$$

A search of the literature for a wide variety of materials and for nonrolling-element fatigue reveals that most stress criteria exponent c vary from 6 to 12. Zaretsky [25] assumes that the value of the stress criteria exponent $c = 9$ and Weibull slope $e = 1.11$. Thus, depth exponent h is equal to 2.61.

Cheng [20] found that the fatigue life of roller bearings under the pure rolling condition can be predicted by simply knowing the Hertzian contact pressure and the contact width, which avoids complicated calculation of the subsurface stresses. Under the conditions of non-frictional Hertzian line contact, maximum orthogonal shear stress τ_0 (Gpa) is related to the maximum Hertzian contact pressure p_0 through the following equation:

$$\tau_0 = 0.250 p_0 \quad (8)$$

Furthermore, the depths of the critical stresses under the condition of non frictional Hertzian line contact can be calculated as:

$$z_0 = 0.496b \quad (9)$$

Therefore, Equation (6) can be rewritten as:

$$\ln\frac{1}{S} = 2.359 \times 10^{-5} \frac{p_0^9 N^{1.11} bl}{b^{1.61}} \quad (10)$$

In beginning, the contact mechanical model was introduced to derive deformation curve on plate surface. Then, Modified Hertzian contact pressure and the contact width were calculated by using the relation between the plate deformation with finite and infinite depth. Due to the fatigue life of roller bearings under the pure rolling condition can be predicted by simply knowing the Hertzian contact pressure and the contact width which avoid complicated calculation of the subsurface stresses, the Lundberg-Palmgren life theory is introduced to predict the fatigue life for the arbitrarily crowned roller compressed between raceways

4. Results and Comparisons

Firstly, in order to verify the correctness of the theoretical solutions, a steel single roller compressed between two flat plates shown in **Figure 2**, was investigated. The comparison of maximum contact pressure of roller between the computation results using the above theoretical expressions and FEM software (ANSYS workbench), is conducted. In the CAD software the solid model is created, shown in **Figure 4**. The finite element model is simplified in order to compare with the theoretical results conveniently. Because of model symmetry, one-eighth of the model is taken to be simulated, shown in **Figure 5**. The dimension of the simplified LGT model of rollers is plotted in **Figure 2**, and **Figure 3** as well as **Table 1**. After the model building completes, the mesh is created shown in **Figure 6**, the full model has 22803 elements, and 41,283 nodes. In the contact setting, the contact between the roller and the plate is set to nonlinear contact [26] shown in **Figure 7**. The type of contact is defined as frictionless, so as to closer to the actual situation. If the contact is set to be a linear contact, there will be a great gap between the theoretical solution and the simulation results. Because of the one-eighth model, the blue face in **Figure 8** is set to be frictionless support in order to meet the symmetry of the set. Moreover, set the bottom face of the plate as fixed support, shown in **Figure 9**.

The displacement is applied on the roller shown in **Figure 10**. The displacement is set to be −0.01 mm, the direction is shown as the red arrow. The total deformation and equivalent strain are shown in **Figures 11** and **12**, respectively. The direction of the reaction force between

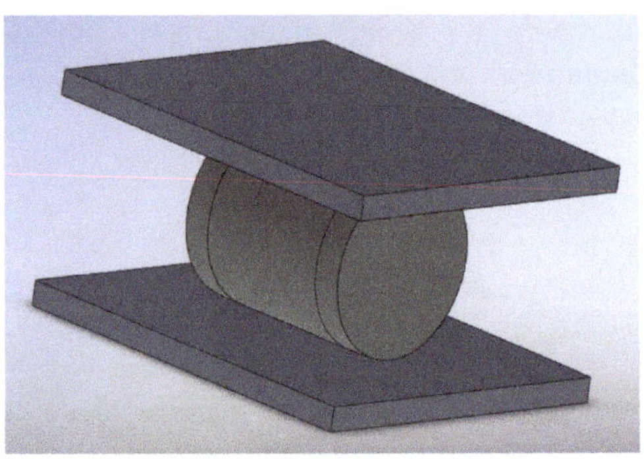

Figure 4. Simplified LGT model.

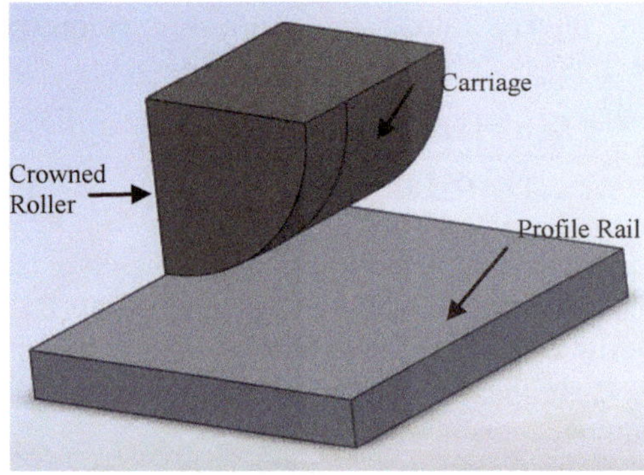

Figure 5. One-eighth of the simplified LGT model.

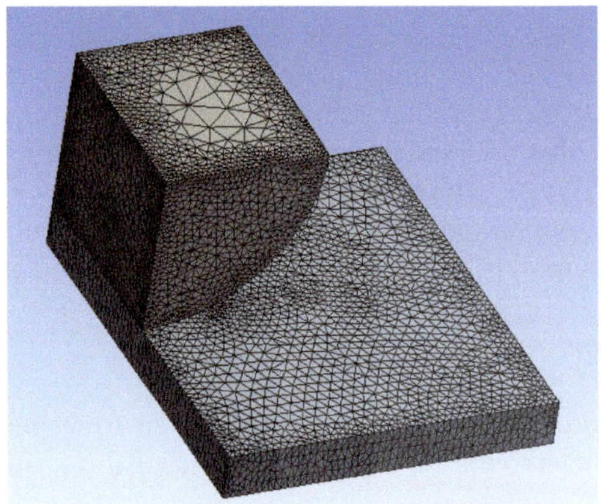

Figure 6. Finite element mesh.

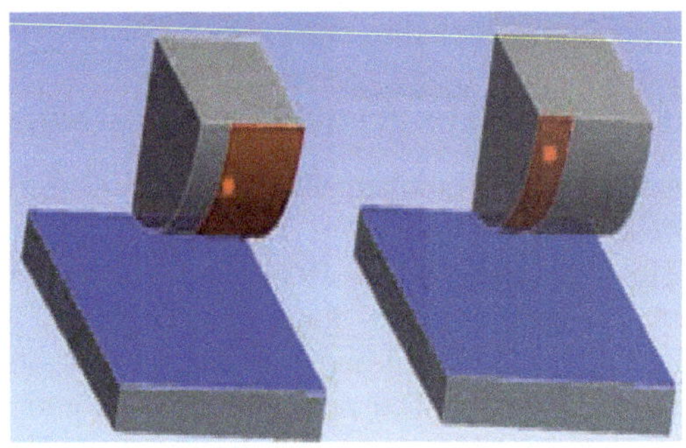

Figure 7. Contact setting.

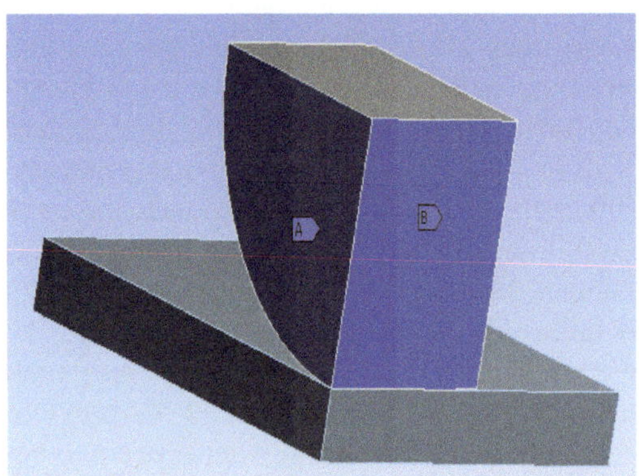

Figure 8. Frictionless support setting.

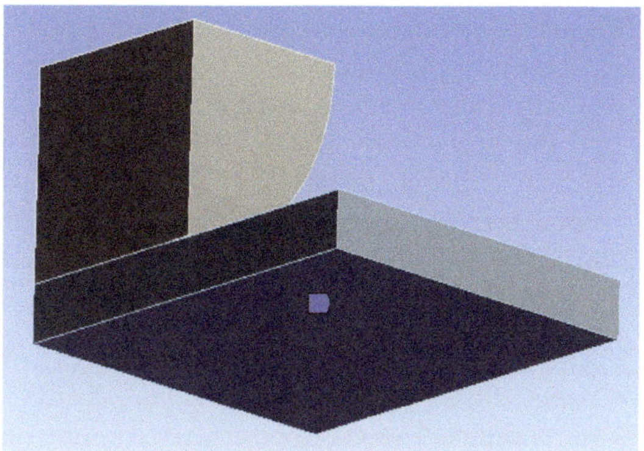

Figure 9. Fixed boundary conditions.

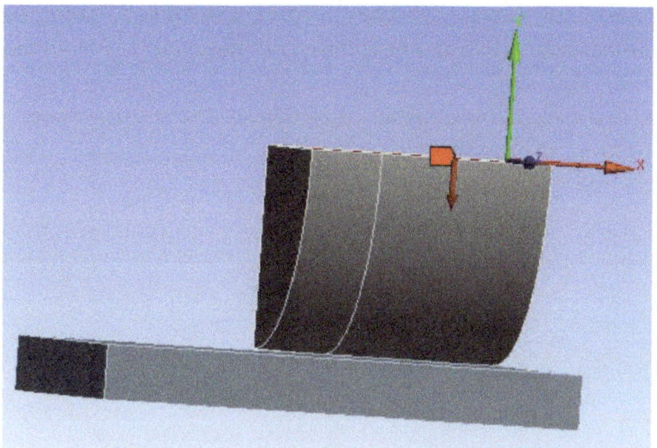

Figure10. Displacement in the model.

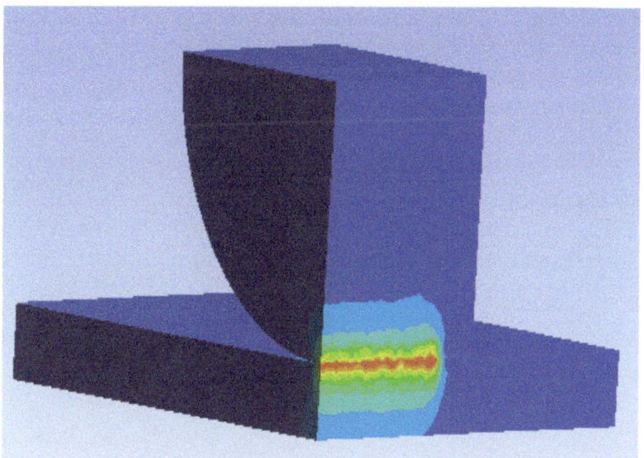

Figure 11. Total deformation.

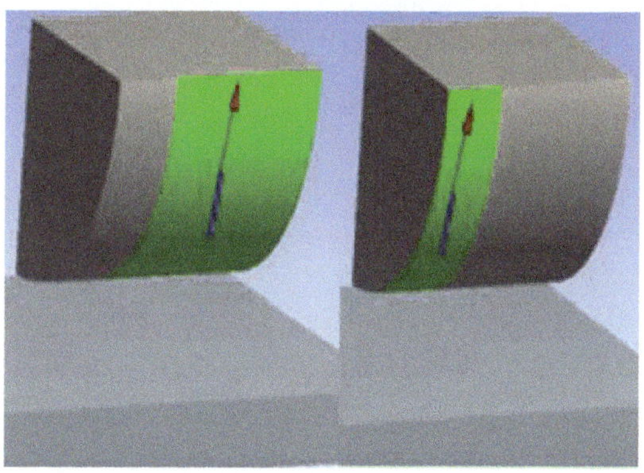

Figure 12. Equivalent strain.

Table 1. Test linear bearing specifications.

Total length of a roller, $2c$	26 mm
Effective depth of the carriage/profile rail, H	20 mm
Radii of roller, r_{cx}	3 mm
Crowned length of roller, ds	1.4506 mm
Crowned depth of roller, dt	0.04 mm
Young's modulus of the carriage/profile rail, E_1	206 Gpa
Young's modulus of the roller, E_2	314 Gpa
Poisson's ratio of the carriage/profile rail, ν_1	0.29
Poisson's ratio of the roller, ν_2	0.3
Number of circuits of recirculating rollers	4

the two contact bodies is shown in **Figure 13**, and the respective value is 1967.5 N and 224.17 N. The contact pressure on roller is 1131.7 Mpa, shown in **Figure 14**. The contact pressure on plate is 539.24 Mpa, shown in **Figure 15**. According to the instructions in the ANSYS Help [27], the actual contact stress should be the summation of the above two stress, so the maximum contact pressure is $1131.7 + 539.24 = 1670.94\text{ Mpa}$, and is very close to the theoretical results obtained by the Equation (10), denoted by 1583.33 Mpa. Therefore, the accuracy of theoretical expressions for the maximum contact pressure is acceptable.

Using the above maximum contact pressure expressions, we calculated the fatigue life of a carriage on LGT recirculating rollers. In the calculation of the fatigue life, the dimensions and material properties of LGT recirculating rollers shown in **Table 1** were used. A steel roller compressed between two flat plates is shown in **Figure 1**, with the material properties being the same as mentioned previously. Firstly, we wish to find the maximum contact pressure p_0 and the half Hertzian contact width b at edge point. Then, the Lundberg-Palmgren life theory is introduced to predict the fatigue life for the arbitrarily crowned roller compressed between raceways. Probability of survival was plotted for fatigue life analyses. **Figure 16** shows a comparison of calculated probability of survival for LGT recirculating rollers with circular crowned profiles subjected to the light displacement load. **Figure 16** indicates that the effect of number of stress circle of LGT recirculating rollers to probability of survival, is not apparent subjected to the extremely light displacement load. Moreover, the probability of survival decrease with increasing applied displacements under the same number of stress circle. This is because of the larger contact pressure for the increasing displacement load. **Figure 17** shows a comparison of calculated probability of survival for LGT recirculating rollers with circular crowned profiles subjected to the heavy displacement

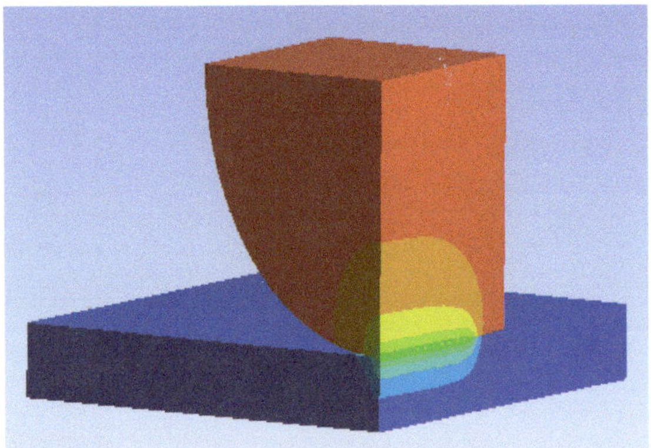

Figure 13. Reaction force.

load. **Figure 17** indicates that the number of stress circle of LGT recirculating rollers significantly affects probability of survival subjected to the heavier displacement load, and the probability of survival decrease abruptly with the heavy displacement load. This is because the stress concentration effect becomes large near the edge when the heavy displacement load is applied to the LGT.

5. Conclusion

In this paper, linear guideway type (LGT) recirculating rollers were modeled as crowned rollers compressed between two plates. Then, an analytical solution of the the maximum contact pressure between roller and plates was

Figure 14. Contact pressure (Roller).

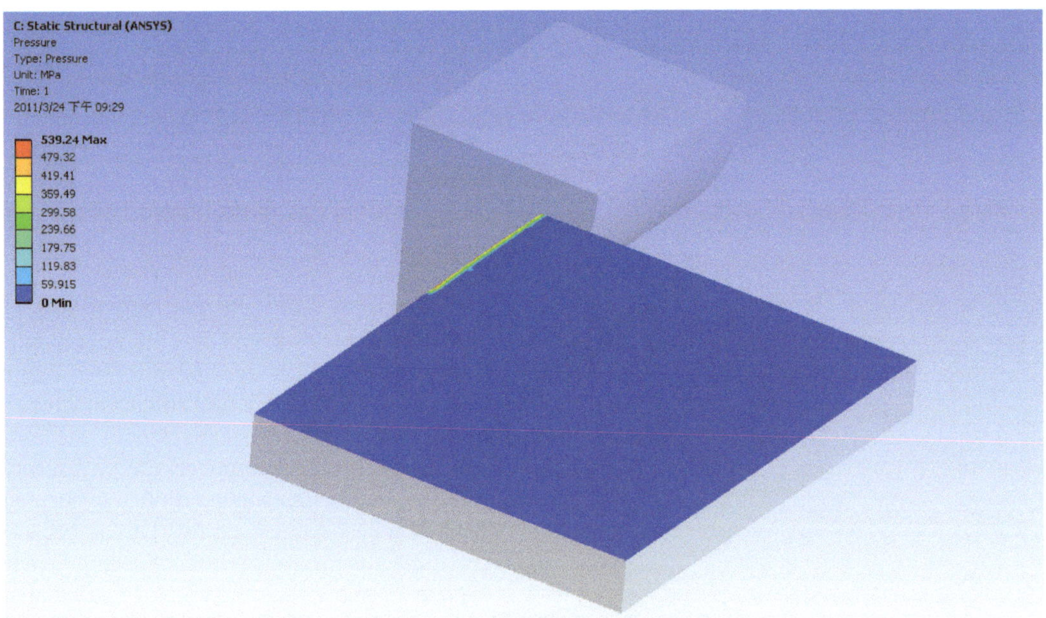

Figure 15. Contact pressure (Plate).

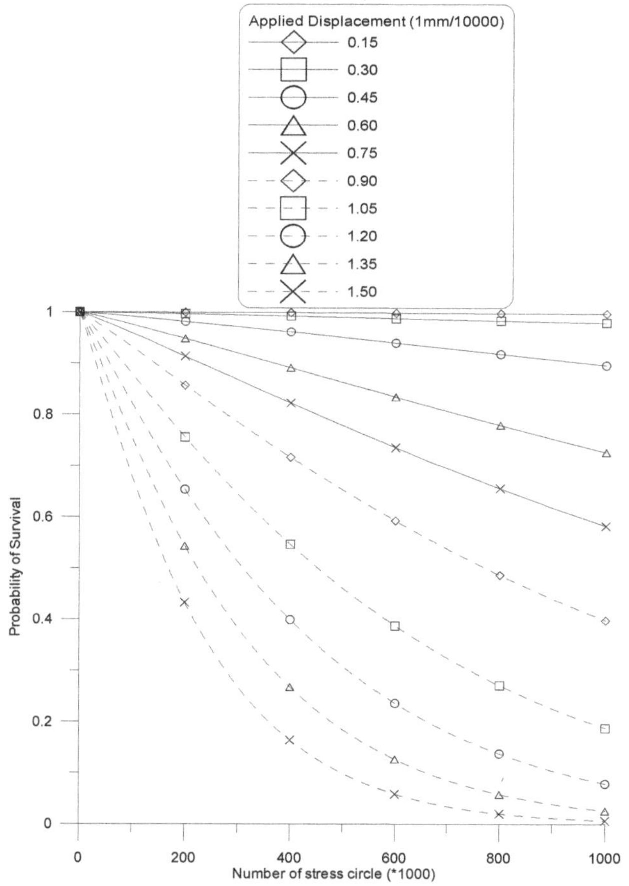

Figure 16. Comparison between the probability of survival subjected to light displacement load.

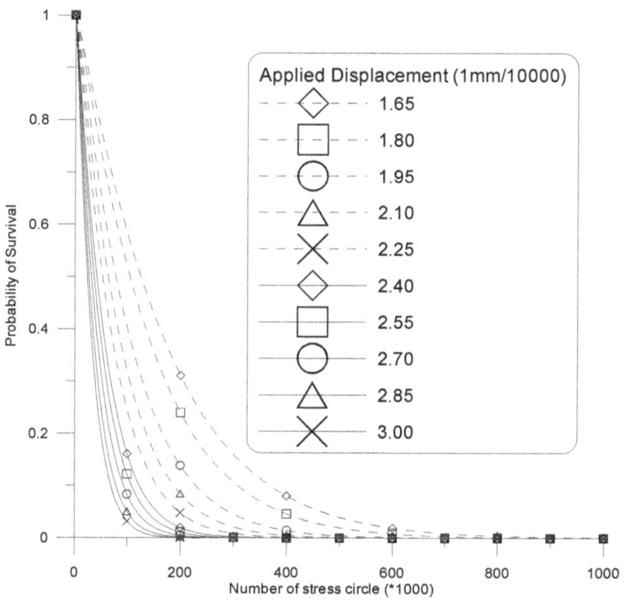

Figure 17. Comparison between the probability of survival subjected to heavy displacement load.

developed. From the comparison with FEM software (ANSYS workbench) result, the calculation model for normal contact pressure can be a reasonable estimate for a crowned roller being compressed between two plates. Finally, the system stress distribution of linear guideways will be obtained by using the superposition method, and the bearing life theory proposed by Lundberg and Palmgren is adopted to describe the contact fatigue life. The fatigue life prediction results reveal that the effect of number of stress circle of LGT recirculating rollers to probability of survival, is not apparent subjected to the extremely light displacement load, and the probability of survival decrease with increasing applied displacements under the same number of stress circle. This is because of the larger contact pressure for the increasing displacement load. Moreover, a number of stress circles of LGT recirculating rollers significantly affects the probability of survival subjected to the heavier displacement load, and the probability of survival decrease abruptly with the heavy displacement load. This is because the stress concentration effect becomes large near the edge when the heavy displacement load is applied to the LGT.

6. Acknowledgements

The authors would like to thank the National Science Council of the Republic of China in Taiwan for providing financial support for this study under grant NSC 99-2221-E-168-021.

REFERENCES

[1] HIWIN, Ltd., "Technical Information of Linear Guideway," HIWIN, Ltd., Taichung, 2007.

[2] H. Schmitz and G. G. Lyon, "Picking a Better Linear Bearing," *Machine Design*, Vol. 66, No. 6, 1994, pp. 63-65.

[3] NSK Ltd., "Precision Machine Parts," NSK Ltd., Fujisawa City, 1989.

[4] S. Kasai, T. Tsukada and S. Kato, "Linear Guides for Machine Tools (in Japanese)," *NSK Technical Journal*, Vol. 647, 1987, pp. 39-50.

[5] S. Kasai, T. Tsukada and S. Kato, "Recent technical Trends of Linear Guides (in Japanese)," *NSK Technical Journal*, Vol. 649, 1988, pp. 27-36.

[6] J. Ye, N. Iijima, F. Tashiro, S. Hagiwara and S. Yamada, Vibration of Linear Motion Bearing (in Japanese)," *Proceedings of Spring JSPE Meeting*, 1988, pp. 199-200.

[7] M. Schneider, "Statisches und Dynamisches Verhalten Beim Einsatz, Liearer Schienfuhrungen auf Walzlager-Basis im Werkzeugmaschinenbau," Carl Hanser Verlag, Munchen, Wien, 1991.

[8] T. L. Horng and S. H. Ju, "Stiffness of Arbitrarily Crowned Roller Compressed between Two Plates," *Journal of Engineering Tribology*, Vol. 217, No. 5, 2003, pp. 375-384.

[9] T. L. Horng, "Analyses of Stiffness in an Arbitrarily Crowned Roller Compressed between Raceways," *Journal of the*

Chinese Society of Mechanical Engineering, Vol. 24, No. 3, 2003, pp. 267-275.

[10] T. L. Horng, "Analyses of Stress Components for a Circular Crowned Roller Compressed Between Two Flat Plates," *Proceedings of the Institution of Mechanical Engineers Part Journal—Journal of Engineering Tribology*, Vol. 221, No. 5, 2007, pp. 581-589.

[11] T. L. Horng, "Analytical Solution of the Stiffness Equation for Linear Guideway Type Recirculating Rollers with Arbitrarily Crowned Profiles," *Proceedings of the Institution of Mechanical Engineers Part C—Journal of Mechanical Engineering Science*, Vol. 223, No. 6, 2009, pp. 1351-1358.

[12] T. L. Horng, "Analytical Solution of the Frequency Expressions for Rigid-body Natural Vibration of a Carriage on Linear Guideway Type Recirculating Rollers," *JSME International Journal—Journal of System Design and Dynamics*, Vol. 3, No. 2, 2009, pp. 215-226.

[13] G. Lundberg and A. Palmgren, "Dynamic Capacity of Rolling Bearings," *Acta Polytechnica Scandinavica. Mechanical Engineering Series*, Vol. 1, No. 3, 1947.

[14] P. Paris and F. Erdogan, "A Critical Analysis of Crack Propagation Laws," *ASME Journal of Basic Engineering*, Vol. 85, No. 4, 1963, p. 528.

[15] L. M. Keer and M. D. Bryant, "A Pitting Model for Rolling Contact Fatigue," *ASME Journal of Lubrication Technology*, Vol. 105, No. 2, 1983 pp. 198-205.

[16] T. E. Tallian, "Simplified Contact Fatigue Life Prediction Model. Part I: Review of Published Models. Part II: New Model," *ASME Journal of Tribology*, Vol. 114, No. 2, 1992, pp. 207-220.

[17] E. V. Zaretsky, R. J. Parker and W. J. Anderson, "A Study of Residual Stress Induced during Rolling," *ASME Journal of Lubrication Echnology*, Vol. 91, No. 2, 1969, pp. 314-319.

[18] N. G. Popinceanu, E. Diaconescu and S. Cretu, "Critical Stresses in Rolling Contact Fatigue," *Wear*, Vol. 71, No. 3, 1981, pp. 265-282.

[19] R. S. Zhou, "Surface Topography and Fatigue Life of Rolling Contact Bearings," *Tribology Transactions*, Vol. 36, No. 3, 1993, pp. 329-340.

[20] W. Cheng and S. Cheng Herbert, "Effect of Roller Profile on Cylindrical Roller Bearing Life Prediction," *ASME Journal of Tribology*, Vol. 119, No. 2, 1997, pp. 233-240.

[21] G. Lundberg and A. Palmgren, "Dynamic Capacity of Rolling Bearings," *Ada Polytechnica Mech. Eng. Series*, Vol. 2, No. 4, 1952.

[22] A. Palmgren, "The Service Life of Ball Bearings," *Zeitschrift des Vereines Deutscher Ingenieure*, Vol. 68, No. 14, 1924, pp. 339-341.

[23] V. A. Thomas and H. R. Hoersch, "Stresses Due to the Pressure of One Elastic Solid upon Another with Special Reference to Railroad Rails," Technical Reports, University of Illinois, Engineering Experiment Station, Bulletin No. 212, 1930.

[24] W. Weibull, "A Statistical Theory of the Strength of Materials," *Royal Swedish Academy of Engineering Sciences*, Vol. 151, No. 151, 1939.

[25] E. V. Zaretsky, J. V. Poplawski and S. M. Peters, "Comparison of Life Theories for Rolling-Element Bearings," *Tribology Transactions*, Vol. 39, No. 2, 1996, pp. 237-248.

[26] E. Madenci and I. Guve, "The Finite Element Method and Applications in Engineering Using Ansys," The University of Arizona, Springer, Berlin, 2006.

[27] ANSYS, Inc., "ANSYS 12.1 HTML Online Documentation ANSYS Structural Nonlinearities Reference," SAS IP, Inc., Canonsburg, 2009.

9

Vibration Control of a Gym Floor Using Tuned Mass Dampers: A Numerical Analysis

Marcos Daniel Souza dos Santos, Daniel Valença Fiuza Lima, Jorge Eliécer Campuzano Carmona, Suzana Moreira Avila, Graciela Nora Doz de Carvalho

Departamento de Engenharia Civil e Ambiental, Faculdade de Tecnologia, Universidade de Brasilia, Brasilia, Brasil

ABSTRACT

This work presents a study on excessive vibration problem occurring on concrete slabs, usually used on residential and commercial building floors. Even well designed slabs, according to ultimate and serviceability limit states criteria, can be vulnerable to undesirable vibrations that lead to user discomfort. A gym floor, that presented real excessive vibrations, located in a commercial building situated in the city of Brasilia, Brazil, was analyzed via Finite Element Method using ANSYS software. The first step in this analysis was to obtain natural frequencies and vibration modes, the structure presented low natural frequencies representing its flexible behavior. Then it was simulated a dynamic loading of people jumping, characteristic of this type of building occupation. Since it was observed the occurrence of excessive vibrations also in the numerical analysis, a Tuned Mass Damper (TMD) control system was proposed, looking for the best set of dampers to improve the control performance. The parameters for the best vibration reduction were obtained via a parametric study considering four different slabs varying dimensions and support conditions. Different models considering one and more TMDs, varying its placements and parameters, besides the frequency reference value to tune the damper were considered. An efficient control solution to this practical problem is presented to reduce its undesirable vibrations.

Keywords: Structural Dynamics; Structural Control; Concrete Slabs; Tuned Mass Damper

1. Introduction

Human induced loads can cause excessive vibration leading to discomfort and even compromising structural safety. Real occurrences of this problem in structures subjected to loads originated from people movement are worldwide known, a famous example is the London Millenium Footbridge [1]. Floor vibration studies have intensified in the last years. A number of factors cause this effect; the more important are: the use of more resistant materials that lead to more slender and flexible slabs, and the occurrence of non-provided activities in the original structural design. Those things have contributed to increasing emergence of excessive vibration on structures, especially on slabs [2,3]. This problem is directly linked to the structural control devices study. Structural control provide changes on stiffness and damping properties of the system, installing external devices or applying external forces, aiming to reduce original structure vibrations.

Among several structural control devices there is the Tuned Mass Damper (TMD), a damper applied in many practical cases, such as high buildings, footbridges, towers, etc. [4]. It is a mass-spring-dashpot system connected to the structure in a way to vibrate out of phase with the main structure, if tuned to the appropriated frequency. In this way the energy is transferred to the TMD reducing amplitude vibration on the main system. The use of TMD to reduce floor vibration has been studied by many researchers lately [5-9]; however there are still many topics to further development, such as the determination of appropriate TMD parameters to optimize control performance. This work presents a floor vibration study to a practical case of a gym. The slab analyzed presented excessive vibration problems when subjected to typical human induced loading of rhythmic exercise practicing. A control system consisting of various TMD was proposed. This proposal was based on a parametrical study to improve the performance of a slab set with typical dimensions of structures like residential and commercial buildings.

2. Human Induced Vibrations

Nowadays the design of building floors with low natural frequencies, with values near of those from human induced loads, can cause undesirable excessive vibrations.

Thus the understanding of structural dynamic behavior and the characteristics of the dynamic load is very important, particularly to floor slabs. Dynamic loading can produce high vibration levels than can cause discomfort to users and even compromise structural safety. Occurrence's of these problems are well known worldwide, like the already mentioned London Millenium Footbridge [1]. The stadium grandstands collapse in the 1920 years lead the American Standards Association (ASA, atual ANSI—American National Standards Institute) to nominate a committee to establish safety standards for this type of structure. Decades later, researchers described mathematically human loading arising from typical activities using a Fourier series. Faisca [10] performed an experimental study focusing on this type of load, describing human induced dynamic load considering loss of contact with the structure. According to her, these loads can change depending on structure's flexibility, namely they depend on people-structure interaction. Varela [5] studied human induced loading correlating the mathematical model with experimental results.

3. Structural Control: Tuned Mass Damper (TMD)

Tuned Mass Damper reduces the energy dissipation request of structural members when subjected to dynamic loads. This reduction occurs by transferring a portion of the vibratory energy to the TMD that in its simplest conception is a mass-spring-dashpot system connected to the structure, like shown on **Figure 1**. Considering the main system like a single-degree-of-freedom model, with mass *M*, damping *C* and stiffness *K*, on which is applied a dynamic load *f*(t) and attached a TMD with mass *m*, damping *c* and stiffness *k*, the equations of motion are:

$$M\ddot{y}(t)+C\dot{y}(t)+Ky(t)=c\dot{z}(t)+kz(t)+f(t) \quad (1)$$

$$m\ddot{z}(t)+c\dot{z}(t)+kz(t)=-m\ddot{y}(t) \quad (2)$$

where:

$\ddot{y}(t)$: main system acceleration;

$y(t)$: main system velocity;

$y(t)$: main system displacement;

$z(t)$: relative displacemente between the TMD and the main system.

When installed, the TMD tries to bring resonance peaks to lower values; this effect is desirable occurring on a wide frequency range.

For this to happen, TMD optimum parameters should be obtained, improving the control system performance [11]. A large number of these devices have been installed in high buildings, bridges, towers and industrial chimneys to control the dynamic response due to strong winds mostly [2]. In recent years studies were developed to

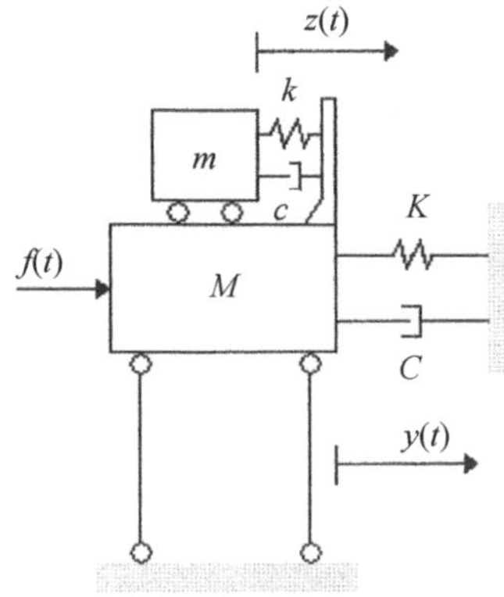

Figure 1. Structural system model with a TMD installed [4].

harness the potential of TMD on floor vibration control. Battista & Varella [12] presented a practical case of using TMDs in commercial building with metallic beams and composite floor decks that presented excessive vibration levels according to ISO 2631/1, ISO 2631/2 and NBR 8800 ABNT codes. Setareh *et al.* [8] describe the application of a passive pendulum working as a TMD for two cases with excessive slab vibrations. In this study they emphasize men-structure interaction on experimental results obtained.

4. TMD Parametric Study

The four slab structures studied in this work are shown in **Figure 2**. The loading applied simulated people on the move.In all cases analyzed it was verified vibration levels higher than those recommended on standard codes. Thus it was performed a study analyzing different TMD configurations in a way to verify the best control proposal for each one of the slabs. This study also analyzed the influence of TMD parameters on the control strategy performance. Because of the absence of design criteria to set parameters to TMD installed on floors, it was performed a parametrical study varying the mass ratio μ, the frequency ratio α and the damping ratio ξ, and verifying for each set of parameters the TMD efficiency. The four models presented natural frequencies near of values that characterize human dynamic loads. On the first slab studied, simply supported at the four edges, the best configuration obtained was a single TMD placed at the midspan, it reduced on 75% displacement amplitude.

Table 1 presents the bests reductions for each slab, applying harmonic loading and human-induced loading, associated frequency ratios are also presented. It can be verified that best reductions were obtained considering the theoretical harmonic loading compared to human-induced loading. In the case of Slab 1, the best verified value of α was 0.95, namely tuning the TMD very close to the corresponding slab natural frequency. For damping ratio it was concluded that increasing its value can worsen TMD

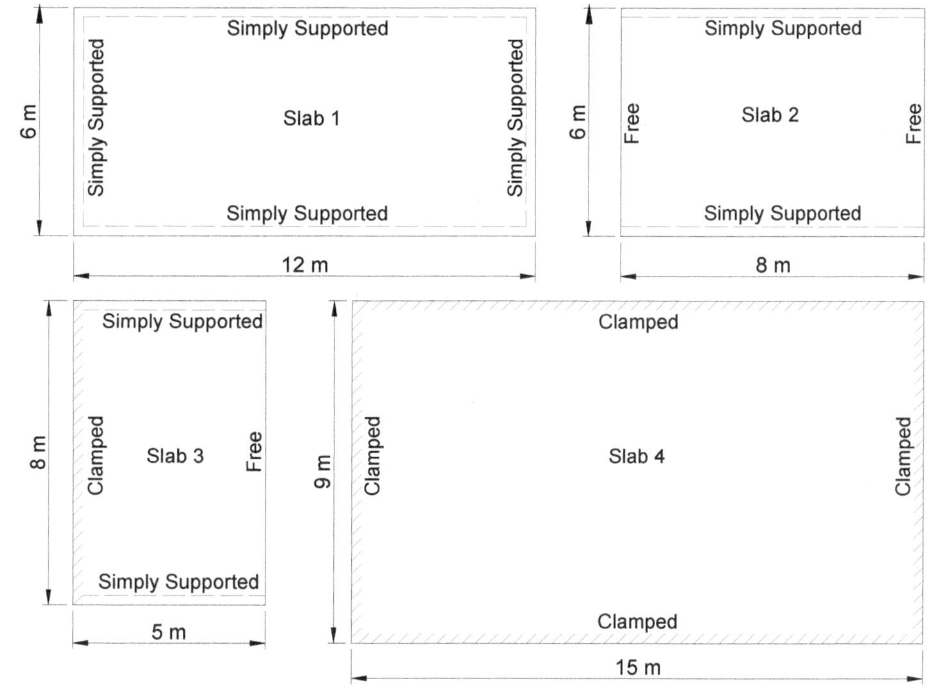

Figure 2. Dimensions and support conditions of the four slab models analyzed.

Table 1. Parametrical study results: frequency ratio and displacement reduction.

	Type of loading	Number of TMDs	Position	α	Displ. reduction (%)
Slab 01	Harmonic	1	Midspan	0.95	74.58
	Human-induced 12 people	1	Midspan	0.98	45.70
Slab 02	Harmonic	2	1 m from the border	1.03	73.12
	Human-induced 8 people	2	At the border	1.10	54.65
Slab 03	Harmonic	1	0.8 m from the border	1.00	79.90
	Human-induced 8 people	1	At the border	1.25	67.6
Slab 04	Harmonic	1	Midspan	0.95	72.09
	Human-induced 8 people	1	Midspan	1.00	56.00

performance. The more increase applied to mass ratio, best is the efficiency. However, this parameter requires special attention because the damper weight increase can lead to excessive static deformations, compromising structural safety. The parametric studies performed allowed to observe the influence of the frequency ratio (α) change in the displacement reducing, shown in **Figure 3**, as well as mass ratio (μ) and damping ratio influence (ξ), represented on **Figures 4** and **5**. Similar studies were performed by in the other slabs shown in **Figure 2**, they are well detailed on the work of Santos [13]. Numerical simulations indicated that the parameter which design should be more careful, since it has a great influence on final results, is the frequency ratio α. Damping ratio ξ and mass ratio μ didn't show the same influence.

5. Analysis of a Gym Slab

A gym floor, that presented excessive vibrations, located in a commercial building situated in the city of Brasilia, Brazil, was analyzed via Finite Element Method using ANSYS software. The commercial building has five floors: a basement, downstairs, first and second floor and roof. All floors were built as concrete solid slabs with 15 MP a resistance, 7 cm of thickness, supported by beams and columns of rectangular section.

A rectangular portion of the floor situated in downstairs floor with 14.074 m × 6.49 m and 3.15 m of height, showed on **Figure 6** was considered here. In this sector of the floor is located the ballroom gymnastics. The physical properties attributed to the numerical model

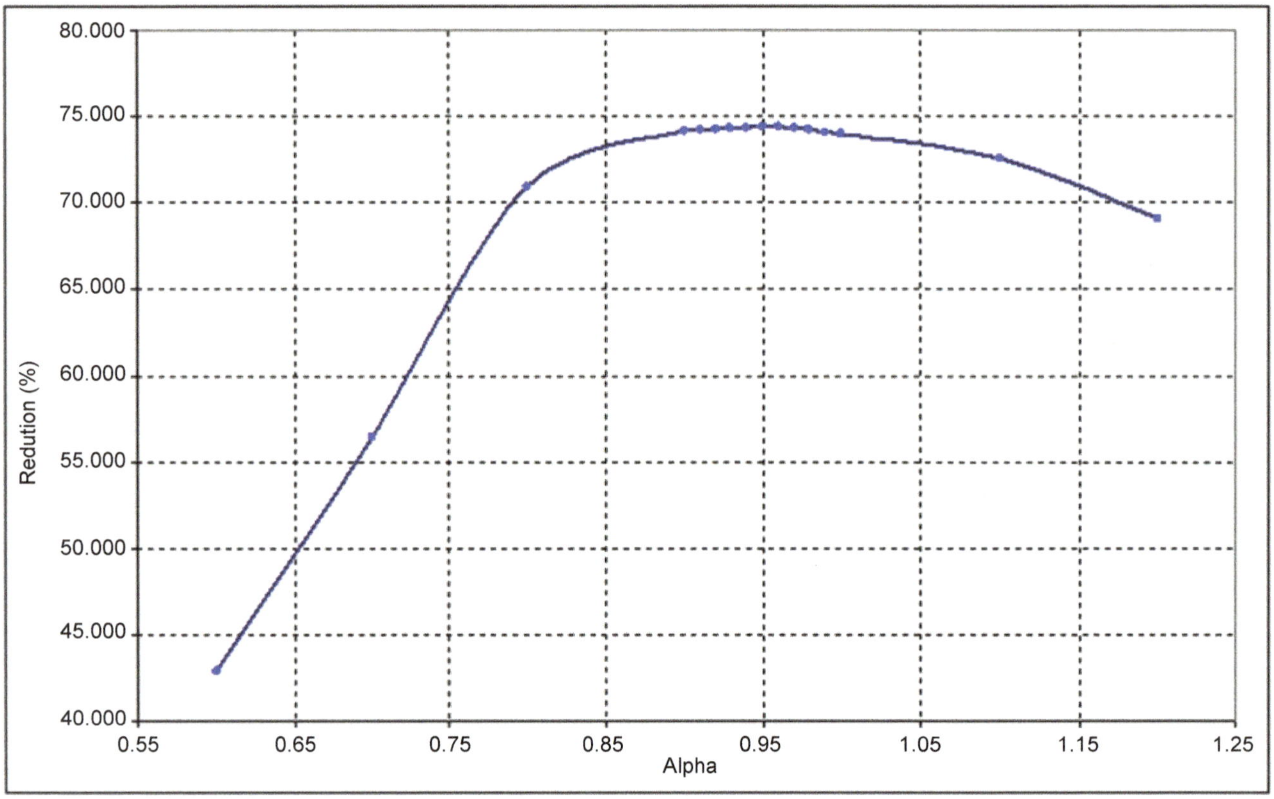

Figure 3. Frequency ratio α influence on displacement reduction ($\mu = 0.10$ and $\xi = 0.40$) Slab 1.

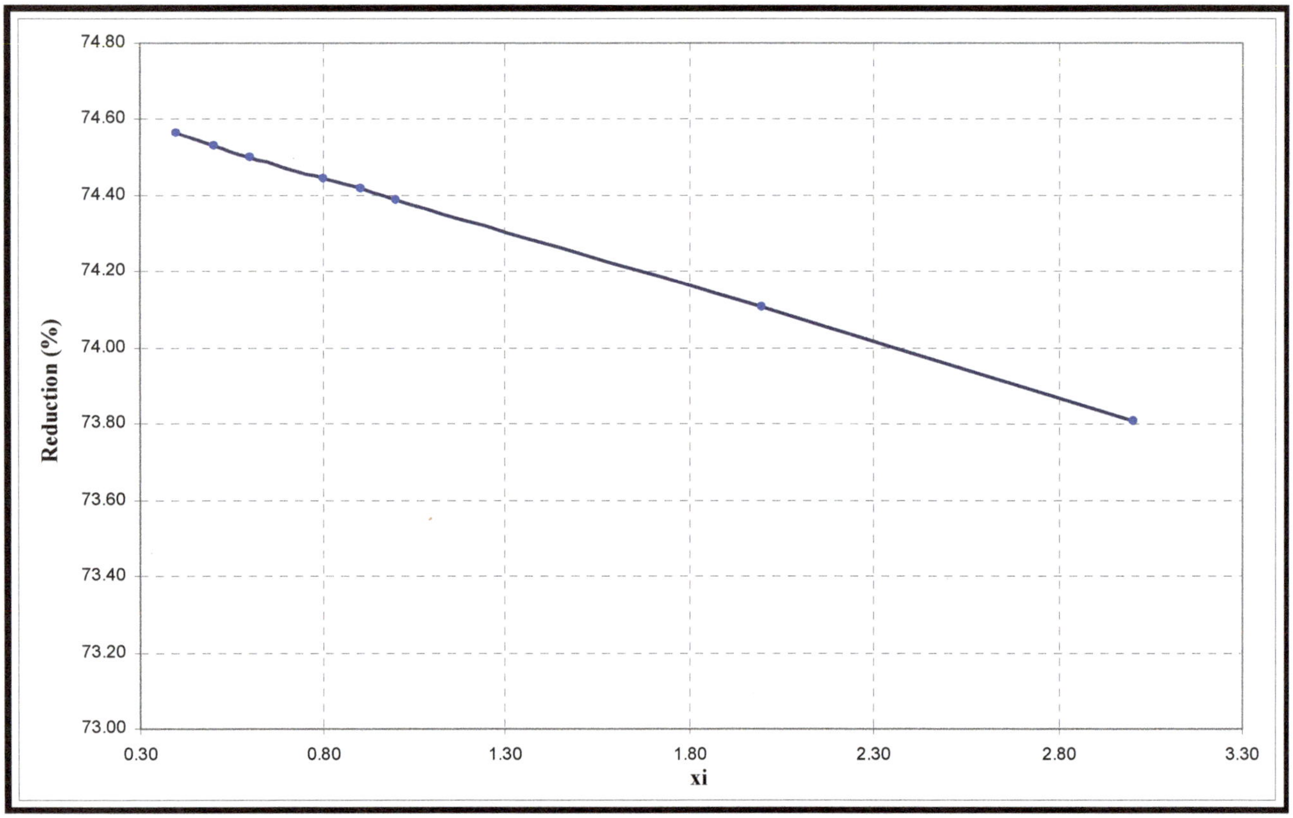

Figure 4. Damping ratio ξ influence on displacement reduction ($\mu = 0.10$ and $\alpha = 0.96$) Slab 1.

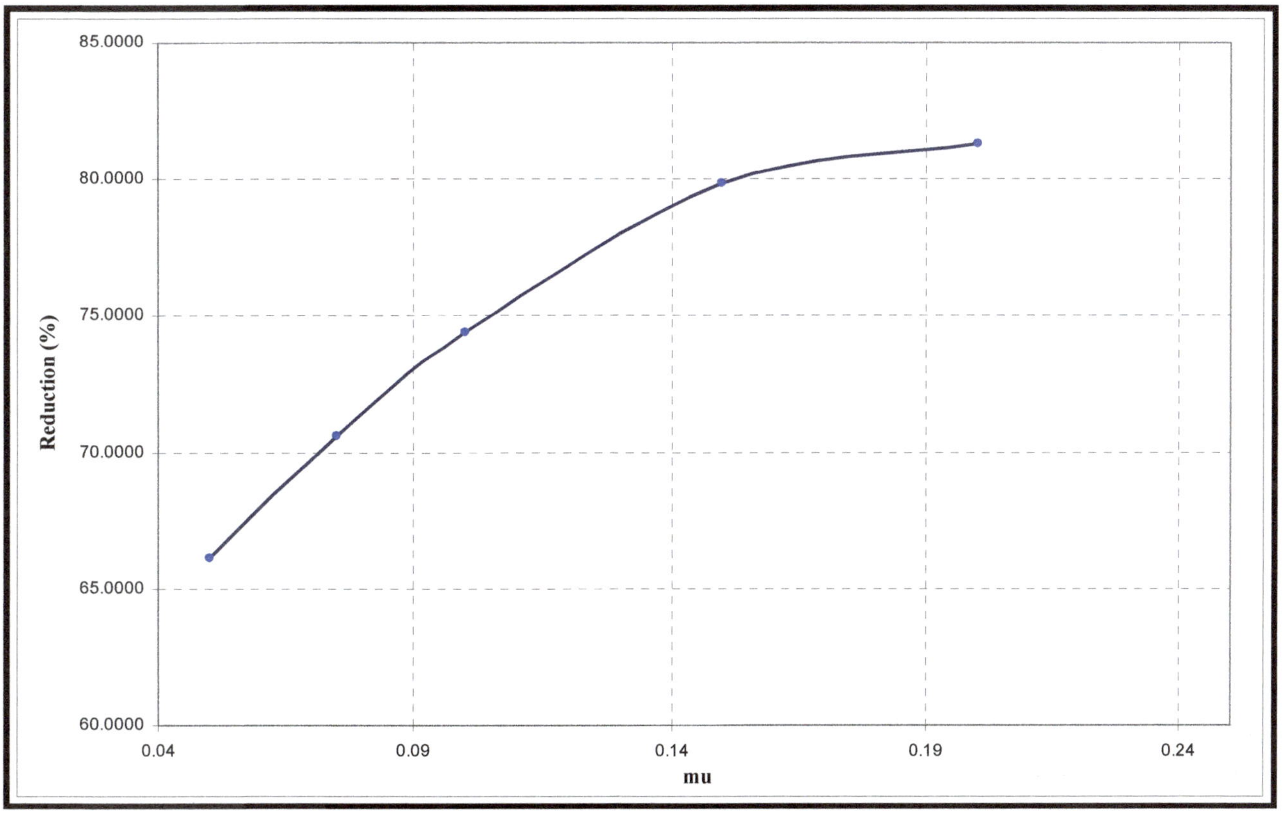

Figure 5. Mass ratio influence μ on displacement reduction ($\xi = 0.4$ and $\alpha = 0.96$) Slab 1.

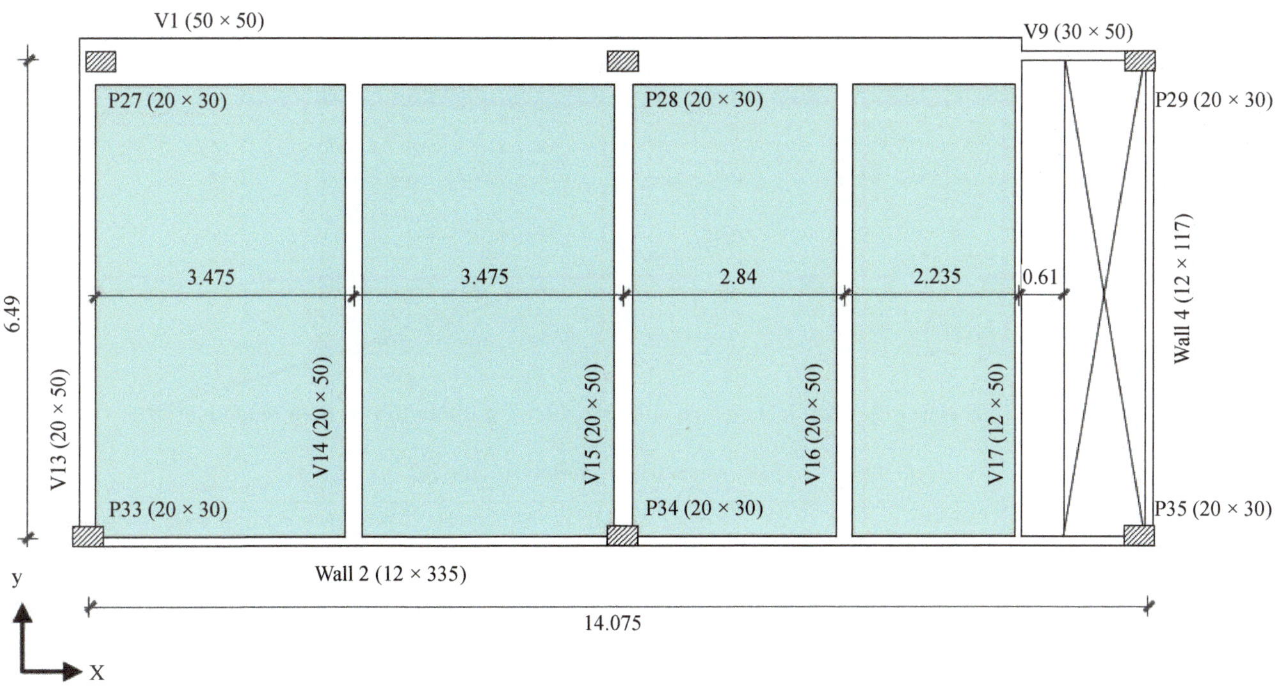

Figure 6. Dimensions and structural sketch of the gym floor portion studied.

were: Modulus of Elasticity of 29 Gpa; density of 2500 kg/m^3 and Poisson ratio of 0.3. The studied slab was modeled with a finite element mesh in the XY plane, using SHELL63 elements. Beams were modeled using BEAM4, as well as columns.

Modal analysis showed natural frequencies values near

of those from human induced loads. The first three mode shapes are in the horizontal direction with bending of the columns in the first mode and column bending and torsion in the second and third modes. From the fourth mode arise vertical displacements on the structure [13]. A transient analysis was also performed simulating the loading of people jumping. Nineteen people were considered placed at a distance of 1.0 m between one another, covering all the slab area.

Table 2 present the maximum response value on the node where it was verified the largest vertical displacement on the transient analysis of nineteen people jumping.

To attenuate these excessive vibrations it was proposed a control strategy using TMDs. In the absence of design TMD criteria, initially Jangid and Den Hartog [4,14] parameters were considered. It is noteworthy that these above parametric suggestions were based on different type of structures like single-degree-of-freedom model and shear buildings. The best performance was achieved using 6 TMD where three of them were tuned to the first natural frequency (2.56 Hz) and working on Y direction (horizontal) and the other ones tuned in the fourth natural frequency (9.35 Hz) and working on Z direction (vertical). The Z directions dampers were placed individually on the midspan of beams V14, V15 and V16. The Z direction dampers were located in the midspan of the three slabs; these nodes presented the largest displacements in the analysis. However, those were not good results, since it was achieved a 20% acceleration reduction, but also a 10% increase on displacements.

Therefore, a new analysis, based on the parametric studies presented earlier in this work, was performed.The nodes chosen to attach the TMDs were the slab's midspan nodes, which presented the largest displacement amplitudes. Three proposals with different numbers of TMDs were studied (1, 2 and 3 dampers). The best proposal showed to be the one using a single TMD.

The TMD was placed on the node in the slab midspan between beams V13 and V14, as shown in **Figure 7**. The adopted TMD parameters were $\alpha = 0.96$; $\mu = 0.10$ and $\xi = 0.4$. The TMD was tuned to the first and fourth frequencies (A1 and A2 models) as shown on **Table 3**. Maximum responses are presented on **Table 4**. Model

Table 2. Uncontrolled largest node response.

No.	Uz (m)	zV (s/m)	zA (s^2/m)
472	0.790	0.0172	0.6694

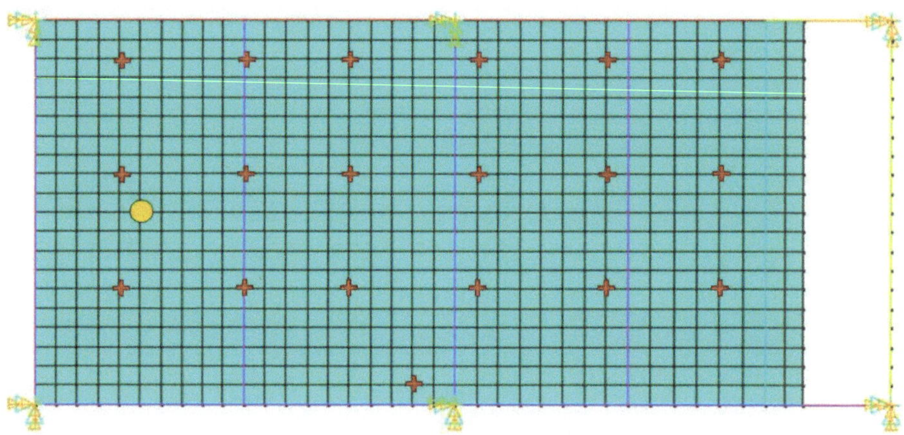

Figure 7. Gym slab with TMD installed on node 386, better position to vibration reduction.

Table 3. TMD properties to models A1 and A2.

Model	No. AMS	f_{est} (Hz)	M_{est} (kg)	μ	ξ	α	ω_{est} (rad/s)	ω_{AMS} (rad/s)	k_{AMS} (N/m)	m_{AMS} (kg)	C_{AMS} (N·s/m)
A1	1	2.56	45034.7	0.10	0.004	0.96	16.08	15.44	1073751.52	45034.7	556.31
A2	1	9.35	45034.7	0.10	0.004	0.96	16.08	56.40	14323431.69	45034.7	2031.83

Table 4. Maximum displacements, accelerations and correspondigreductionr to A1 and A2 models.

Model	Node		Undamped	Damped	Reduction Δ%
A1	386	U_z (mm)	0.712	0.638	**10.364**
		A_z (m/s²)	0.6214	0.541	12.955
A2	386	U_z (mm)	0.712	0.827	−16.15
		A_z (m/s²)	0.6214	0.494	**20.56**

Table 5. TMD properties to models A3 and A4.

Model	No. AMS	f_{est} (Hz)	M_{est} (kg)	μ	ξ	α	ω_{est} (rad/s)	ω_{AMS} (rad/s)	k_{AMS} (N/m)	m_{AMS} (kg)	C_{AMS} (N·s/m)
A3	1	2.56	45034.7	0.10	0.004	0.82	16.08	13.19	783409.85	4503.47	475.18
A4	1	9.35	45034.7	0.10	0.004	0.82	48.17	48.17	10450385.71	4503.47	1735.52

Table 6. Maximum displacements, accelerations and corresponding reductions to A3 and A4 models.

Model	Node		Undamped	Damped	Reduction Δ%
A3	386	U_z (mm)	0.712	0.6585	**7.514**
		A_z (m/s²)	0.6214	0.6500	9.881
A4	386	U_z (mm)	0.712	0.7817	−9.789
		A_z (m/s²)	0.6214	0.4376	**29.578**

Table 7. Comparative results of displacements and accelerations.

Proposals	No. AMS	Uz (mm) Δ%	Az (m/s²) Δ%
A1 model	01	10.449	12.955
6 TMD proposal	06	−8.421	23.887

A1 was the better alternative with a 10.449% displacement reduction and a 12.995% acceleration reduction. Despite presenting a greater acceleration reduction, model A2 presented displacement amplification.

Still considering a single TMD placed at the same node, a new set of parameters was attributed based on work of Santos [13] parametric studies of a 3D model considering beams and columns, similar to the present work structure. The parameter values adopted were α = 0.82, μ = 0.10 and ξ = 0.4, as indicated on **Table 5**, where A3 is the model tuning the TMD on the first frequency and A4 the TMD tuned on the fourth frequency. The dynamic response in both cases is presented on **Table 6**.

It is verified that, as well as in the last proposal, a better performance with the TMD tuned on the first frequency presenting a 7.514% displacement reduction and a 9.881% acceleration reduction. Nevertheless, model A1 showed better performance.

Based on all presented results it can be concluded that the displacement reduction is better achieved tuning the TMD to the first frequency. In acceleration reduction the best performance happened when tuning the TMD to the fourth frequency, but in these cases an undesirable increase on displacements was observed. In this way, the best proposal to control the structure is model A1, using a single TMD tuned to the first natural frequency of the structure. A comparison between model A1 and the 6 TMD initial proposal is shown in **Table 7**. Model A1 was the more indicated design in a way that despite not reducing so well accelerations, no increase in displacement is noticed. A displacement increase is not recommendable in a structural point of view.

6. Conclusion

This paper presents the study of the vibration problem of a commercial building floor used as a gym. The concrete slab analyzed presented excessive vibrations when subjected to typical exercise rhythmic loading. A control system was proposed using the Tuned Mass Damper device. Also a parametric study to design the TMD was performed considering four different concrete slabs with varying dimensions, support conditions and loads. The control system showed to be effective reducing by 10% displacement amplitudes of the gym floor analyzed.

7. Acknowledgements

This work has been supported by CNPq and CAPES brazilian agencies which are gratefully acknowledged.

REFERENCES

[1] P. Dallard, A. I. Fitzpatrick, A. Flint, S. Le Bourva, A. Low, R. Smith and M. R. Willford, "The London Millennium Footbridge," *The Structural Engineer*, Vol. 79, No. 22, 2001, pp. 17-33.

[2] I. Saidi, "Development of a Viscoelastic Tuned Mass Damper to Reduce Walking Induced Vibrations in Building Floors," Thesis of Doctor of Philosophy, Swinburne University of Technology, Melbourne, 2012.

[3] J. E. Campuzano, R. Castro, S. M. Avila and G. Doz, "Platform for Dynamic Tests: Preliminary Studies, De-

sign and Construction," *RevistaIbracon de Estruturas e Materiais*, Vol. 6, No. 1, 2013, pp. 55-74.

[4] T. T. Soong, G. F. Dargush, "Passive Energy Dissipation in Structural Engineering," John Wiley & Sons, Chichester, 1997.

[5] W. D. Varela, "Modelo Teórico Experimental para Análise de Vibrações Induzidas por Pessoas Caminhando Sobre Lajes de Edifícios," Ph.D. Thesis, COPPE/UFRJ, Rio de Janeiro, 2004.

[6] D. V. F. Lima, "Controle de Vibrações Induzidas em uma Laje de Academia de Ginástica com a Utilização de Amortecedores de Massa Sintonizados," Master Thesis, Universidade de Brasília, Brasília, 2007.

[7] M. Setareh, J. K. Ritchey, A. J. Baxter and T. M. Murray, "Pendulum Tuned Mass Dampers for Floor Vibration Controll," *Journal of Performance of Constructed Facilities, ASCE*, Vol. 20, No. 1, 2006, pp. 64-73.

[8] M. Setareh, J. K. Ritchey, T. M. Murray, J. H. Koo and M. Ahmadian, "Semiactive Tuned Mass Dampers for Floor Vibration Control," *Journal of Structural Engineering, ASCE*, Vol. 133, 2007, pp. 242-250.

[9] W. D. Varela and R. C. Battista, "Control of Vibration Induced by People Walking on Large Span Composite Floor Decks," *Engineering Structures*, Vol. 33, No. 9, 2011, pp. 2485-2494.

[10] R. G. Faisca, "Caracterização de Cargas Dinâmicas Geradas por Atividades Humanas," Ph.D. Thesis, COPPE/UFRJ, Rio de Janeiro, 2003.

[11] S. M. Avila, "Controle Híbrido para Atenuação de Vibrações em Edifícios," Ph.D. Thesis, Pontifícia Universidade Católica do Rio de Janeiro, Rio de Janeiro, 2002.

[12] R. C. Battista and W. D. Varela, "Medidas Corretivas para Vibrações de Painéis Contínuos de Lajes de Edifícios," *Proceedings of the* 30*th Jornadas Sul-Americanas de Engenharia Estrutural*, Brasilia, 2002.

[13] M. D. S. Santos, "Análise Numérica Do Controle de Vibrações em Lajes de Edifícios Utilizando Amortecedores de Massa Sintonizados," Master Thesis, Universidade de Brasília, Brasília, 2009.

[14] R. Jangid, "Optimum Multiple Tuned Mass Dampers for Base Excited Undamped System," Earthquake Engineering and Structural Dynamics, Vol. 28, No. 9, 1999, pp. 1041-1049.

10

Semi-Automatic Modeling Technique of Torque Converter Flow Passage

Shiping Liu, Shujuan Zheng
North China Institute of Water Conservancy and Hydro-Electric Power, Zhengzhou, China

ABSTRACT

The modeling technique of hydrodynamic torque converter flow passage was investigated. The semi-automatic modeling technique of torque converter flow passage was proposed. The flow passage model of each converter wheel is considered as a revolution entity sliced by two curved surfaces. In order to generate the revolution entity, a new approximation method, condition optimum arc approximation, was proposed. The method was used to approximate the meridional streamlines of the inner and outer wall. As a result, the three-dimensional revolution entity can be conveniently generated. In order to create slice surfaces, the central stream surface of flow passage was approximated with a quadric surface. The normal vector of the quadric surface and the thickness/thickness-function of bade were used to calculate the discrete point coordinates of blade surfaces. Via the rotation transformation to the coordinates, the discrete point coordinates of slice surfaces were obtained. A parameterized program code used for the hydrodynamic torque converter design and semi-automatic modeling was developed. Modeling errors were calculated and analyzed. The flow passage model was generated in several minutes with the help of the program code, Auto CAD and Solidworks software. Finally, the model was inputted into Gambit, and the pre-processing task used for the numerical simulation of torque converter flow field was successfully completed. The investigation results show that the semi-automatic modeling not only can ensure the accuracy of modeling, but also librates the research and design workers of torque converter from the time-consuming modeling work, which paves the way for the numerical simulation of the complex flow field of the hydrodynamic torque converter.

Keywords: Hydrodynamic Torque Converter; Flow Field Simulation; Semi-Automatic Modeling; Quadric Surface; Condition Optimum Approximation

1. Introduction

Semi-automatic modeling refers to programming and, with the help of a computer, to directly generating the model for numerical simulation. In the research and design of the torque converter, the flow field numerical simulation has become an indispensable procedure. Many scholars have done a lot of work in numerical simulation (Masatoshi Yamada *et al.* [1], R. R. By *et al.* [2], Hyuk Jae Chang *et al.* [3], Pradeep Attibele *et al.* [4], Marco Cigarini *et al.* [5], Zhao Dingxuan *et al.* [6], Yan Peng *et al.* [7], Tian Hua *et al.* [8], H. Schulz *et al.* [9], Won Sik Lim *et al.* [10]). In general, the flow passage model of a converter wheel is a hexahedron which includes 6 curved surfaces. For such a complex shape, modeling is a challenging task. On the one hand, the existing modeling methods are uneasy to ensure the accuracy of the model (Won Sik Lim *et al.* [11]). Obviously, the accuracy of Mode not only affects the accuracy of solution, but also affects the convergence of the iterative process. On the other hand, the existing modeling methods also spend too much time. Even if some modeling software is used for modeling in a human-machine-dialogue manner, the modeling is also a very time-consuming work.

Fluent is the most widely used CFD software, and its pre-processing software Gambit can be used to create a two-dimensional or three-dimensional model. However, for such a complex model of converter wheel passage, it is not very effective for Gambit to be used for modeling. Therefore, some scholars completed modeling with the help of the other software in a human-machine-dialogue manner. Some examples state that the communication between Gambit and the other modeling software is successful. On the other hand, References [12] and [13] established an analytic research and design system of torque converter which laid the mathematic foundations for the semi-automatic modeling. If a program code is developed which is used for the design and modeling of a torque converter, and an interface between high-level programming language and the modeling software is created, the semi-automatic modeling technique is feasible.

The flow passage model of a torque converter wheel includes four revolution surfaces (inner wall surface, outer wall surface, entrance surface and exit surface) and two freeform surfaces (point-based surfaces) which are blade external surfaces. Therefore, the flow passage model of torque converter can be considered as a revolution entity sliced by two curved surfaces.

2. Approximation of the Meridional Streamlines of Inner and Outer Wall

In order to facilitate modeling, the meridional streamlines of inner and outer wall can be approximated with circular arcs. In general, according to evaluation indexes, there are three kinds of curve approximation methods. The first method is square approximation. The advantage of square approximation is that each coefficient of the fitted equation can directly be obtained by solving linear equations. The disadvantage is that the deviation of some points may be substantially large. The second method is that the residual serves as the objective function and is mathematically called optimum approximation or uniform approximation. The optimum approximation makes the maximum error as small as possible. Therefore, the optimum approximation is more reasonable. However, in fact it is very difficult to directly obtain coefficients of the fitted equation. In addition, the relative error can serve as the objective function, which can be called relative approximation. The relative approximation can give a quantitative feeling, but it is difficult to directly find the coefficients of the fitted equation as well. For a converter wheel, the starting and ending point parameters of meridional streamlines serve as original parameters and the parameter errors at the two points should not be allowed. After the design conditions of torque converter is taken into account, a new approximation method, condition optimum arc approximation, is put forward. The basic idea of the approximation is as follows: first, construct a circular arc by using three points, and then optimize it by adjusting the coordinates of the intermediate point. In this manner, the parameters of the arc equation can be obtained.

For any converter wheel, according to the [12], take on $\xi = -1$, thus, the meridional streamline equation of inner wall is

$$\rho = \frac{h\left[2(\delta-\sigma)+\delta^2\sin\theta\right]}{1+\sqrt{(1+\delta\sin\theta)^2-2\sigma\sin\theta}} \tag{1}$$

The x-coordinate at a point located on the meridional streamline of inner wall is

$$x = \frac{h\left[2(\delta-\sigma)+\delta^2\sin\theta\right]\cos\theta}{1+\sqrt{(1+\delta\sin\theta)^2-2\sigma\sin\theta}} \tag{2}$$

The y-coordinate at a point located on the meridional streamline of inner wall is

$$y = h\sqrt{(1+\delta\sin\theta)^2-2\sigma\sin\theta} \tag{3}$$

In order to approximate the meridional streamline of the inner wall, the polar angle θ needs to be discretized. Take on $\theta_1,\theta_2,\cdots,\theta_i,\cdots,\theta_n$, where θ_1 is the polar angle at the starting point and located on the meridional streamline of the inner wall, while θ_2 is the polar angle at the ending point and located on the meridional streamline of the inner wall.

By using (2) and (3), the discrete point x-coordinates $x_1,x_2,\cdots,x_i,\cdots,x_n$ and y-coordinates $y_1,y_2,\cdots,y_i,\cdots,y_n$ can be obtained.

Assume that the circular arc equation takes the form of

$$(x-x_C)^2+(y+y_C)^2=r^2 \tag{4}$$

The condition optimum arc approximation of the meridional streamline of the inner wall can be described as follows:

Under the condition of $(x_1-x_C)^2+(y_1+y_C)^2=r^2$ and $(x_n-x_C)^2+(y_n+y_C)^2=r^2$, the radius residual approaches its minimum, that is

$$\max_{1<i<n}\left|r-\sqrt{(x_i-x_C)^2+(y_i-y_C)^2}\right| \to \min \tag{5}$$

where x_C, y_C and r are used to represent the x-coordinate of arc center, y-coordinate of arc center and arc radius, respectively. x_1 and y_1 are the x-coordinate and y-coordinate of the first discrete point, while x_n and y_n are the x-coordinate and y-coordinate of the last discrete point.

Because there are both positive and negative bias, and there are always

$$\begin{cases}\max\limits_{1<i<n}\left\{r-\sqrt{(x_i-x_C)^2+(y_i+y_C)^2}\right\}>0\\ \min\limits_{1<i<n}\left\{r-\sqrt{(x_i-x_C)^2+(y_i+y_C)^2}\right\}<0\end{cases}$$

Consequently, Equation (5) can be rewritten as

$$\begin{aligned}&\max_{1<i<n}\left\{r-\sqrt{(x_i-x_C)^2+(y_i+y_C)^2}\right\}\\ &=\left|\min_{1<i<n}\left\{r-\sqrt{(x_i-x_C)^2+(y_i+y_C)^2}\right\}\right|\end{aligned} \tag{6}$$

To obtain the arc equation parameters x_C, y_C and r, the starting point (x_1,y_1), the ending point (x_n,y_n)

and the intermediate point (x_m, y_m) can be used to construct an arc ($m=(1+n)/2$, rounding). Substituting the coordinates of the three points into (4), we have

$$\begin{cases}(x_1-x_C)^2+(y_1-y_C)^2=r^2\\(x_m-x_C)^2+(y_m-y_C)^2=r^2\\(x_n-x_C)^2+(y_n-y_C)^2=r^2\end{cases}\quad(7)$$

The first formula of (7) is used to subtract the second and the third formula of (7) respectively, and then they are rearranged. We can obtain

$$\begin{bmatrix}a_{11} & a_{12}\\a_{21} & a_{22}\end{bmatrix}\begin{bmatrix}x_C\\y_C\end{bmatrix}=\begin{bmatrix}b_1\\b_2\end{bmatrix}\quad(8)$$

where

$$\begin{cases}a_{11}=x_1-x_m\\a_{12}=y_1-y_m\\a_{21}=x_1-x_n\\a_{22}=y_1-y_n\\b_1=\left[(x_1-x_m)(x_1+x_m)+(y_1-y_m)(y_1+y_m)\right]/2\\b_2=\left[(x_1-x_n)(x_1+x_n)+(y_1-y_n)(y_1+y_n)\right]/2\end{cases}\quad(9)$$

Solving (8) by using Cramer's Rule, the central coordinates and radius of the circular arc are

$$\begin{cases}x_C=(a_{22}b_1-a_{12}b_2)/(a_{11}a_{22}-a_{21}a_{12})\\y_C=(a_{11}b_2-a_{21}b_1)/(a_{11}a_{22}-a_{21}a_{12})\\r=\sqrt{(x_m-x_C)^2+(y_m-y_C)^2}\end{cases}\quad(10)$$

The maximum error of the arc approximation is

$$E_{\max}=\max_{1<i<n}\left\{r-\sqrt{(x_i-x_C)^2+(y_i-y_C)^2}\right\}\quad(11)$$

The minimum error of the arc approximation is

$$E_{\min}=\min_{1<i<n}\left\{r-\sqrt{(x_i-x_C)^2+(y_i-y_C)^2}\right\}\quad(12)$$

If $|E_{\max}|=|E_{\min}|$, the arc equation determined by the above parameters is the condition optimum arc approximation of the meridional streamline of inner wall. Otherwise, the above parameters need to be corrected by optimization.

If the vector from arc center (x_C, y_C) to intermediate point (x_m, y_m) is expressed as $(x_m-x_C)i+(y_m-y_C)j$, the correction values of the intermediate point coordinates take the form of

$$\begin{cases}x'_m=x_m+K\left[(x_m-x_C)/r\right](E_{\max}+E_{\min})\\y'_m=y_m+K\left[(y_m-y_C)/r\right](E_{\max}+E_{\min})\end{cases}\quad(13)$$

where K is a correction factor less than 1 and can take on 0.7 generally.

Substitute x'_m for x_m and substitute y'_m for y_m, iterating by using (9)-(13). After several iteration steps, the parameters of the arc equation hardly vary. Thus, the condition optimum approximation arc parameters of the meridional streamline of inner wall can be obtained.

Similarly, the condition optimum approximation arc parameters of the meridional streamline of outer wall can be obtained as well.

3. Calculation of Slice Surface

Slice surfaces are the two lateral surfaces of flow passage. One lateral surface is the pressure surface of a blade, while another lateral surface is the suction surface of the adjacent blade.

3.1. Basic Idea of Slice Surface Calculation

From [12] and [13], it can be found that the central stream surface (or blade camber surface) cannot be expressed as a rectangular coordinate equation. Therefore, it is difficult to directly derive the equation of blade surface or the equation of flow passage lateral surface. A feasible option is to calculate the coordinates of discrete points by numerical method.

Discrete parameter ξ and polar angle θ. In other words, take on $\xi_1, \xi_2, \cdots, \xi_m$, and $\theta_1, \theta_2, \cdots, \xi_n$.

In order to calculate the point coordinates on blade surfaces, an imaginary blade is arranged just in the central stream surface of flow passage. The relation among flow passage central stream surface, the surfaces of the imaginary blade and the flow passage lateral surfaces is shown in **Figure 1**.

First, calculate the coordinates of Point P located on central stream surface of the flow passage. Then draw a

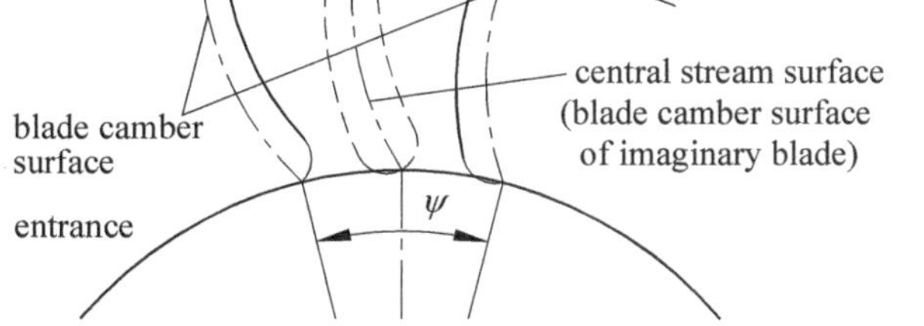

Figure 1. Relation among passage lateral surface, imaginary blade surfaces and flow passage central stream surface.

normal line of the central stream surface passing through point P. In this manner, intersection Points A and B are obtained. They are located on two external surfaces of the imaginary blade, respectively. Finally, rotate the Point A and B a half of flow passage angle $\psi/2$ around x-axis in different directions, respectively. Points C and D can be obtained. The two points are located on the lateral surfaces of the flow passage.

According to [12], the coordinate calculation formula of the flow passage central stream surface is

$$\begin{cases} x = \dfrac{h\left[2(\delta-\sigma)+\delta^2\sin\theta\right]\cos\theta}{1+\sqrt{(1+\delta\sin\theta)^2-2\sigma\sin\theta}} \\ y = r\cos\dfrac{s}{r} \\ z = r\sin\dfrac{s}{r} \end{cases} \tag{14}$$

where

$$r = \sqrt{(1+\delta\sin\theta)^2+\sigma\xi\sin\theta}$$

$$s = C_1(\theta-\theta_1)+C_2(\theta^2-\theta_1^2) + C_3(\theta^3-\theta_1^3)+C_4(\theta^4-\theta_1^4)$$

For the given ξ_i and θ_j, the coordinates of any discrete point on the flow passage central stream surface can be calculated.

3.2. Approximation of Central Stream Surface

As the equation of central stream surface is substantially complex, it is difficult to obtain the normal equation of the central stream surface. Therefore, a quadratic surface is used to approximate the central stream surface. For a given point $(x_{i,j}, y_{i,j}, z_{i,j})$, the point and its surrounding 8 points are used to construct a quadric surface. Assume that the equation of the quadric surface takes the form of

$$ax^2+by^2+cz^2+fxy+gxz +hyz+ux+vy+wz=1000 \tag{15}$$

where a, b, c, f, g, h, u, v and w are all coefficients to be determined.

The constant item of (15) takes on 1000 rather than 1 so as to avoid too small coefficients.

Inserting the coordinates of the above 9 points into (15), we have

$$\boldsymbol{AX} = \boldsymbol{B} \tag{16}$$

where

$$\boldsymbol{X} = [a \quad b \quad c \quad f \quad g \quad h \quad u \quad v \quad w]^{\mathrm{T}}$$

$$\boldsymbol{B} = 1000\times[1 \quad 1 \quad 1 \quad 1 \quad 1 \quad 1 \quad 1 \quad 1 \quad 1]^{\mathrm{T}}$$

$$\boldsymbol{A} = \begin{bmatrix} a_{11} & a_{12} & a_{13} & a_{14} & a_{15} & a_{16} & a_{17} & a_{18} & a_{19} \\ a_{21} & a_{22} & a_{23} & a_{24} & a_{25} & a_{26} & a_{27} & a_{28} & a_{29} \\ a_{31} & a_{32} & a_{33} & a_{34} & a_{35} & a_{36} & a_{37} & a_{38} & a_{39} \\ a_{41} & a_{42} & a_{43} & a_{44} & a_{45} & a_{46} & a_{47} & a_{48} & a_{49} \\ a_{51} & a_{52} & a_{53} & a_{54} & a_{55} & a_{56} & a_{57} & a_{58} & a_{59} \\ a_{61} & a_{62} & a_{63} & a_{64} & a_{65} & a_{66} & a_{67} & a_{68} & a_{69} \\ a_{71} & a_{72} & a_{73} & a_{74} & a_{75} & a_{76} & a_{77} & a_{78} & a_{79} \\ a_{81} & a_{82} & a_{83} & a_{84} & a_{85} & a_{86} & a_{87} & a_{88} & a_{89} \\ a_{91} & a_{92} & a_{93} & a_{94} & a_{95} & a_{96} & a_{97} & a_{98} & a_{99} \end{bmatrix}$$

$$a_{11} = x_{i-1,j-1}^2, a_{12} = y_{i-1,j-1}^2, a_{13} = z_{i-1,j-1}^2$$

$$a_{14} = x_{i-1,j-1}\cdot y_{i-1,j-1}, a_{15} = x_{i-1,j-1}\cdot z_{i-1,j-1}$$

$$a_{16} = y_{i-1,j-1}\cdot z_{i-1,j-1}$$

$$a_{17} = x_{i-1,j-1}, a_{18} = y_{i-1,j-1}, a_{19} = z_{i-1,j-1}$$

$$a_{21} = x_{i,j-1}^2, a_{22} = y_{i,j-1}^2, a_{23} = z_{i,j-1}^2$$

$$a_{24} = x_{i,j-1}\cdot y_{i,j-1}, a_{25} = x_{i,j-1}\cdot z_{i,j-1}$$

$$a_{26} = y_{i,j-1}\cdot z_{i,j-1}$$

$$a_{27} = x_{i,j-1}, a_{28} = y_{i,j-1}, a_{29} = z_{i,j-1}$$

$$a_{31} = x_{i+1,j-1}^2, a_{32} = y_{i+1,j-1}^2, a_{33} = z_{i+1,j-1}^2$$

$$a_{34} = x_{i+1,j-1}\cdot y_{i+1,j-1}, a_{35} = x_{i+1,j-1}\cdot z_{i+1,j-1}$$

$$a_{36} = y_{i+1,j-1}\cdot z_{i+1,j-1}$$

$$a_{37} = x_{i+1,j-1}, a_{38} = y_{i+1,j-1}, a_{39} = z_{i+1,j-1}$$

$$a_{41} = x_{i-1,j}^2, a_{42} = y_{i-1,j}^2, a_{43} = z_{i-1,j}^2$$

$$a_{44} = x_{i-1,j}\cdot y_{i-1,j}, a_{45} = x_{i-1,j}\cdot z_{i-1,j}$$

$$a_{46} = y_{i-1,j}\cdot z_{i-1,j}$$

$$a_{47} = x_{i-1,j}, a_{48} = y_{i-1,j}, a_{49} = z_{i-1,j}$$

$$a_{51} = x_{i,j}^2, a_{52} = y_{i,j}^2, a_{53} = z_{i,j}^2$$

$$a_{54} = x_{i,j}\cdot y_{i,j}, a_{55} = x_{i,j}\cdot z_{i,j}, a_{56} = y_{i,j}\cdot z_{i,j}$$

$$a_{57} = x_{i,j}, a_{58} = y_{i,j}, a_{59} = z_{i,j}$$

$$a_{61} = x_{i+1,j}^2, a_{62} = y_{i+1,j}^2, a_{63} = z_{i+1,j}^2$$

$$a_{64} = x_{i+1,j}\cdot y_{i+1,j}, a_{65} = x_{i+1,j}\cdot z_{i+1,j}$$

$$a_{66} = y_{i+1,j}\cdot z_{i+1,j}$$

$$a_{67} = x_{i+1,j}, a_{68} = y_{i+1,j}, a_{69} = z_{i+1,j}$$

$$a_{71} = x_{i-1,j+1}^2, a_{72} = y_{i-1,j+1}^2, a_{73} = z_{i-1,j+1}^2$$

$$a_{74} = x_{i-1,j+1}\cdot y_{i-1,j+1}, a_{75} = x_{i-1,j+1}\cdot z_{i-1,j+1}$$

$$a_{76} = y_{i-1,j+1}\cdot z_{i-1,j+1}$$

$$a_{77} = x_{i-1,j+1}, a_{78} = y_{i-1,j+1}, a_{79} = z_{i-1,j+1}$$

$$a_{81} = x_{i,j}^2, a_{82} = y_{i,j}^2, a_{83} = z_{i,j}^2$$

$$a_{84} = x_{i,j+1} \cdot y_{i,j+1}, a_{85} = x_{i,j+1} \cdot z_{i,j+1}$$

$$a_{86} = y_{i,j+1} \cdot z_{i,j+1}$$

$$a_{87} = x_{i,j+1}, a_{88} = y_{i,j+1}, a_{89} = z_{i,j+1}$$

$$a_{91} = x_{i,j}^2, a_{92} = y_{i,j}^2, a_{93} = z_{i,j}^2$$

$$a_{94} = x_{i+1,j+1} \cdot y_{i+1,j+1}, a_{95} = x_{i+1,j+1} \cdot z_{i+1,j+1}$$

$$a_{96} = y_{i+1,j+1} \cdot z_{i+1,j+1}$$

$$a_{97} = x_{i+1,j+1}, a_{98} = y_{i+1,j+1}, a_{99} = z_{i+1,j+1}$$

It should be noteworthy that the existence and uniqueness of the equation solution need to be investigated before solving (16). According to the linear algebra theory, the necessary and sufficient condition for the existence and uniqueness of solution is that the coefficient matrix of the equation is a non-singular matrix, or $\det(A) \neq 0$. The solution error of the simultaneous equations is closely related with the ill-posed problem of the equation. The condition number of equation needs to be used to determine if an equation is ill-posed (Xi Meicheng [14]). Obviously, it is difficult to prove whether the matrix is a non-singular matrix or not in advance. Similarly, it is uneasy to determine if an equation is ill-posed in advance. However, for the given practical engineering example, it should be seen that the central stream surface is a smooth surface. No dramatic curvature change occurs and there is no singularity either. It should be said that the existence and uniqueness of the equation solution is undoubted.

There are many methods of solving linear equations. Column pivot Gauss-Jordan elimination method is used to solve (16). The ill-posed problem taken into account, some pre-processing measures are taken and double-precision format data are used for program design. By solving the linear equations, each coefficient of (15) can be obtained.

3.3. Numerical Calculation of Passage Lateral Surface

Let

$$F(x,y,z) = ax^2 + by^2 + cz^2 + fxy + gxz + hyz + ux + vy + wz - 1000$$

Take partial derivatives of the above expression, there resulting

$$\begin{cases} F_x(x,y,z) = \partial F/\partial x = 2ax + fy + gz + u \\ F_y(x,y,z) = \partial F/\partial y = 2by + fx + hz + v \\ F_z(x,y,z) = \partial F/\partial z = 2cz + gx + hy + w \end{cases}$$

The normal vector of the quadric surface passing through point $(x_{i,j}, y_{i,j}, z_{i,j})$ is

$$\begin{aligned} v = \{&2ax_{i,j} + fy_{i,j} + gz_{i,j} + u, 2by_{i,j} + fx_{i,j} \\ &+ hz_{i,j} + v, 2cz_{i,j} + gx_{i,j} + hy_{i,j} + w\} \end{aligned} \tag{17}$$

With the normal vector of the quadric surface, assume that the thickness of the blade is $2t$. For a punched blade $t = \text{constant}$; for a casted blade $t = t(\theta)$. The coordinates of Point A shown in **Figure 1** are

$$\begin{cases} x_A = x_{i,j} - \dfrac{F_x(x_{i,j}, y_{i,j}, z_{i,j})}{|v|} t \\ y_A = y_{i,j} - \dfrac{F_y(x_{i,j}, y_{i,j}, z_{i,j})}{|v|} t \\ z_A = z_{i,j} - \dfrac{F_z(x_{i,j}, y_{i,j}, z_{i,j})}{|v|} t \end{cases} \tag{18}$$

The coordinates of point B are

$$\begin{cases} x_B = x_{i,j} + \dfrac{F_x(x_{i,j}, y_{i,j}, z_{i,j})}{|v|} t \\ y_B = y_{i,j} + \dfrac{F_y(x_{i,j}, y_{i,j}, z_{i,j})}{|v|} t \\ z_B = z_{i,j} + \dfrac{F_z(x_{i,j}, y_{i,j}, z_{i,j})}{|v|} t \end{cases} \tag{19}$$

With coordinates of Point A and B, by using the coordinate rotation formula, coordinates of Point C are

$$\begin{cases} x_C = x_A \\ y_C = y_A \cos(-\psi/2) - z_A \sin(-\psi/2) \\ z_C = y_A \sin(-\psi/2) + z_A \cos(-\psi/2) \end{cases} \tag{20}$$

Coordinates of Point D are

$$\begin{cases} x_D = x_B \\ y_D = y_B \cos(\psi/2) - z_B \sin(\psi/2) \\ z_D = y_B \sin(\psi/2) + z_B \cos(\psi/2) \end{cases} \tag{21}$$

Coordinates of Points C and D are stored in file Converter.txt. With coordinates of Points C and D, the offset angles corresponding to the above two points are

$$\begin{cases} \phi_C = \arcsin\left[z_C / \left(y_C^2 + z_C^2\right)^{1/2}\right] \\ \phi_D = \arcsin\left[z_D / \left(y_D^2 + z_D^2\right)^{1/2}\right] \end{cases} \tag{22}$$

It can be inferred that offset angle ϕ_C and ϕ_D are the function of parameter ξ and θ, naturally varying with i and j. By means of the program, the maximum offset angle ϕ_{max} and minimum offset angle ϕ_{min} of the flow passage can be obtained conveniently. They are

$$\begin{cases} \phi_{max} = \max\{\phi_{Cmax}, \phi_{Dmax}\} \\ \phi_{min} = \min\{\phi_{Cmin}, \phi_{Dmin}\} \end{cases} \quad (23)$$

4. Generation of Flow Passage Mode

4.1. Automatic Generation of Revolution Entity

If meridional streamlines of inner and outer wall of torque converter are approximated with arcs, the torus of each converter wheel consists of two arcs (inner and outer wall) and two straight line segments (entrance and exit). Pump torus is shown in **Figure 2**.

Coordinates of arc endpoints, center coordinates of inner and outer arc, and arc radii have been calculated. The central angle of the outer arc (from Point 3 to 4) is

$$\alpha = 2\arcsin\frac{\sqrt{(x_4 - x_3)^2 + (y_4 - y_3)^2}}{2R_2} \quad (24)$$

By using high-level language programming for the above calculation, the parameters of all graphic entities (straight line segments and circular arcs) can be obtained. With the graphic entity parameters, the torus of each converter wheel can be generated automatically by using a drawing program.

In order to generate a three-dimensional revolution entity with the above torus, the "revolve" command of Auto CAD needs to be applied. The command need specify a revolution angle. Theoretically, the revolution angle should be equal to the difference between the maximum offset angle ϕ_{max} and minimum offset angle ϕ_{min} of the flow passage. In fact, after considering 10° margin and rounding, the revolution angle takes on $\phi_{rev} = \text{int}(\phi_{max} - \phi_{min} + 10)$.

The "revolve" command of Auto CAD is prescribed to revolve an angle from 0° to ϕ_{rev}. In general, the revolution angle of model is often not in the range. Therefore, it is necessary to adjust the space position of the revolution entity. Thus, the "rotate3d" command of Auto CAD needs to be applied. Similarly, after considering 5° margin and rounding, the rotation angle of the "rotate3d" command takes on $\phi_{rot} = \text{int}(\phi_{min} - 5)$.

The developed high-level language program can output a file converter.lsp which is a drawing program and includes various drawing commands. By using Auto CAD, the three-dimensional revolution entity can be generated automatically and stored in file RevolutionEntity.sat. Next, with the help of Solidworks, transform the file into RevolutionEntity.sldprt file.

4.2. Generation of Slice Surfaces

Run Solidworks, open the file Converter.txt, read in the point cloud data, and then the points on the slice surfaces of the flow passage will appear in Solidworks graphic window, as shown in **Figure 3**.

In order to transform these discrete points into smooth surfaces, the ScanTO3D Wizard tool can be used to accomplish this function. The generated slice surfaces are shown in **Figure 4**.

4.3. Inserting Revolution Entity

In the drop-down menu of Solidworks, select "Insert Part" command, select the RevolutionEntity.sldprt. Thus, the revolution entity is inserted into the graphic window of SolidWorks, as shown in **Figure 5**.

4.4. Slicing the Revolution Entity

By using "Insert" command as well as its sub-command, the entity slice is completed. After slicing, the entity is shown in **Figure 6**.

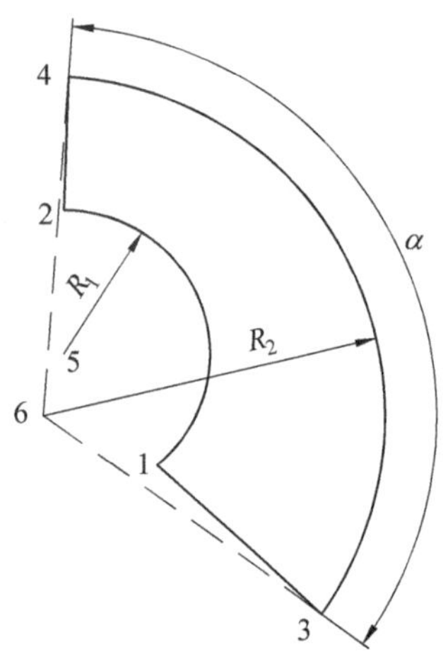

Figure 2. Pump torus of hydrodynamic torque converter.

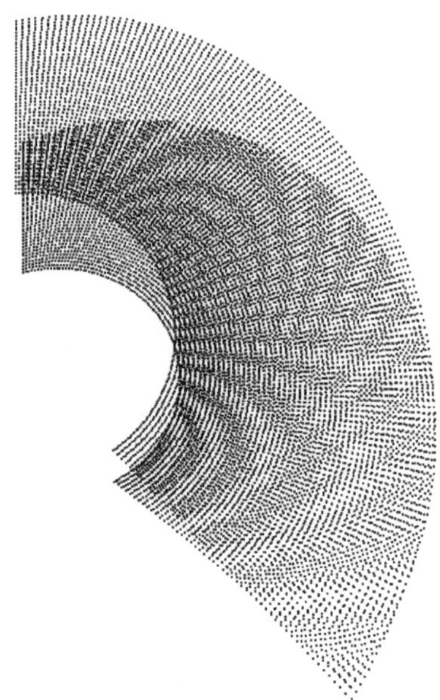

Figure 3. Discrete points on slice surfaces.

Figure 4. Two generated slice surfaces.

Figure 5. A revolution entity and two slice surfaces.

Figure 6. Revolution entity after slicing.

From **Figure 6**, it can be seen that the two slice surfaces still exist after completing slice operation. This is because Solidworks software was design in history base. The two surfaces cannot be deleted simply. Otherwise the model will be destroyed. The simplest solution is to insert the entity into a part drawing.

Something also needs to be said about the file format used to store the model. Parasolid format can be a correct selection. Enter the file name Converter.x_t, and save the file. At this stage, the model of pump flow passage is shown in **Figure 7**.

From **Figure 7**, it can be found that each lateral surface of flow passage model is not a single surface, but consists of a number of small surfaces. Moreover, the number of small surfaces is not fixed.

4.5. Further Process of Flow Passage Model

In order to facilitate the automatic generation of mesh, by using Gambit, these small surfaces of each lateral surface can be merged into a large surface, so as to make the model become a surface hexahedral. After merging surfaces, the model is shown **Figure 8**.

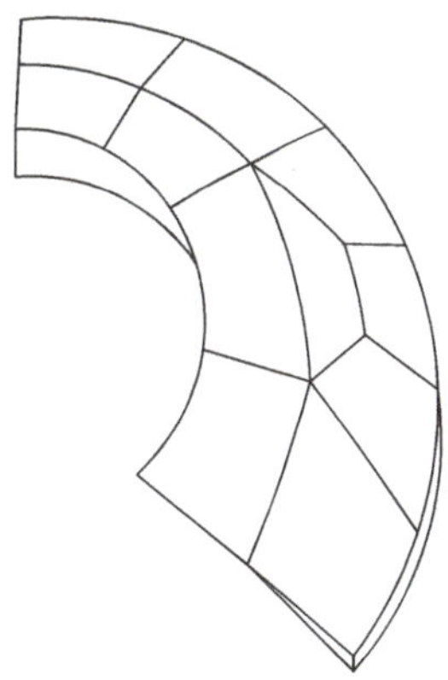

Figure 7. Pump flow passage model created by using one revolution entity and two slice surfaces.

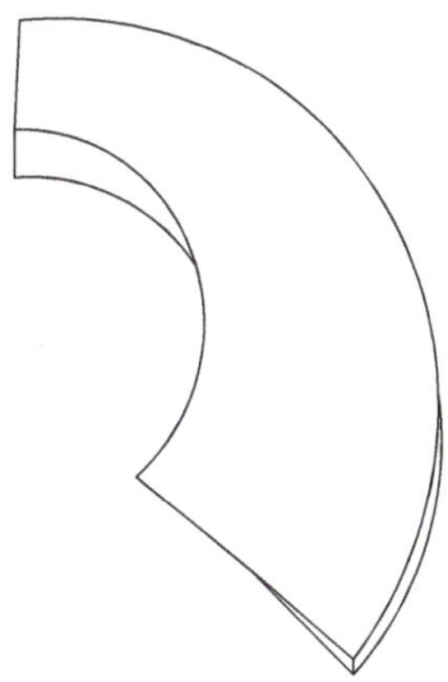

Figure 8. Pump flow passage model after merging small curved surfaces.

5. Parameterized Program

In order to achieve modeling automation or semi-automation, a parameterized program code is developed. Various design and modeling tasks, such as, design calculation, graphic drawing, file output, etc., are accomplished with the program code. The program's input and output interface is shown in **Figure 9**.

From the figure, it can be seen that the program interface includes 10 input parameters and 2 output parameters. After the input parameters are given, click on the Design button. It can be observed the 2 output parameters. If the 2 output parameters are satisfactory, click on the Exit button to end the program. After running, the program will output two important files, Converter.lsp and Converter.txt. The former is used to generate a three-dimensional revolution entity, while the later is used to generate two slice surfaces.

Practice has stated that the various design and modeling tasks can be completed within several minutes with the help of the program code.

6. Model Error Analysis

Of the six surfaces of the torque converter flow passage model, entrance surface and exit surface are accurate, while the inner wall surface and the outer wall surface are not very accurate. In addition, the two lateral surface of the flow passage model are not very accurate, either. The errors of the inner surface and the outer wall surface result from the approximation of meridional streamline, while the errors of two lateral surfaces of the flow passage model result from the approximation of central stream surface.

6.1. Meridional Streamline Approxi-Mation Error

As meridional streamlines of inner and outer wall are approximated with circular arcs, which will result in model error. The error can be expressed as the arc radius

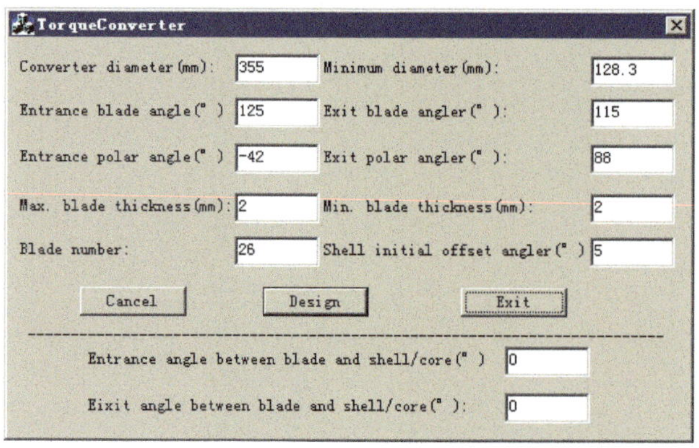

Figure 9. Input and output interface of program.

error. With the help of the program code, the absolute error and relative error of arc radius can be obtained directly. As an example, model error of YB355 torque converter is calculated.

For the pump and turbine, the maximum absolute error of inner wall arc radius is 0.19376853 mm, and the maximum relative error is 0.7983%. The maximum absolute error of outer wall arc radius is 0.43897597 mm, and the maximum relative error is 0.7703%.

For the stator, the maximum absolute error of inner wall arc radius is 0.02159697 mm, and the maximum relative error is 0.080558%. The maximum absolute error of outer wall arc radius is 0.1617693 mm, and the maximum relative error is 0.326664%. These data indicate that the condition optimum arc approximation possesses substantially high approximation accuracy and can meet the requirements of engineering practice.

6.2. Central Stream Surface Approxi-Mation Error

As the central stream surface is approximated with a quadratic surface, the node coordinates are accurate, but the other point coordinates are not very accurate. It is very difficult to derive the error formula used for the quadratic surface approximation. A feasible selection is to directly calculate the errors by using numerical method.

First, take on $\xi = (\xi_i + \xi_{i+1})/2$ and $\theta = (\theta_j + \theta_{j+1})/2$. Next, by using (14), theoretical coordinate (x_t, y_t, z_t) and theoretical revolution radius r_t can be obtained. After that, substitute $y = r_t \cos\gamma$ and $z = r_t \sin\gamma$ into (15). Finally solve the equation for γ. Thus, approximate coordinates are $y_a = r_t \cos\gamma$ and $z_a = r_t \sin\gamma$.

The absolute error expressions of *y*-coordinate and *z*-coordinate are

$$\begin{cases} e_y = |y_t - y_a| \\ e_z = |z_t - z_a| \end{cases} \quad (25)$$

The relative error expressions of *y*-coordinate and *z*-coordinate are

$$\begin{cases} \varepsilon_y = |(y_t - y_a)/y_t| \times 100\% \\ \varepsilon_z = |(z_t - z_a)/z_t| \times 100\% \end{cases} \quad (26)$$

With the help of the program code, the error values can be obtained.

For the pump, the absolute error $e_{ymax} = 6 \times 10^{-6}$ mm and $e_{zmax} = 23 \times 10^{-6}$ mm. The relative error $\varepsilon_{ymax} < 10^{-6}\%$ and $\varepsilon_{zmax} < 10^{-6}\%$.

For the turbine, the absolute error $e_{ymax} = 6 \times 10^{-6}$ mm and $e_{zmax} = 52 \times 10^{-6}$ mm. The relative error $\varepsilon_{ymax} < 10^{-6}\%$ and $\varepsilon_{zmax} < 10^{-6}\%$.

For the stator, the absolute error $e_{ymax} = 22 \times 10^{-6}$ mm and $e_{zmax} = 44 \times 10^{-6}$ mm. The relative error $\varepsilon_{ymax} < 10^{-6}\%$ and $\varepsilon_{zmax} < 10^{-6}\%$.

The above data clearly show that the quadric surface approximate is substantially accurate and able to meet the requirements of engineering practice.

7. Conclusions

The modeling technique of torque converter is investigated. The main results of this paper are as follows:

1) The semi-automatic modeling technique of torque converter is put forward and implemented;

2) A new approximation method, condition optimum approximation, is proposed. And, the method was used for the arc approximation of the meridional streamlines of inner and outer wall. In this manner, the problem that the three-dimensional revolution entity is automatically generated is solved;

3) The central stream surface of flow passage is approximated with a quadratic surface. The coordinates of the blade surface points and the flow passage lateral surface points are substantially accurately calculated with numerical method. The problem of automatic generation of slice surface is fairly well solved;

4) The various tasks (design calculation, graphic drawing, file output, etc.) are accomplished with a parameterized program code, achieving semi-automatic modeling of torque converter passage, and greatly reducing the modeling time-consumption;

5) By means of the program code, the errors of flow passage model are calculated and analyzed with numerical method. The error estimation of modeling was solved.

In short, the semi-automatic modeling technique is of engineering application value.

8. Acknowledgements

The authors wish to thank the financial support of Henan Provincial Tackle Key Program of China. In addition, the authors would like to thank Prof. L. Quan, for his help and advice in the research.

REFERENCES

[1] M. Yamada and K. Imai, *et al.*, "Numerical Analysis of the Torque Converter Stator Blade by the Boundary Element Method," SAE Paper No. 921692, 1992.

[2] R. R. By and R. Kunz, *et al.*, "Navier-Stokes Analysis of the Pump Flow Field of an Automotive Torque Converter," *Journal of Fluids Engineering*, Vol. 117, No. 1, 1995, pp. 116-122.

[3] H. J. Chang, M. Athavale and S. Shin, "Numerical Investigation of the Pump Flow in an Automotive Torque Converter," SAE Paper No. 1999-01-1056, 1999.

[4] P. Attibele and V. Korivi, *et al.*, "Torque Converter CFD Engineering-Part I: Torque Ratio and K Factor Improvement through Stator Modifications," SAE Paper No. 2002-01-0883, 2002.

[5] M. Cigarini and S. Jonnavithula, "Fluid Flow in an Automotive Torque Converter: Comparison of Numerical Results with Measurements," SAE Paper No. 950673, 1995.

[6] D. X. Zhao and X. Z. Shi, *et al.*, "Method of Internal 3-D Flow Field Numerical Simulation for Hydrodynamic Torque Converter," *Journal of Jilin University* (*Engineering and Technology Edition*), Vol. 36, No. 3, 2006, pp. 199-203.

[7] P. Yan and G. Q. Wu, "Performance Investigation of Torque Converter," Journal of Tongji University(Natural Science), Vol. 32, No. 11, 2004, pp. 1504-1507.

[8] H. Tian and A. L. Ge, *et al.*, "Numerical Analysis on Internal Flow Field in Pump Wheel of a Torque Converter," *Journal of Jilin University* (*Engineering and Technology Edition*), Vol. 34, No. 3, 2004, pp. 378-382.

[9] H. Schulz and R. Greim, "Calculation of Three-Dimensional Viscous Flow in Hydrodynamic Torque Converters," *Journal of Turbomachinery*, Vol. 118, No. 6, 1996, pp. 578-589.

[10] Y. Liu and Y. X. Pan, *et al.*, "Numerical Analysis on Three-Dimensional Flow Field of Turbine in Torque Converter," *Chinese Journal of Mechanical Engineering*, Vol. 20, No. 2, 2007, pp. 94-96.

[11] W. S. Lim and C. Lee, *et al.*, "Three- Dimensional Flow Field Simulation to Estimate Performance of a Torque Converter," SAE Paper No. 2000-01-1146.

[12] S. P. Liu and L. Quan, "Mathematical Model of Hydrodynamic Torque Converter and Analytic Description of Streamline," *Chinese Journal of Mechanical Engineering*, Vol. 22, No. 1, 2009, pp. 70-77.

[13] S. P. Liu and L. Quan, "Torus Streamline Method Design of 3-Element Centripetal-Turbine Hydraulic Torque Converters," *Transactions of the Chinese Society for Agricultural Machinery*, Vol. 40, No. 9, 2009, pp. 20-24, 44.

[14] M. C. Xi, "Numerical Analysis Methods," University of Science and Technology of China Press, Hefei, 2003.

Nomenclature

h = distance from rotation axis line to the geometric center of design streamline;
θ = polar angle at a point located on the meridional streamline;
ρ = polar radius at a point located on the meridional streamline;
σ = constant;
$\delta = R/h$;
R = geometric radius of design streamline;
r = arc radius or revolution radius;
S = offset arc length;
t = one half of blade thichness;
ϕ = offset angle;
α = arc central angle;
ψ = flow passage angle (2π divided by blade number);
x, y, z = Cartesian coordinates;
ξ = parameter used to specify blade span direction, $-1 \le \xi \le 1$. For the inner wall $\xi = -1$; for the outer wall $\xi = 1$; for the design streamline $\xi = 0$.

Subscription

C = arc center;
1 = first discrete point or pump/turbine/stator inlet;
t = theoretical value;
a = approximation calculation value;
n = last discrete point or pump/turbine/stator exit;
m = intermediate discrete point.

Investigation of a Pressure Sensor with Temperature Compensation Using Two Concentric Wheatstone-Bridge Circuits

Chi-Chang Hsieh[1], Chih-Ching Hung[2], Yan-Huei Li[3]
[1]Department of Mechanical and Automation Engineering, National Kaohsiung First University of Science and Technology, Taiwan
[2]Department of Product Design, Shu-Te University, Taiwan
[3]Department of Mechanical and Automation Engineering, National Kaohsiung First University of Science and Technology, Taiwan

ABSTRACT

This study presents a silicon-based pressure sensor with temperature compensation. The eight piezoresistors were designed on the polycrystalline silicon membrane and constructed by two concentric Wheatstone-bridge circuits to perform two sets of sensors. The sensor in the central circuit measures the membrane deflection caused by the combined effects of pressure and temperature, while the outer one measures only the deflection caused by the working temperature. From this arrangement, it is reliable and accurate to measure the pressure by comparing the output signals from the two concentric Wheatstone-bridge circuits. The optimal positions of the eight piezoresistors were simulated by simulation software ANSYS. The investigated pressure sensor was fabricated by the micro electro-mechanical systems (MEMS) techniques. The measuring performance and an indication of the conventional single Wheatstone-bridge pressure sensor is easily affected under variation of different working temperature and causes a maximum absolute error up to 45.5%, while the double Wheatstone-bridge pressure sensor is able to compensate the error, and reduces it down to 1.13%. The results in this paper demonstrate an effective temperature compensation performance, and have a great performance and stability in the pressure measuring system as well.

Keywords: Pressure Sensor; Temperature Compensation; Wheatstone-Bridge Circuit; MEMS

1. Introduction

Pressure sensors have been extensively applied in the automobile industries, environmental monitoring systems, process controls and biomedical fields [1-4]. The pressure sensor was traditionally manufactured by large-scale mechanical devices, while recently the micro-electro-mechanical systems (MEMS) technologies have the advances and great improvements to fabricate highly-sophisticated micro-sensors with excellent repeatability characteristics and ultra-low-pressure detection capabilities [5-7]. In such devices, the pressure is generally measured by piezoelectric thin films [8,9] or the form of capacitance structures [10]. Since the capacitive-type pressure sensors have better sensitivity to the influences of the parasitic capacitance, this pressure sensors do not interfere with the test results of pressure by the outer temperature. By contrast, the sensitivity, linearity and repeatability characteristics of the piezoresistive-type pressure sensors have been greatly improved, and can be easily manufactured by mass production using standard MEMS-based techniques to reduce the cost [11,12]. Most of piezoresistive pressure sensors are easy to produce the error or deformation for test result of pressure when the outer temperature yields variations. In the conventional piezoresistive pressure sensors, the sensing element has four diffused or ion-implanted strain gauges which compose a single Wheatstone-bridge circuit on a thin silicon membrane [13]. Such devices have the drawback of zero-pressure offset error due to the temperature [14]. It was shown that the zero-pressure offset error in the single Wheatstone-bridge piezoresistor was primarily caused by the thermal residual stresses due to different thermal expansion coefficients of the monocrystalline silicon substrate and the polycrystalline silicon membrane [15,16]. In

Reference [15], the dual bridge pressure sensor only considers the different pressure variables will lead to what influences for the test result of the pressure sensor. However, the different pressure and temperature variables should be included in the measurement of the pressure sensor.

The purpose of this paper is to present a piezoresistive pressure sensor which composed by two concentric Wheatstone-bridge circuits on a polycrystalline silicon membrane, shown in **Figure 1**. This piezoresistive pressure sensor has accurately compensation of temperature, high-sensitivity, and high-reliability characteristics. The proposed pressure sensor was fabricated by conventional MEMS-based techniques, including low-pressure chemical vapor deposition (LPCVD), light-boron and heavy-boron ion implantation, dry etching, electrochemical etch-stop processing, and anodic bonding. In this investigation, the eight piezoresistors were designed on the polycrystalline silicon membrane, see **Figure 1**, and this design is to find out the best arrangement of the two central and the outer Wheatstone-bridge circuits to compensate the errors. For this reason, this study is to combine the pressure sensor circuit (inner piezoresistors) and temperature compensation circuit (outer piezoresistors) on the same sensor. It can not only enhance the performance, but also shrink the volume of the pressure sensor system.

The four piezoresistors of the Wheatstone-bridge circuit in the central region on the membrane measure the membrane deflection caused by the combined effects of the pressure and temperature, while the four outer ones measure only the deflection caused by the thermal load. A reliable indication of the pressure measurement can therefore be obtained by combining the output signals of the two Wheatstone-bridge circuits under different temperature surroundings. Here, the stress distributions caused by the working temperature and pressure respecttively on the membrane was simulated by finite element analysis (FEA) method and ANSYS software tool in order to establish the optimal locations of the eight piezoresistors in each Wheatstone-bridge circuits. The experiment conditions of the introduced pressure sensor are under the pressures ranging from zero to 150 psi and the temperatures ranging from −10°C to 75°C.

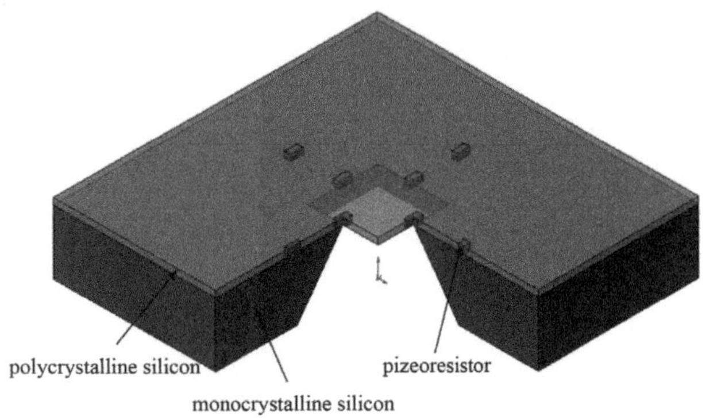

Figure 1. Illustration of the two concentric Wheatstone-bridge circuits on the polycrystalline silicon membrane.

2. Numerical Simulation and Analysis

In the practical applications of pressure sensors, the membrane deformations are caused by both the working temperature and pressure. The variations of temperature will affect the measuring result of pressure. The phenomena call thermal stress when the variations of pressure lead to the membrane deformations. Therefore, the thermal stress is a major pressure course for the membrane. And, the different materials have different values of thermal expansions and contractions that will bring about different thermal stress. Similarly, different locations of the piezoresistors will lead to different result of membrane deformations. As a result, by using the simulation to analysis the influence of pressure is absolutely to do when establish the model in the initial phase. Here, this study used the FEA as the simulation and analysis method. And, by applying the ANSYS software as the simulation and analysis tool to obtain the maximum pressure distribution, and thus, the optimal locations of the piezoresistors on the membrane will be got.

Figures 2(a) and **(b)** illustrate the top and bottom views of the FEA model respectively. The settings and definitions of the materials are basically monocrystalline and polycrystalline silicon. Recently, the integrated circuit trends to have more and more complex with multi-functions but smaller and smaller in size. It needs much more exact calculation and designation for electrical characteristics. Poor design and process-induced mechanical stress effect will cause serious problem during fabrication stages, especially for the packaging process in the front-end process, which make more variation to electrical characteristics and affect the quality and performance of the device in the end. Mechanical stress induced unstable and variation of electrical characteristics is one of the major factors to affect the entire performance of a device seriously. And, the different type of silicon has the different direction of etch in the process of fabrication. The different direction of etch will affect the electrical characteristics of sensor. Therefore, in this study, the monocrystalline silicon is design as the substrate, where the polycrystalline silicon is the membrane of the pressure sensor. The polycrystalline silicon membrane will deform when the pressure is loaded, and the deformation analysis due to the thermal stresses can be calculated by different thermal expansion coefficients of the monocrystalline and polycrystalline silicon materials. The properties of the materials are shown in **Table 1**.

Considering the boundary conditions in the ANSYS software, the displacement of each node along the X, Y, Z axes at the bottom of the substrate respectively are set

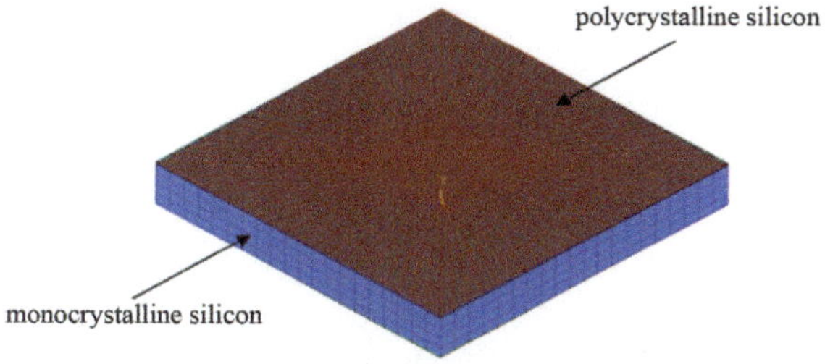

(a)

(a)

Figure 2. The (a) top view, and (b) bottom view of the ANSYS FEA model.

Table 1. Material properties of the monocrystalline substrate and the polysilicon membrane.

	monocrystalline siliconsubstrate	polycrystalline siliconmembrane
Young's modulus (Pa)	190×10^9	160×10^9
Poisson's ratio	0.23	0.23
density (kg/cm^3)	2330	2330
thermal expansion coefficient (1/°C)	2.6×10^{-6}	2.8×10^{-6}

to be zero. The types of the loadings are considered as: pressure loadings and temperature loadings. The loading conditions are divided into three parts: applying temperature without pressure; applying pressure at room temperature (25°C); and applying both pressure and temperature loadings. Under the above loading conditions, the effects and influences about the pressure and temperature to the membrane deformations can be analyzed. The values of the loading conditions are ranging from zero to 150 psi, and −10°C to 75°C for pressure and temperature respectively. Since the stress distributions caused by simply pressure loading, simply temperature loading, or both pressure and temperature loadings, the pressure and temperature distributions on the membrane of the pressure sensor can be known after the simulation. Furthermore, the best locations of the four inner piezoresistors can be obtained based on the maximum stress distribution to perform an effective output measurement indication.

3. Fabrication Process and Measurement

In this study, the fabrication of the proposed double Wheatstone-bridge pressure sensor is in comparison with the conventional single Wheatstone-bridge pressure sensor. Both the dimensions are $1600 \times 1600 \times 425$ μm in length, width, and height respectively.

3.1. Fabrication of the MEMS-Based Pressure Sensors

In the fabrication process, the two types of sensors were patterned using the mask designs shown in **Figures 3(a)** and **(b)** respectively. Note that the patterning process was performed using cheaper film-photo masks with 30 μm line width and only two quartz masks with 1 μm line width respectively. As a result, it was found that the application of combining the two types of masks above reduced nearly 40% of the fabricating cost.

Figure 4 illustrates a schematic overview of the fabrication processes of the two types of pressure sensors, and the only difference is the choices of the mask patterns. In the beginning, a thin (20 μm) layer of polycrystalline silicon was deposit by the LPCVD technique on the surface of monocrystalline silicon substrate. The substrate was then located in the oxidation furnace at temperature 1092°C and pressure 1 Pa for 30 hours to form a thin (800 Å) silicon oxide layers on the top and bottom surfaces of it shown in **Figure 4(a)**. The substrate was then patterned using a photolithography technique, and immersed in a Buffered Oxide Etch (**BOE**) **solution** to remove the silicon oxide layer from the bottom surface. Partly of the silicon oxide layer from the top surface was also removed at the same time in order to create the designed basic outline of the piezoresistor structures shown in **Figure 4(b)**. An ion-implantation method was then utilized to implant the polycrystalline silicon layer with boron to create the piezoresistor structures. In order to establish a resistance gradient within the structure, the light-ion implantation process was initially performed with 3.2×10^{14} atom/cm^3 boron implant intensity to create a thin and high-resistance boron layer shown in **Figure 4(c)**. After this procedure, the heavy-ion implantation process was performed with 1.5×10^{16} atom/cm^3 boron implant intensity to create a thicker and low-resistance layer shown in **Figure 4(d)**. Note that in both cases of the plating processes are performed by 80 KeV implantation energy. In the following procedure, the LPVCD technique was utilized to deposit the silicon nitride layers with a thickness of 1400 Å on the top and bottom surfaces of the substrate shown in **Figure 4(e)**. A reactive ion etching (RIE) process was then performed to remove the silicon nitride layer from the top surface shown in **Figure 4(f)**. After this, an aluminum layer with a thickness of 900 Å was sputtered on the top surface, patterned lithographically using AZ5214N photoresist, and etched at a temperature of 65°C in H_3PO_4 acid to create the electrical Wheatstone-bridge circuits shown in

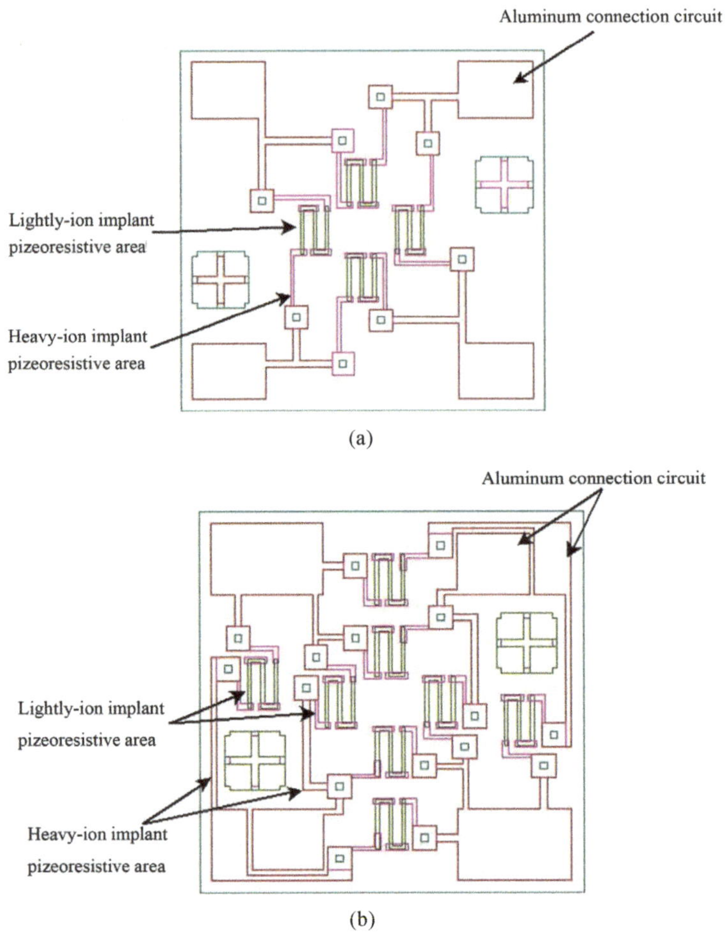

Figure 3. Mask designs of (a) the single Wheatstone-bridge, and (b) the double Wheatstone-bridge pressure sensors.

Figure 2. The silicon nitridelayer on the bottom surface of the substrate was then removed using a RIE process in order to produce an etching mask shown **Figure 4(g)**, and the substrate was etched by KOH chemical composition to create a cavity shown in **Figure 4(h)**. Eventually, the remaining silicon nitride layer was removed from the bottom surface of the substrate by an etching process shown in **Figure 4(i)**, and the substrate was finally fused with 7740 glass by an anodic bonding technique at 470˚C temperature shown in **Figure 4(j)**.

The substrate cavity was etched by an electrochemical etch-stop process to accurately control the etching depth, and not to destroy the polycrystalline silicon membrane. The criterion of etching stop depends on the etching depth. In this study, a small current was initially detected after 150 minutes etching, and a large amount of silicon substrate was removed. It was observed that the current increased radically after 250 minutes etching, and this indicates that the etching process was completed and stopped, and the polycrystalline silicon membrane was not destroyed yet. **Figure 5** illustrates the variations of currents with time in the electrochemical etch-stop process. **Figure 6** shows an optical microscope image of the etched cavity, and **Figure 7** shows a cross-sectional scanning electron microscope (SEM) image of the pressure sensor bonded with 7740 glass.

3.2. Packaging of the Pressure Sensor

After the following fabrication process described in the above section, a total number of 1600 pressure sensor dies were manufactured. The individual pressure sensor die was then ready to be separated using a diamond cutter and assembled to a printed circuit board (PCB) by epoxy glue at temperature 80˚C for 25 minutes shown in **Figure 8(a)**. Thereafter, the gold wires were bonded between the sensor die and the PCB, and then packaged by a thermo-setting polyimide resin at a temperature 100˚C for 30 minutes. Finally, an input pressure nozzle and an O-ring were assembled shown in **Figure 8(b)**.

3.3. Measurement of the Pressure Sensor

Figure 9 illustrates the layouts and equipments for the measurement of the pressure sensor. Two types of the pressure sensors were tested at temperatures ranging from −10˚C to 75˚C and pressures ranging from zero to

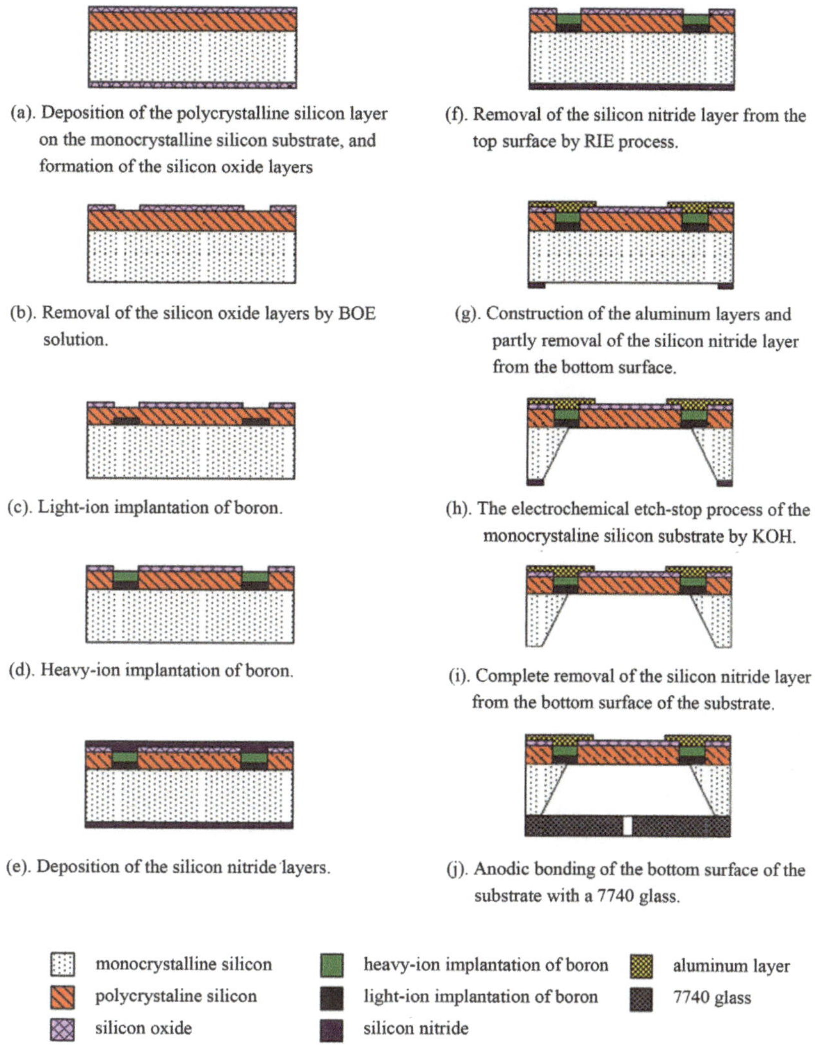

Figure 4. Overview of the pressure sensor fabrication process.

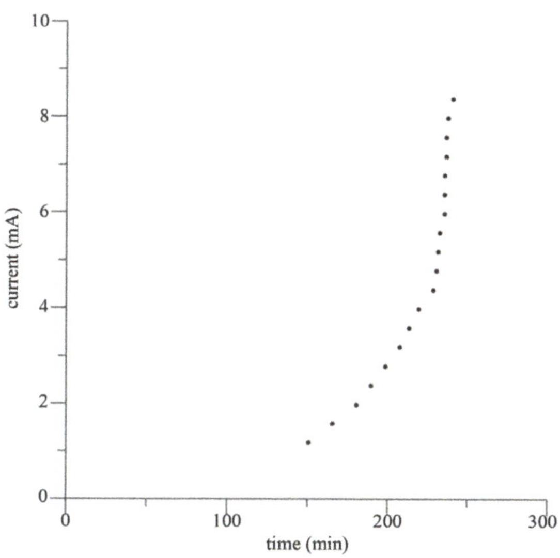

Figure 5.The variations of currents with time in the electrochemical etch-stop process.

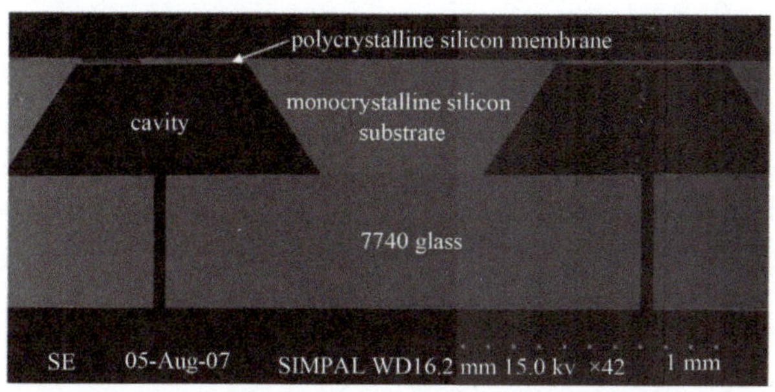

Figure 6. An optical microscope image of the etched cavity.

150 psi in the GH800 GIANT FORCE chamber, and the pressure was supplied by a high pressure gas cylinder which was controlled by the DRUCK DPI515 system. The input voltage was supplied by Agilent-E3633A DC power supply with 50V maximum output voltage and 4A current. The output signal of the pressure sensors was measured by GW-DIGITAL MULTIMETER (GDM-8055) with a minimum 1 μV measuring accuracy.

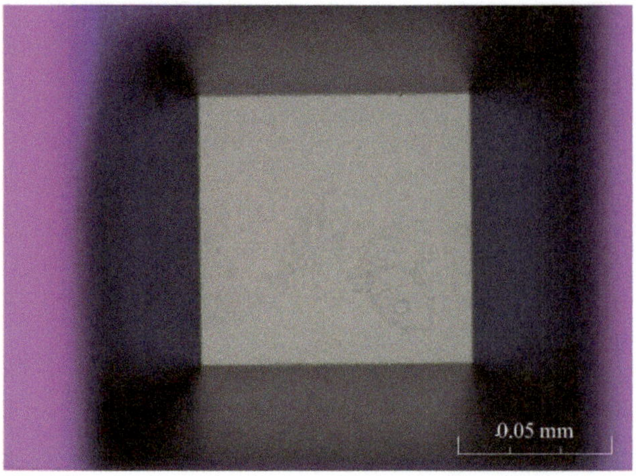

Figure 7. A cross-sectional SEM image of the pressure sensor bonded with a 7740 glass.

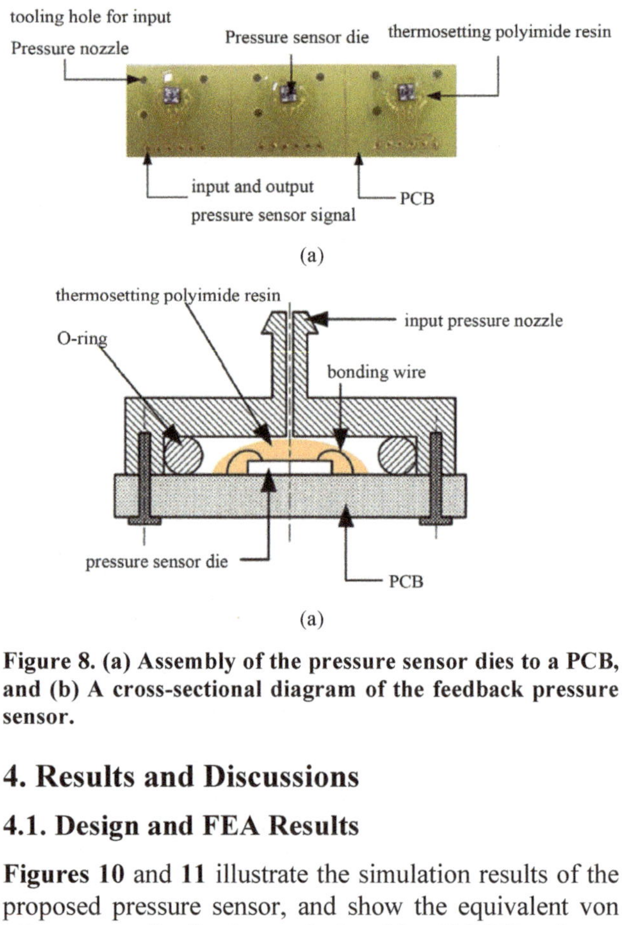

Figure 8. (a) Assembly of the pressure sensor dies to a PCB, and (b) A cross-sectional diagram of the feedback pressure sensor.

4. Results and Discussions

4.1. Design and FEA Results

Figures 10 and **11** illustrate the simulation results of the proposed pressure sensor, and show the equivalent von Mises stress distributions calculated by ANSYS software. Since the temperature has an obvious effect to the membrane and lead to it yields the deformation, the membrane deformed locations can be used to set the four outer piezoresistors in order to perform the temperature compensation. **Figure 10** illustrates the simulation resultunder 75˚C thermal load without pressure in order to estimate the locations of estimate the four outer piezoresistors, and the temperature should take into consideration if the working temperature is above the 25˚C room temperature. When the pressure sensor has been pressured, the edge of middle of the membrane will be produced maximum stress. The location with the edge of middle of the membrane can be used to set the four inner piezoresistors. **Figure 11** shows the equivalent von Mises stress distribution of the pressure sensor under the combination of 75˚C thermal load and 150 psi pressure load and it estimates the locations of the four inner piezoresistors. Therefore, this study in order to find out the optimal designing location of the two concentric Wheatstone-bridge circuits, it is necessary to know the maximum stress distributions which caused by the thermal and pressure loads individually, shown in the **Figures 10** and **11**. As a result, the centric Wheatstone-bridge circuit was selected to design in the location where the maximum von Mises stress appeared due to the pressure load. And the outer one is located where the maximum von Mises stress appeared due to the thermal load.

4.2. Analysis of the Output Voltage Affected by Temperature and Pressure

In the following subsections demonstrate that the temperature has a considerable effect to the output voltage in the conventional single Wheatstone-bridge pressure sensor. Nevertheless, the proposed pressure sensor is able to compensate the error affected by the working temperature using the double Wheatstone-bridge circuits design.

4.2.1. The Relation of the Output Voltage to the Pressure Load under Fixed Temperature

The experiment input voltage was set to be 5 V for the single Wheatstone-bridge pressure sensor at 25˚C room temperature, and the pressure values were set to be zero psi, 50 psi, 100 psi, and 150 psi as the input parameters respectively. Therefore, the 0.358 V, 0.270 V, 0.191 V, and 0.110 V output voltages were measured respectively. The diagram of the measured output voltages $\left(V_{\text{output}}\right)$ to the input pressures (P) are shown in **Figure 12**, and their relation can be obtained as an idealized equation below:

$$V_{\text{output}} = -0.0016P + 0.359 \tag{1}$$

The idealized output voltage at 25 psi, 75 psi, and 125 psi respectively can then be calculated in advance from Equation (1) shown in **Figure 13**, and the 0.32%, 0.66%, and 2.63% absolute error in comparison with the measured values were analyzed respectively. It demonstrated that the pressure sensor with a single Wheatstone-bridge circuit has its accuracy and approximately linear output properties at room temperature under zero to 150 psi input pressure load.

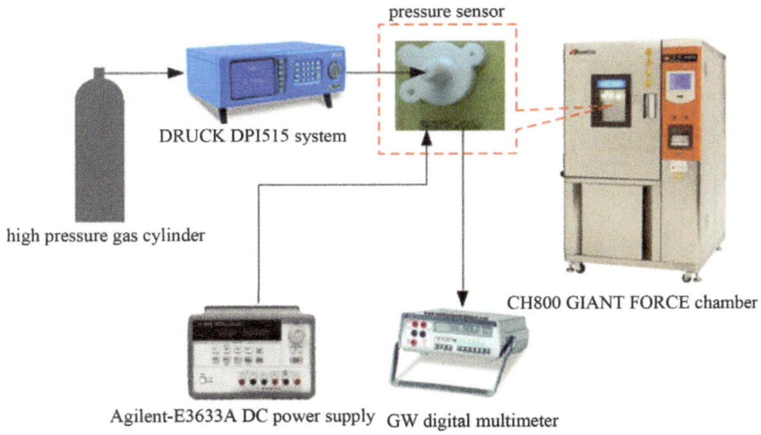

Figure 9. A schematic illustration of the layouts and equipments for the measurement.

Figure 10. Results of the ANSYS FEA simulationunder 75˚C thermal load without pressure.

Figure 11. Results of the ANSYS FEA simulation under the combination of 75˚C thermal.

4.2.2. Analysis of the Output Voltage to the Input Pressure under Different Temperature

The output voltages of the single Wheatstone-bridge pressure sensor were then experimented and measured under different temperature circumstances. It was found in **Figure 14** that the output voltage varied from 0.335V to 0.390 V under zero pressure setting with a temperature variation from −10˚C to 75˚C. The values of the output voltage without pressure loadings are the so-called zero-pressure offset errors. These values are very critical and will lead to further errors. **Figure 14** illustrates that the zero-pressure offset errors under different temperature conditions exist a maximum 8.2% absolute error in comparison with the standard output voltage at 25˚C room temperature. Furthermore, **Figure 14** also illustrates the output voltages of the single Wheatstone-bridge pressure sensor varied from 0.060 V to 0.178 V at 150 psi pressure loading under different temperature conditions, and there exists a maximum 45.5% absolute error in comparison with the standard output voltage at 25˚C room temperature. The experiment proved that the accuracy of the single Wheatstone-bridge pressure sensor is easily affected by the surrounding temperature that leads to zero-pressure offset errors.

4.3. Figures and Tables

According to the above section, the single Wheatstone-bridge pressure sensor is easily to produce the zero-pressure offset errors by the different surrounding temperature. The proposed pressure sensor will improve this disadvantage. From the measurement of experiments, it was found that the zero-pressure offset errors can be reduced by means of the proposed design under different working temperature shown in **Figure 15**. The values of the measured output voltage varied from 0.0766 V to 0.0768 V under zero pressure setting with a temperature variation from −10˚C to 75˚C. The zero-pressure offset error of the proposed design has only a maximum 0.06% absolute error in comparison with the standard output voltage at 25˚C room temperature, and has a great compensation improvement up to 99.94% caused by the surrounding temperatures. Furthermore, **Figure 15** alsoillustrates the output voltages varied from 0.0745 V to 0.0747 V at 150 psi pressure loading under different temperature conditions, and there only exists a maximum 1.13% absolute error in comparison with the standard output voltage at 25˚C room temperature. It was found that the error will not increase with the increasing pres-

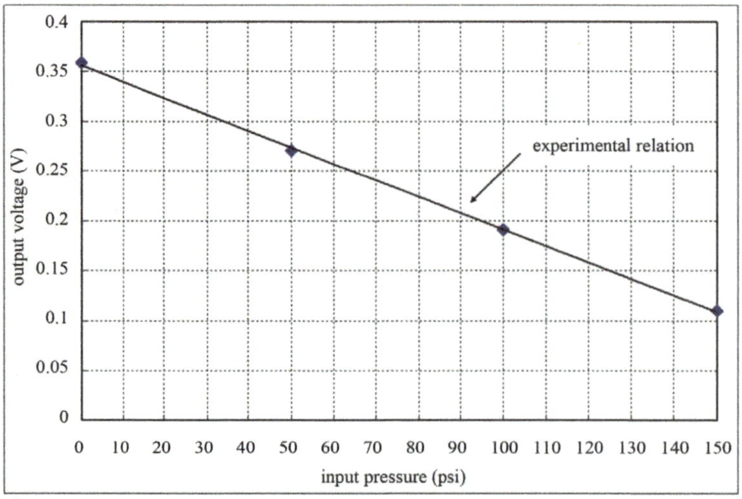

Figure 12. The relation between the measured output voltages and the input pressures under 0 psi, 50 psi, 100 psi, and 150 psi pressure loads respectively.

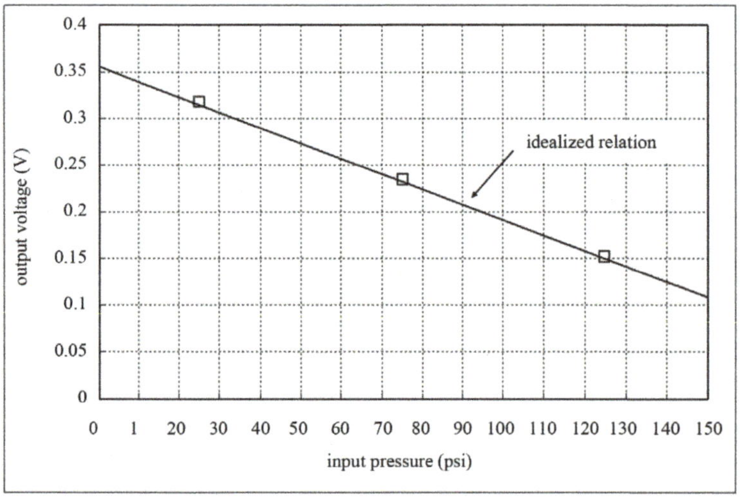

Figure 13. The relation between the measured output voltages and the input pressures under 25 psi, 75 psi, and 125 psi pressure loads respectively.

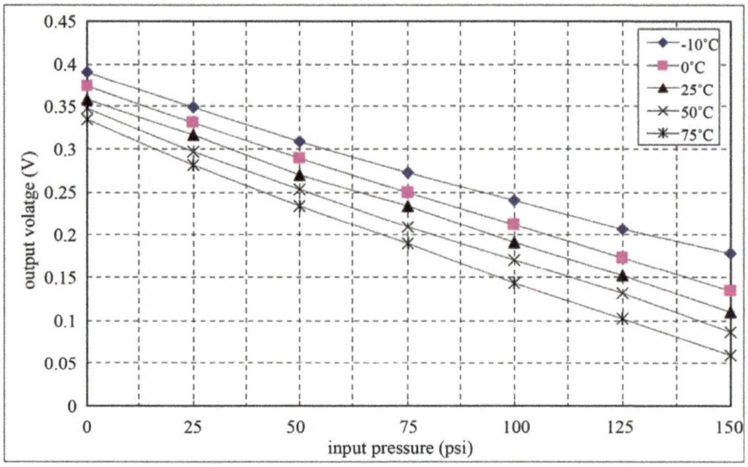

Figure 14. The relation between the measured output voltages and the input pressures of the single Wheatstone-bridge pressure sensor.

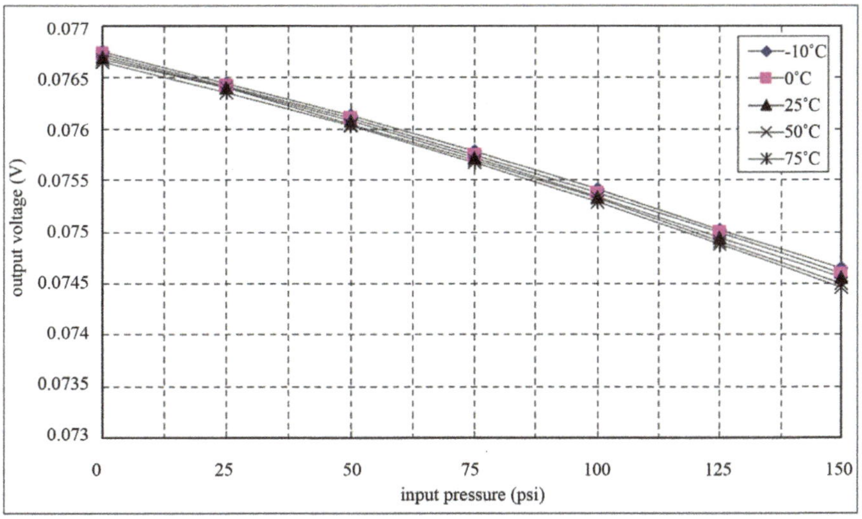

Figure 15. The relation between the measured output voltages and the input pressures of the proposed double Wheatstone-bridge pressure sensor.

sure loading. The proposed pressure sensor is better than the single Wheatstone-bridge pressure sensor when compared the results between the **Figures 14** and **15**. Consequently, it has been proved that the proposed double Wheatstone-bridge pressure sensor has its accuracy and stability for the application to the pressure measuring system under temperatures ranging from −10°C to 75°C.

5. Conclusion

The double Wheatstone-bridge pressure sensor with temperature compensation that proposed in this paper is fabricated by MEMS techniques. And, it is based on FEA method and ANSYS tool to simulate and analyze the optimal locations of the inner and outer piezoresistors in the membrane. The proposed pressure sensor with temperature compensation has been proved to have a better accuracy and stability in comparison with the conventional single Wheatstone-bridge pressure sensor. And, it enables to enhance the test result of pressure effectively and exactly when the pressure sensor faces the bad surrounding temperatures. The fabrication process proposed in this study can reduce nearly 40% of the manufacturing cost by using only two quartz masks. The maximum zero-pressure offset error was found to be only 0.06% in comparison with the standard output voltage at 25°C room temperature, and had a great improvement of temperature compensation up to 99.94%. Furthermore, maximum absolute error of the output voltage in comparison with the standard one at room temperature reduces down to 1.13%. As a result, the proposed pressure sensor with temperature compensation can be applied to the pressure measuring systems, high-tech industry, or vehicle safety systems, and it will benefit to increase the production yield rates for these industries.

6. Acknowledgements

This research was partly sponsored by the National Science Council of Taiwan under contracts NSC 98-2622-E-327-005-CC3 and NSC 98-2221-E-132-004.

REFERENCES

[1] S. A. Wright, H. Z. Harvey and Y. B. Gianchandani, "A Microdischarge-Based Deflecting-Cathode Pressure Sensor in a Ceramic Package," *Journal of Microelectromechanical Systems*, Vol. 22, No. 1, 2013, pp. 80-86

[2] A. L. Kang, W. B. Wang, Y. Q. Liu and T. Han, "A Wireless Pressure Sensor Based On Surface Transverse Wave," *IEEE Conference Publications*, Shenzhen, 9-11 December 2011, pp. 231-235.

[3] Y. Nam and J. W. Park, "Child Activity Recognition Based on Cooperative Fusion Model of a Triaxial Accelerometer and a Barometric Pressure Sensor," *IEEE Journal of Biomedical and Health Informatics*, Vol. 17, No. 2, 2013, pp. 420-426.

[4] X. Meng, K. D. Browne, S. M. Huang, C. Mietus, D. K. Cullen, M. R. Tofighi and A. Rosen, "Dynamic Evaluation of a Digital Wireless Intracranial Pressure Sensor for the Assessment of Traumatic Brain Injury in a Swine Model," *IEEE Transactions on Microwave Theory and Techniques*, Vol. 61, No. 1, 2013, pp. 316-325.

[5] M. M. Sisto, S. García-Blanco, L. L. Noc, B. Tremblay, Y. Desroches, J.-S. Caron, F. Provencal and F. Picard, "Pressure Sensing in Vacuum Hermetic Micropackaging in MOEMS-MEMS," *Journal of Micro/Nanolithography, MEMS, and MOEMS*, Vol. 9, No. 4, 2012, Article ID: 041109.

[6] L. Sujatha, V. S. Kale and E. Bhattacharya, "Critical Study of High-Sensitivity Pressure Sensors with Silicon/Porous Silicon Composite Membranes," *Journal of Micro/Nanolithography, MEMS, and MOEMS*, Vol. 8, No. 3,

2009, Article ID: 033070.

[7] Z. Wen, Z. Wen, G. Chen and S. Xu, "Fabrication of a Novel Vacuum Microelectronic Pressure Sensor," *Journal of Micro/Nanolithography, MEMS, and MOEMS*, Vol. 3, No. 4, 2004, pp. 574-578.

[8] P. Gonzalez, B. Guo, M. Rakowski, K. D. Meyer and A. Witvrouw, "CMOS Compatible Polycrystalline Silicon-Germanium Based Pressure Sensors," *Sensors and Actuators A*: *Physical*, Vol. 188, 2012, pp. 9-18.

[9] J. T. Huang and S. C. Cheng, "Study of Injection Molding Pressure Sensor with Low Cost and Small Probe," *Sensors and Actuators A*: *Physical*, Vol. 101, No. 3, 2002, pp. 269-274.

[10] G. Chitnis, T. Maleki, B. Samuels, L. B. Cantor and B. Ziaie, "A Minimally Invasive Implantable Wireless Pressure Sensor for Continuous IOP Monitoring," *IEEE Transactions on Biomedical Engineering*, Vol. 60, No. 1, 2013, pp. 250-256.

[11] L. Kaabi, A. Kaabi, J. Sakly and M. F. Abdel Malek, "Modeling and Analysis of MEMS Sensor Based on Piezoresistive Effects," *Materials Science and Engineering*: *C*, Vol. 27, No. 4, 2007, pp. 691-694.

[12] Z. Dibi, A. Boukabache and P. Pons, "Effect of the Membrane Flatness Defect on the Offset Voltage of a Silicon Piezoresistive Pressure Sensor," *ICECS Proceedings of the 10th IEEE International Conference*, United Arab Emirates, 14-17 December 2003, pp. 902-905.

[13] J. Wang, X. Xia and X. Li, "Monolithic Integration of Pressure plus Acceleration Composite TPMS Sensors with a Single-Sided Micromachining Technology," *Journal of Microelectromechanical Systems*, Vol. 21, No. 2, 2012, pp. 284-293.

[14] M. J. Hu, B. C. Sang and H. L. Jong, "Design of Smart Piezoresistive Pressure Sensor, Science and Technology," *The 5th Korea-Russia International Symposiumon Science and Technology*, Tomsk, 26 June-31 July 2001, pp. 202-205.

[15] C. Pedersen, S. T. Jespersen, J. P. Krog, C. Christensen and E. V. Thomsen, "Combined Differential and Static Pressure Sensor Based on a Double-Bridged Structure," *Sensors Journal for IEEE*, Vol. 5, No. 3, 2005, pp. 446-454.

[16] T. L. Young, D. S. Hee, K. Akihisa, Y. Tetsuhiro, M. Yoshinori, I. Makoto and N. Tetsuro, "Compensation Method of Offset and Its Temperature Drift in Silicon Piezoresistive Pressure Sensor Using Double Wheatstone-Bridge Configuration," *8th International Conference on Solid-State Sensors and Actuators*, Stockholm, 25-29 June 1995, pp. 570-573.

12

A Survey of Modeling and Control of Piezoelectric Actuators

Jingyang Peng, Xiongbiao Chen
Department of Mechanical Engineering, University of Saskatchewan, Saskatoon, Canada

ABSTRACT

Piezoelectric actuators (PEAs) have been widely used in micro- and nanopositioning applications due to their fine resolution, fast responses, and large actuating forces. However, the existence of nonlinearities such as hysteresis makes modeling and control of PEAs challenging. This paper reviews the recent achievements in modeling and control of piezoelectric actuators. Specifically, various methods for modeling linear and nonlinear behaviors of PEAs, including vibration dynamics, hysteresis, and creep, are examined; and the issues involved are identified. In the control of PEAs as applied to positioning, a review of various control schemes of both model-based and non-model-based is presented along with their limitations. The challenges associated with the control problem are also discussed. This paper is concluded with the emerging issues identified in modeling and control of PEAs for future research.

Keywords: Piezoelectric Actuators; Modeling; Control

1. Introduction

Piezoelectric actuators (PEAs) and PEA-driven positioning systems have been widely used in the fields of micro- and nanopositioning such as atomic force microscopes [1-3], adaptive optics [4], computer components [5], machine tools [6], aviation [7], internal combustion engines [8], micromanipulators [9], and synchrotron-based imaging systems [10] due to their high displacement resolution (sub nanometer) and large actuating force (typically a few hundreds of N). PEA-driven positioning systems have also been developed with various configurations, for example, 1-degree-of-freedom (1-DOF) positioning systems with flexure hinge mechanisms [11,12], stick-slip actuators [13], multiple PEAs-driven inchworms [14], or walking actuators [15]; multi-DOF positioning systems with series mechanism [16], parallel mechanism [17], or stick-and-clamping actuators [18]. In all positioning control applications, the hysteresis and creep effects of PEAs have shown to be able to significantly degrade the system performance and even system stability [19]. For improvement, models are desirable to represent these effects and on this basis, tracking controllers can be designed and implemented to achieve desired positioning requirements.

This paper reviews the modeling and control methods of PEAs in micro- and nanopositioning applications and some remaining issues to be solved, and is arranged as follows. Firstly, the working principles and behaviors of PEAs are briefly introduced in Section 2. In Section 3 various models and modeling techniques for PEAs are examined, along with their limitation; and in Section 4 various control schemes for PEAs are reviewed with associated challenges. Section 5 concludes with emerging issues identified in modeling and control of PEAs for future research.

2. Piezoelectric Actuators and Their Behaviors

2.1. Working Principles of Piezoelectric Actuators

PEAs utilize the converse piezoelectric effect of piezoelectric materials to generate displacement and force, *i.e.* a piece of piezoelectric material will be mechanically strained if subject to an electric field (by placing it into the electric field or applying voltages to its surfaces) [20]. This property is resulted from the microscopic structure of piezoelectric materials as explained in the following. Most piezoelectric materials used in PEAs, e.g. lead zirconate titanate, undergo a structural phase transition as its temperature drops through the so-called Curie temperature, during which their structurally and electrically symmetric cubic unit cells deform into structurally and electrically asymmetric tetragonal unit cells, resulting in spontaneous strain and polarization [21]. Groups of adjoining unit cells with uniformly oriented spontaneous polarization are called Weiss [20] or ferroelectric [21] domains. The directions of spontaneous polarization of different domains thus developed are random, so in this state the piezoelectric materials exhibit no overall piezoelectric behavior. To fabricate PEAs from such piezoelectric materials, they are further exposed to a strong electric field (10^6 to 10^7 V/m) at a temperature just below the Curie temperature, forcing the directions of spontaneous polarization of the domains to align with the electric field. Such alignment can be approximately preserved after the removal of the electric field. This process is referred to as poling [21,20]. After poling, due to the approximate alignment of the spontaneous polarization of the domains, the deformations of the domains in the direction of their respective spontaneous polarization resulting form the application of an voltage to the piece of piezoelectric material (which generates an electric field that either enhance or suppress the spontaneous polarization of the domains, causing their dimension to change in the direction of the spontaneous polarization) can accumulate and causes an overall deformation or displacement. Thus the piece of piezoelectric material now possesses overall piezoelectric property including the converse piezoelectric effect and can be used as a PEA [20].

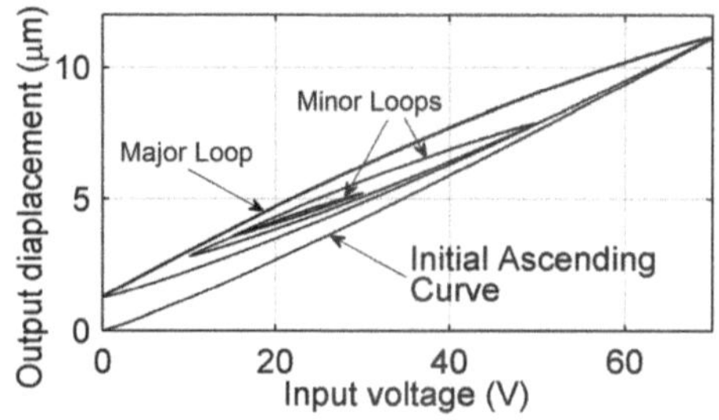

(a)

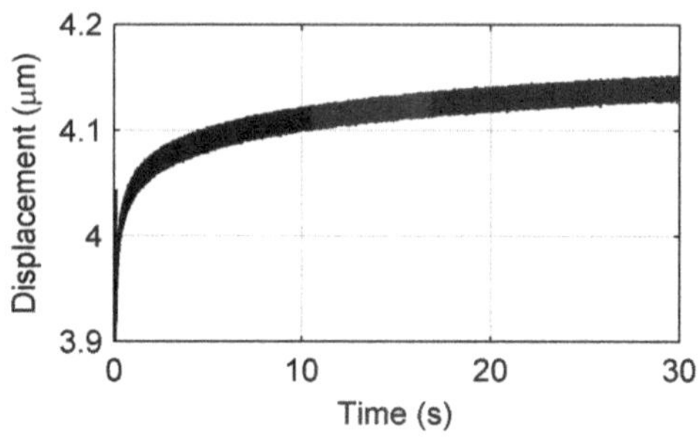

(b)

Figure 1. (a) Hysteresis of a PEA and (b) creep of a PEA subject to a 30 V step input.

2.2. Behaviors of Piezoelectric Actuators

In micro- and nanopositioning applications, typical behaviors of PEAs concerned include hysteresis, creep, and vibration dynamics.

Hysteresis is the nonlinear dependence of a system not only on its current input but also on its past input. In PEAs, hysteresis exists in both the electric field (voltage)-polarization relationship and the electric field (voltage)-strain (deformation or displacement) relationship, with the latter being mostly concerned in micro- and nanopositioning, and it is caused by the nonlinearities in the converse piezoelectric effect of the unit cells and the switching and movement of domain walls [21]. According to [22], the relationship between the actuating force exerted on the PEA and the resulting displacement of the PEA is linear, as such the electric field (voltage)-strain (displacement) hysteresis can also be treated as the result of the voltage-internal actuating force hysteresis.

The strain (deformation or displacement)-electric field (voltage) hysteresis, which is usually also the input-output hysteresis relationship, of a typical PEA is shown in **Figure 1(a)**. It can be seen in **Figure 1(a)** that the hysteresis trajectory of a PEA can be treated as being composed of three types of components: 1) the major loop which is the hysteresis loop that spans the whole input (voltage) range, 2) the minor loops which are the hysteresis loops that only span portions of the input range, and 3) the initial ascending curve which is the input-output trajectory traversed whenever the magnitude of the input applied to the PEA exceeds the maximum magnitude in the input history.

Hysteresis in PEAs occurs in both relatively static operating conditions (*i.e.* with constant or slow varying input) and dynamic operating conditions (*i.e.* with fast varying input). If the influence of the rate of change of the input on the hysteresis can be ignored, then the hysteresis is referred to as rate-independent, otherwise rate-dependent. The latter is usually treated as a combined effect of the rate-independent hysteresis and the vibration dynamics.

As hysteresis being the major nonlinearity of PEAs and possessing detrimental effects on the positioning accuracy and stability margins of feed-back control systems [23], compensation of hysteresis has always been a major concern in modeling and control of PEAs.

Creep is the slow variation in the PEA displacement that occurs without any accompanying change in the input voltage [24] as shown in **Figure 1(b)**. It is caused by the same piezoelectric material properties as PEA hysteresis [20]. Being a slow and a small effect (on the order of 1% of the last displacement per time decade [20]), creep is sometimes neglected in closed loop and high frequency operations, e.g. in [25,26]. However, for slow and open-loop operations of PEAs, creep must be considered to avoid large positioning error [27].

Finally, PEAs also exhibit linear vibration dynamics, which is the dynamic relationship between the total force exerted on a PEA and the displacement of the PEA, and it resembles a distributed parameter system as the mass of the PEA is not concentrated at certain points [22]. However, if the positioning mechanism attached to the PEA, which is usually much more massive than the PEA, is also taken into account, the resulting overall vibration dynamics will approach that of a lumped parameter system [28].

3. Modeling of Piezoelectric Actuators

A large number of PEA models have been developed to mathematically represent the behaviors of PEAs mentioned in Section 2.2, and they can be generally classified into macroscopic models, which models a PEA as a whole, microscopic models, which models a PEA as a combination of a series of ferroelectric domains or discretized cells [21,29-39], and hybrid models, which combines the ideas behind these two categories has also been reported in [40]. Due to the requirement of using finite element methods, which is computationally expensive, to obtain the overall PEA behaviors, microscopic models and hybrid models are not suitable for the use in micro- and nanopositioning applications. As such, the following discussion only concerns macroscopic models.

The linear electromechanical model reported in [41] is an early example of macroscopic PEA models. But it can represent neither the nonlinear behaviors (hysteresis and creep) nor the linear vibration dynamics in PEAs due to its linear and static nature. To solve this, various sub-models have been developed with each representing one or two of the linear/nonlinear behaviors of PEAs mentioned in Section 2.2, and then connected in appropriate manners to form (or used alone as) macroscopic PEA models. These sub-models, the methods of constructing macroscopic PEA models based on such sub-models, and the methods of inverting such sub-models for the use in model-based open-loop feedforward control of PEAs are reviewed as follows.

3.1. Hysteresis Sub-Models

3.1.1. Models for the Major Hysteresis Loop

At the early stage, hysteresis models were developed only to represent the major loop in the hysteresis trajectories of PEAs (refer to **Figure 1**). For example, in [42] polynomials were used to represent the major loop. The Jiles-Atherton hysteresis model, which was initially developed for representing ferromagnetic hysteresis [43] and afterwards adopted for the ferroelectric hysteresis in piezoelectric materials [44], is also limited to represent the major loop [45]. The inability of such models to represent minor loops in the hysteresis curves of PEAs limits their applications.

3.1.2. Rate-Independent Hysteresis Models

To model both the major loops and the minor loops in the hysteresis of PEAs without concerning the influence of the rate of chance of the input, in other words, to model the rate-independent hysteresis of PEAs, both existing and newly developed rate-independent hysteresis models have been adopted. Among which the Preisach hysteresis model [46] and its modifications [47-58], the Prandtl-Ishlinskii (PI) hysteresis model [25,59-63], and the Maxwell resistive capacitor (MRC) model [64-67] are among the most widely used. These rate-independent hysteresis models are reviewed as follows.

The Preisach hysteresis model represents the hysteresis by the combined effect of an infinite number of Preisach hysteresis operators $\delta_P\left(\alpha,\beta,u(t)\right)$, as shown in **Figure 2**. Two parameters are used to characterize a Preisach hysteresis operator: the up switching value α and the down switching value β, with $\alpha \geq \beta$. Each operator has two saturation values: 0 and 1, and its contribution to the model output is adjusted by $\mu(\alpha,\beta)$, referred to as the Preisach weighting function. As such, the resulting Preisach hysteresis model is expressed as

$$f(t)^{CPM} = \iint_{\alpha\geq\beta} \mu(\alpha,\beta)\delta_P\left(\alpha,\beta,u(t)\right)\mathrm{d}\alpha\mathrm{d}\beta \qquad (1)$$

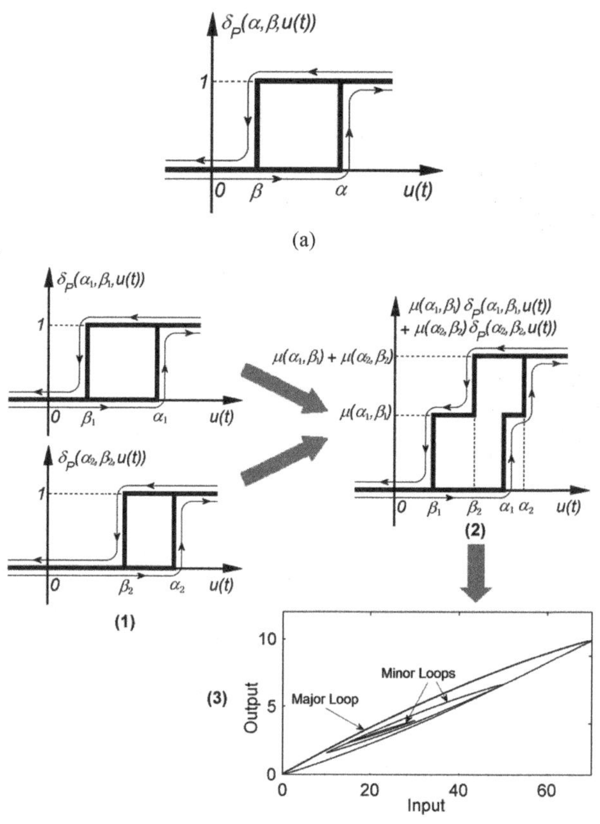

Figure 2. (a) Preisach hysteresis operator and (b) working principle of the Preisach hysteresis model: the sum of two hysteresis operators of different switching values as shown in (1) can represent a simple but unsmooth hysteresis loop as shown in (2); by using the sum of an infinite number of hysteresis operators of different switching values, smooth hysteresis loops including both major and minor loops can then be represented as shown in (3).

where $u(t)$ is the input and $f(t)^{CPM}$ is the output of the hysteretic system [47]. To reduce the computational effort due to the involvement of the double integration, alternative expressions of the Preisach hysteresis model that only involves arithmetic operations have also been derived from Equation (1) and the properties of the Preisach hysteresis operator [68,69].

In PI hysteresis models, hysteresis is represented by the combined effect of a finite number of play or backlash operators as shown in **Figure 3**. Following this idea, the expression of a PI hysteresis model involving n play operators is given as

$$f(t)^{\mathrm{PI}} = \sum_{i=1}^{n} \delta_{\mathrm{PI}}\left(w_i, r_i, u(t)\right) \tag{2}$$

where $f(t)^{\mathrm{PI}}$ is the model output; i denotes the index of the play operator; w_i and r_i are the weight parameter and the threshold parameter of the i-th play operator, respectively.

Similar to the PI hysteresis model, the MRC hysteresis model represents the hysteresis by the combined effect of a finite number of elasto-slide elements or operators as shown in **Figure 4** [64]. Each elasto-slide element or operator (**Figure 4(a)(1)**) is composed of 1) a mass sliding on a surface with a Coulomb friction μN, where μ is the friction coefficient and N is the normal force between the mass and the surface, and 2) a spring of stiffness k with one end connected to the mass whilst

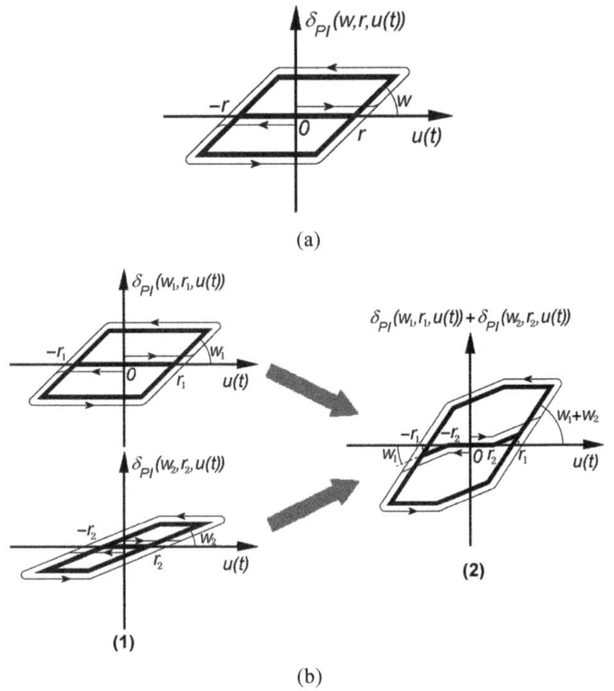

Figure 3. (a) Play operator and (b) working principle of the PI hysteresis model: the sum of two play operators of different parameters as shown in (1) form a simple PI hysteresis model as shown in (2).

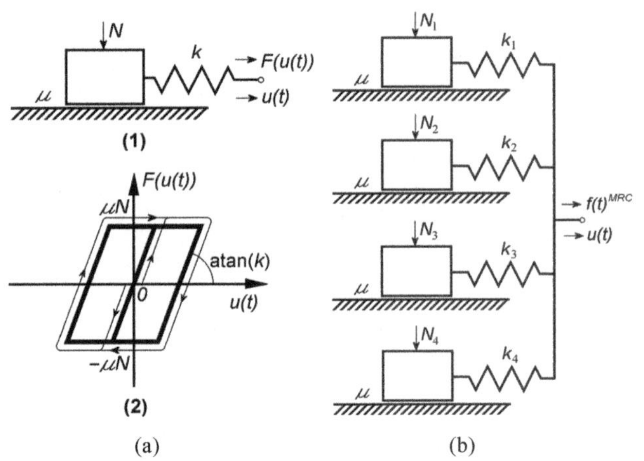

Figure 4. (a) An elasto-slide element and its input-output relationship; (b) The physical interpretation of the MRC hysteresis model [63].

the displacement of the other end (the free end) $u(t)$ can be freely assigned and use as input to the elasto-slide element. As such, hysteresis relationship exists between the (input) displacement of the free end of the spring $u(t)$ and the resulting force in the spring $F(u(t))$, as can be seen in **Figure 4(a)(2)**.

Connecting n elasto-slide elements as in **Figure 4(b)**, the resulting MRC hysteresis model or the relationship between the input displacement of the free ends of the springs $u(t)$ and the total force experienced at the free ends of the springs $f(t)^{\mathrm{MRC}}$ can be expressed as

$$f(t)^{\mathrm{MRC}} = \sum_{i=1}^{n} F_i\left(u(t)\right) \tag{3}$$

where $F_i(u(t))$ is the force in the spring of the i-th elasto-slide element induced by $u(t)$.

3.1.3. Rate-Dependent Preisach and PI Hysteresis Models

The rate-independent hysteresis models discussed in Section 3.1.2 are only capable of representing the hysteretic behavior of a PEA in narrow (a few Hz) frequencies bands. In wide band applications where both hysteresis and vibration dynamics are significant, the rate-independent hysteresis models need either to be combined with vibration dynamics models or to be modified into rate-dependent hysteresis models to represent the behaviors of the PEA. In the literature, both the Preisach hysteresis model and the PI hysteresis model have been modified into rate-dependent hysteresis models.

In the rate-dependent Preisach hysteresis models, the Preisach weighting function $\mu(\alpha,\beta)$ are adjusted to account for the input-rate dependency by using rate-dependent multiplicative modifying factors [68,70] or neural network approaches [69]. In [68], the rate-dependent multiplicative modifying factor for adjusting $\mu(\alpha,\beta)$

is a function of 1) the average voltage input rate between two input extrema and 2) the variation of the voltage between two input extrema. Even though this model can effectively represent the PEA hysteresis up to 800 Hz, it requires the *a priori* knowledge of the input voltage waveform (at least to the next extremum), and thus being not applicable to conventional situations in which the future input voltage is unknown. To avoid this problem, in [70] a multiplicative modifying factor which depends only on the input-rate or the first derivative of the input voltage is used instead. However, the small differences between the input-rate at the input extrema at different input frequencies limits the model accuracy at higher input frequencies (the model is effective typically below 10 Hz). To achieve better accuracy (error of the model prediction being less than 5% of the maximum displacement of the PEA) in operations involving higher frequencies (~30 Hz), in [69] a neural network with rate-dependent output is used to account for the input-rate dependence of the Preisach weighting function. But the applicable frequency range remains low compared to the resonant frequencies of conventional PEAs (typically over 1 KHz).

Similarly, the rate-dependent PI hysteresis models, which are modified from the PI models by modeling the parameters as functions of the input rate, are also limited to represent low frequency (<30 Hz) PEA operations or the error becomes excessively high (over 10% of the peak-to-peak sinusoidal displacement of the PEA) [25,26].

So, in conventional applications where the *a priori* knowledge of the input voltage is not available, the rate-dependent Preisach and the rate-dependent PI hysteresis models are both limited to represent PEA operations over frequency bands only slightly wider than the applicable frequency bands of the rate-independent models.

3.1.4. Inverse Preisach and PI Hysteresis Models

Inverse hysteresis models are essential to the model-inversion-based feedforward hysteresis compensation technique for PEAs. The inverses of the two most widely used hysteresis models, *i.e.* the Preisach hysteresis model and the PI hysteresis model, have been established by employing different methods in the literature.

The Preisach hysteresis model cannot be inversed analytically; hence several methods of approximately inverting the Preisach hysteresis model have been developed instead [71]. Such approximate inverting methods can be classified into two categories.

The first category involves directly identifying the Preisach hysteresis model inversely (using the measured plant output as the input to the Preisach hysteresis model while using the plant input as the output to the Preisach hysteresis model) [1,72,73] or directly identifying the inverses of some intermediate functions that constitute the alternative expressions of the Preisach hysteresis model and then expressing the inverse Preisach hysteresis model based on them [56,71]. The advantage of this category of methods is that the computational effort of the inverse Preisach hysteresis model is the same as the Preisach hysteresis model, not more. However, in applications that also require the Preisach hysteresis model in additional to the inverse one, a different parameter identification procedure has to be performed (assume without using iterative methods) since the former can not be analytically derived from the latter, and the two resulting models are not necessarily accurate inversions of each other due to the different model errors introduced in their respective parameter identification procedures.

The second category involves iterative-algorithms-based methods [71], which find the input to an identified Preisach hysteresis model iteratively until the output of the hysteresis model converges to the desired value. Such methods are developed based on the contraction mapping principle, which are firstly proposed in [74,75], and they have been successfully applied to inverse the Preisach hysteresis modeled hysteresis in a magnetostrictive actuator [71] and a PEA [76]. The advantage of such methods is that both the Preisach hysteresis model and the inverse one can be obtained via a single parameter identification procedure and the inversion can be highly accurate. However, the computational effort involved in the iterative-algorithms-based methods is generally much higher than that of calculating the Preisach hysteresis model output since the Preisach hysteresis model output needs to the calculated in each iteration.

The PI hysteresis model, in the contrary, can be inversed analytically, so good accuracy and low computational effort can be achieved simultaneously, making it much easier to implement than the inverse Preisach hysteresis models. The inversion algorithms for PI hysteresis models can be found, for example, in [63]. In [25,26] the inversion of the rate-dependent PI hysteresis model was also presented. In either case of rate-independent or rate-dependent, the PI hysteresis model inversion algorithms fail if the input frequency is so high that the rounding of the hysteresis loops due to phase lag introduced by other dynamics causes the hysteresis loading and unloading curves to be no longer monotonic. This problem was later solved in [77] by additional mappings to and back from a domain where the inversion of the rate-dependent PI hysteresis model always exists.

3.1.5. Other Hysteresis Models

Other hysteresis models, such as a first-order nonlinear differential equation model [78] and the Bouc-Wen hysteresis model [79-81] have also been used to model hysteresis in PEAs with promising results [22,82-84]. In [85], a non-linear auto-regressive moving average model with exogenous inputs (NARMAX) was employed to represent the hysteresis of a PEA. In [73] a hysteresis model was developed for PEAs by using a polynomial-based linear mapping strategy. In [86], a hysteresis model based on a new hysteresis operator, which in turn is based on two hyperbola functions, are developed for PEAs along with its inverse model. Such hysteresis models have not seen wide application mainly due to their high complexity but limited improvements on accuracy as compared to other widely used hysteresis models discussed before.

3.2. Creep Sub-Models

3.2.1. Frequently Used Creep Models

The creep response in a PEA has a logarithmic shape over time, so it is sometimes referred to as the Log(t)-type creep, and the nonlinear creep model given by Equation (4) has been used to represent such creep behavior [87,88].

$$y(t) = y_0\left[1+\gamma \log_{10}\left(\frac{t}{t_0}\right)\right] \quad (4)$$

where $y(t)$ is the creep model output, y_0 is the creep output at t_0 after the static input voltage is applied, and γ is a coefficient depending on the input voltage. The dependence of γ on the input voltage and the unboundness of such a nonlinear creep model as $t \to 0$ and $t \to +\infty$ pose difficulties in its implementation [19].

To avoid the problem of unboundness encountered in the nonlinear creep model Equation (4) and to avoid the nonlinearity, a linear creep model was proposed as.

$$G_c(s) = 1+\sum_{i-1}^{n_c}\frac{1}{c_i s + k_i} \quad (5)$$

where the creep transfer function $G_c(s)$ is represented by the combined effects of n_c spring (k_i) damper (c_i) systems [1,88]. Besides, an operator-based linear creep model was also reported in [60]. However, these linear creep models have linear and memory free equilibrium values, whist the actual equilibrium value of the creep effect of PEA exhibits hysteresis nonlinearity [89].

To further solve the problems encountered in both the nonlinear and linear creep model above, the PI hysteresis model design approach was adopted in [63] to develop an operator-based nonlinear creep model, whose output is a weighted sum of the values of a series of elementary creep operators.

3.2.2. Inverse Creep Models

Similar to the case of inverse hysteresis models, inverse creep models are essential in feedforward creep compensation.

While the analytic inversion of the nonlinear creep model was not mentioned in the literature, a method of feeding a voltage with an inverse Log(t)-type creep itself to a PEA to compensate for the creep in the output displacement was found to be effective [87].

On the other hand, the inversion of the linear creep model (Equation (5), not operator-based) is usually carried out together with the inversion of the vibration dynamics model since they can be combined into a single linear dynamics model. An optimal inversion approach, which minimizes a quadratic cost function representing the weighted sum of the input voltage energy and the output displacement error energy [90], can be used to inverse the combined vibration dynamics and creep model in [1]. Linear creep models can also be combined with hysteresis models and then inverted, e.g. via iterative methods [88].

The inversion of the operator-based linear and nonlinear creep models together with other dynamics are reported in [60,63], respectively.

3.3. Vibration Dynamics Sub-Models

3.3.1. Frequently Used Vibration Dynamics Sub-Models

According to Section 2.2, for the consideration of accuracy, the vibration dynamics of a PEA should be modeled as a distributed linear system for a stand-alone PEA [22]. But for the convenience of practical uses, the vibration dynamics of a PEA and the attached positioning mechanism is often modeled as a whole as a single second-order system, as in [28,64,91,92], whilst higher order linear dynamical systems are only used in applications requiring higher level of accuracy [1]. Such lumped linear vibration dynamics models of PEAs are usually identified through fitting the measured frequency response of the PEA to that of a specific model structure by using a dynamic signal analyzer [1,93]. On the other hand, the axiomatic design method has also been used to find a higher order linear system representation of the vibration dynamics of a PEA with an attached positioning mechanism [94,95].

3.3.2. Inverse Vibration Dynamics Sub-Models

Inverse vibration dynamics are usually used in open-loop feedforward control of PEAs, especially in the case of controlling piezoelectric tube actuators as in [1]. Methods for inverting the vibration dynamics have been reviewed in Section 3.2.2.

3.4. Model Structures

With different sub-models representing hysteresis, creep, and vibration dynamics of PEAs defined, macroscopic PEA models are constructed by connecting such sub-models according to certain model structures. There are two categories of model structures for macroscopic PEA models.

In the first category, the behavior of a PEA, when subject to a voltage input and an external loading force, is decoupled into several effects such as rate-independent hysteresis, vibration dynamics, and creep based on physical laws. Such effects are then represented by individual sub-models and such sub-models are connected in various manners to form a comprehensive model of the PEA.

The most well known model structure of PEAs belonging to this category is the electromechanical model proposed in [64], which is further refined in [22]. **Figure 5** shows the schematic representation of this model.

There are 3 effects in **Figure 5**, namely 1) the hysteresis, H, between the charge in the PEA, q, and the resulting voltage, u_h; 2) the electromechanical transducer, T_{em}, converts the mechanical deformation of the PEA, y, into charge, (inverse piezoelectric effect), and converts the voltage, u_p, into an actuating force, F_p (piezoelectric effect); and 3) the vibration dynamics, M, relating the deformation of the PEA, y, to the internal actuating force, F_p, and the external load, F_e. Other symbols in **Figure 5** include u_{pea}, which is the voltage across the PEA, L, which is the original length of the PEA, and C, which is the capacitance of the PEA. The model can then be expressed as

$$u_{pea} = u_h + u_p \quad (6)$$

$$q = H(u_h) \quad (7)$$

$$q = Cu_p + q_p \quad (8)$$

$$q_p = T_{em} y \quad (9)$$

$$F_p = T_{em} u_p \quad (10)$$

$$y = M(F_p - F_e) \quad (11)$$

The hysteresis effect, H, and the vibration dynamics, M, can be modeled by different kinds of sub-models. Specifically, H is usually represented by rate-independent hysteresis models, e.g. the MRC model was used in [64] whist the first-order nonlinear differential equation model, which was firstly proposed in [78], was employed in [22]. The vibration dynamics, M, can be modeled by any models discussed in Section 2.1.3. It is noted that the use of this electromechanical model usually requires the measurements of the charge, q, in the PEA for model parameter estimation, which is difficult in practice. So this model is usually employed in applications that the relationship between u_{pea} and F_p, which is difficult to identify, can be approximated by a simple function (e.g. an affine function) with an error term as model uncertainties, such as in [91,92]. Also, creep is usually not considered in this electromechanical model [22,78] expect being treated as model uncertainties [91,92].

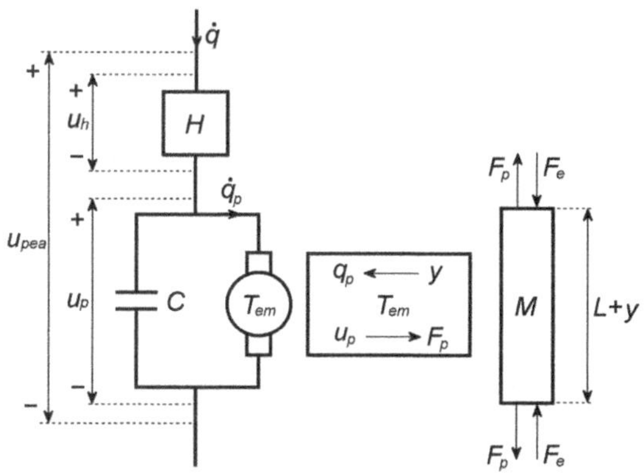

Figure 5. Schematic representation of a electromechanical model of a PEA [22].

Another model structure proposed in [1] models the relationship between u_{pea} and F_p by a rate-independent hysteresis model directly, based on a conclusion drawn from the analysis in [64] that the rate-independent hysteresis of a PEA only exists between the input voltage and the resulting internal actuating force. This hysteresis sub-model is then cascaded to a vibration dynamic sub-model and a creep sub-model to form the complete model of a PEA, as shown in **Figure 6**. Without the involvement of charge measurement for parameter estimation, this model structure is much easier to apply than the electromechanical model. Similar to the case of the electromechanical model, this model structure is highly flexible that each sub-model can be implemented in various ways, e.g. in [1], the Preisach hysteresis model was used for H and a fourth-order and a third-order linear dynamic systems are used to represent M and the creep effect, G_c, whist in [63], a modified PI hysteresis model is used for H and an operator based model of complex Log(t)-type creep for G_c. Also, due to the cascading connection between the sub-models, this model structure is very suitable for implementing control schemes with inverse nonlinearity and/or vibration dynamics compensation as in [1,63,93,96,97]. A similar model structure constructed using a bond-graph representation was also presented in [67].

In the second category, the different behaviors of a PEA are not decoupled. In such cases, rate-dependent hysteresis models such as the rate-dependent Preisach hysteresis model, the rate-dependent PI hysteresis model (Section 3.1.3), and the rate-dependent neural-network-based hysteresis model [98] are used directly as models of the PEA. However, as mentioned in Section 3.1.3,

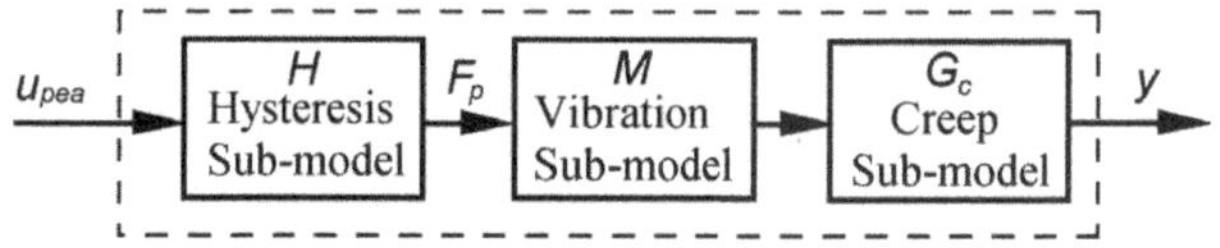

Figure 6. Model structure of a PEA as in [1].

such models are only accurate over small frequency ranges. And, as can be seen above, creep is usually neglected. These disadvantages seriously limit their usage.

4. Control of PEAs

Over the years, a number of control schemes have been reported for positioning control of PEAs. Due to the detrimental effects of the nonlinearities (especially hysteresis) of PEAs on positioning control performances in terms of positioning error and stability [19], the compensation or rejection of such nonlinear effects has been a major consideration in all of these control schemes. In the following, the typical control schemes for PEAs are reviewed.

4.1. Open-Loop Control Schemes

Open-loop control schemes are usually employed in applications in which position feedback are difficult to implement due to mechanical constraints, e.g. atomic force microscopes [1-3]. In such control schemes, inverse model of the PEA to be controlled is found and then cascaded to the PEA. The methods for finding such inverse models have been reviewed in Sections 3.1.4, 3.2.2, and 3.3.2. The inverse model generates an input voltage u_{pea} to the PEA according to the desired displacement y_d, such that the PEA subject to u_{pea} produces an output displacement of PEA y that follows y_d. **Figure 7** shows a typical open-loop control scheme for PEAs.

In some of the open-loop control schemes, all three effects in the PEA shown in **Figure 6** are compensated for by the inverse model of PEA to achieve best tracking accuracy (as the case in **Figure 7**). For example, in [1,67], inverse models of PEA representing hysteresis, vibration dynamics, and creep are used to compensate for these effects; whist in [93] a charge control technique was firstly employed to remove the hysteresis effect and then an inverse model was used to compensate for the linear dynamics.

Other open-loop control schemes only focus on compensating for some of the three effects shown in **Figure 6**. For example, in [72,73,86,99], only hysteresis was compensated for by using the inverses of the Preisach hysteresis model, a hysteresis model developed by using a polynomial-based linear mapping strategy, a hysteresis model developed based on hyperbola-shaped hysteresis operators, and a hysteresis model developed in [100], respectively, since hysteresis is the most significant nonlinearity in PEAs. In [90,101], creep is also compensated for in additional to hysteresis by using operator-based inverse creep models and inverse PI hysteresis models to achieve better tracking performance. However, due to the negligence of the vibration dynamics in these schemes, their performances are only guaranteed in fixed frequency or narrow band operations unless inverse ratedependent hysteresis models are employed, e.g. the use of inverse rate-dependent PI hysteresis models in [25,26,77]. Nonetheless, the applicable frequency ranges (with acceptable tracking error, e.g. <5% of the maximum displacement of the PEA) of the schemes using inverse rate-dependent hysteresis models are still relatively small (less than 50 Hz) compared with those also inverting the vibration dynamics (up to a few hundreds of Hz) [1,25,77].

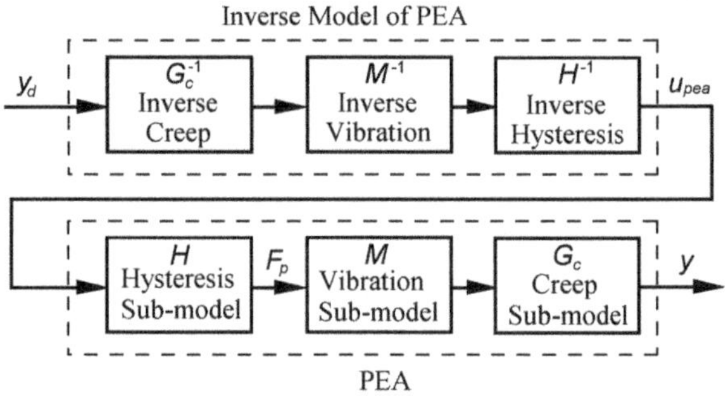

Figure 7. Open-loop control scheme for a PEA [1].

In no load condition the open-loop control schemes have been shown to be highly effective in their applicable frequency ranges [1].

However, one major disadvantage of the open-loop control schemes is that their positioning performances are highly sensitive to unknown effects such as model errors, external loads, and changes in the dynamics of the PEA. To partially alleviate the problem, reference [65] proposed to use a disturbance-observer-based scheme in addition to the inverse MRC hysteresis model to compensate for the hysteresis and unknown effects simultaneously. Specifically, the estimated disturbance input to the PEA, which has the same effect as the aforementioned unknown effects on the PEA output, is subtracted from u_{pea}. However, the effectiveness of this scheme was not examined through experiments. In [102], the inverse hysteresis model is updated adaptively to account for the unknown effects. However, the influence of the unknown effects on the PEA output, though largely suppressed, can still be clearly observed in the experiment results. As such, the best solution to this problem remains introducing feedback into the control schemes, which is to be discussed in the following sections.

4.2. Feedback

Feedback control schemes as shown in **Figure 8** lead to strong suppression of the unknown effects including model errors, external loads, and changes in the dynamics of the PEA on the position control performances, hence they are widely used.

In static or low frequency operations, classical control techniques such as PID control or using multiple integrators for tracking desired displacement profiles are widely used because of their simplicity and capability of elimi-

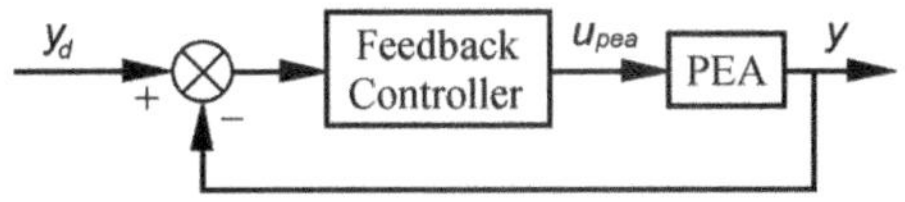

Figure 8. A feedback control scheme for PEAs [19].

nating steady state errors in such applications [20,103,104]. Various PID tuning techniques has been reported in PEA positioning control applications, e.g. by trial and error [20], by grey relational analysis [105], using an optimal linear quadratic regulation method [106], by a semi-automatic tuning technique [107], and by an automatic tuning technique [108]. However, in broadband operations with large system uncertainties including modeling errors, nonlinearities, external loads, etc., advanced control techniques are required because PID control is limited in bandwidth while dealing with uncertainties [109].

Among all of such advanced control techniques, sliding mode control has drawn a lot of attention in the recent years. This is due to the fact that sliding mode controllers can completely reject the effects of the so-call matched uncertainties or uncertainties in the input channel of the plant, resulting in strong robustness [110].

Specifically, in its basic form, the control signal generated by a sliding mode controller can be divided into two parts. The first part is discontinuous, referred to as the switching control, for compensating for the matched uncertainties while driving the plant trajectory to a prescribed geometric entity in the phase space (a space with each dimension corresponding to one state variable in the plant model), referred to as the sliding surface. The second part is continuous, referred to as the equivalent control, for keeping the plant trajectory on the sliding surface after the latter is reached. Since the sliding surface and hence the closed loop system dynamics after reaching the sliding surface can be freely designed, the closed-loop system becomes insensitive to the matched uncertainties after the sliding surface is reached [110]. The major problem of implementing the sliding mode control techniques on PEAs is chattering and the need of the knowledge about the bounds of the matched uncertainties for designing the discontinuous control signal [110], which is to be discussed in more detail in Section 5.2.

Based on the model structure shown in **Figure 6**, hysteresis and creep in a PEA can be treated as (matched) uncertainties in the input channels of the vibration dynamics sub-model of the PEA (given that a linear creep model is used). As such, a sliding mode controller can be designed using the vibration dynamics sub-model as a nominal model of the PEA, and the effects of hysteresis and creep on the PEA output displacement can be strongly suppressed without knowing the exact expression of the hysteresis sub-model and the creep sub-model, thus largely reducing the efforts involved in hysteresis and creep modeling and compensation [95,111-114]. Various modifications of the basic sliding mode control technique have also been reported for controlling PEAs for improved performance. For instance, adaptive methods are combined with sliding mode control to remove the requirement of model parameter estimation or lead to better compensation for the nonlinearities as in [115-117]. And in [118] the nonlinearities of the positioning mechanism driven by a PEA in additional to those of the PEA itself were considered in sliding mode controller design.

Robust control techniques constitute another branch of widely used advanced control techniques for positioning control of PEAs. They try to find the control law via optimizing an objective function that incorporates the robustness objective, for example, minimizing the H_∞ norm [88,119-122] or the H_2 norm [76] of the transfer functions relating the disturbances to the plant output, thus minimizing the effects of the disturbances on the plant output and enhancing robustness. Robust control techniques can be combined with sliding mode control to suppress unmatched uncertainties (which the sliding mode control techniques cannot compensate for), while sliding mode control are considered to be superior in dealing with matched uncertainties because robust control techniques can only minimize the effects of the matched uncertainties on the plant output whilst sliding mode control techniques can completely reject such effects in theory [110].

Other advanced control techniques such as state feedback [123], optimal control [109,124], adaptive control [125-128], and neuron network methods [129-131] have also been applied to control PEAs.

Inverse sub-models of the nonlinearities of PEA are sometimes inserted between the feedback controller and the controller in **Figure 8** to linearize the PEA thus facilitating the use of linear feedback controllers. For example, an inverse Preisach hysteresis model is used to cancel out the hysteresis and then a linear H_2 controller is designed based on the remaining linear dynamics of a PEA in [76], while an inverse PI hysteresis model and a sliding mode controller are used in [113].

One problem with the feedback control schemes for PEAs is that the resulting closed-loop systems have low gain margins, which limit the use of high gain controllers. Such low gain margin is largely due to the quick phase loss near the first resonant peak in the frequency response of a PEA [23]. This problem can be alleviated by using a notch filter to reduce the gain around the first resonant peak [20,132,133].

4.3. Feedback with Feedforward

Feedforward is sometimes used to augment feedback controllers for nonlinearity compensation. A typical control scheme of this type is shown in **Figure 9**. An advantage of this scheme is that the low gain margin problem

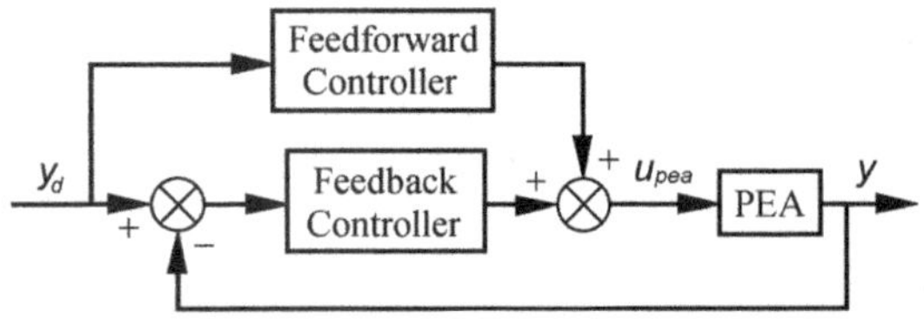

Figure 9. A feedback augmented with feedforward control scheme for PEAs [19].

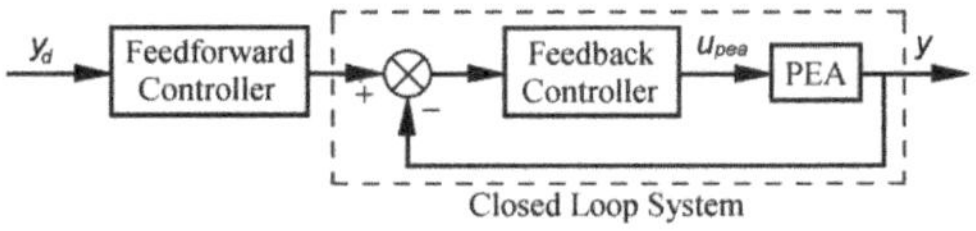

Figure 10. A feedback controller is used for nonlinear suppression whist a feedforward controller is the inverse model of the closed loop system for canceling out the dynamics of the closed loop system, which can be treated as linear [19].

can be alleviated such that high frequency positioning performances can be improved over the feedback control schemes [124,134,135]. Recent researches adopting this scheme can be found, e.g. in [136] a hysteresis compensator based on the inverse Preisach hysteresis model was used as the feedforward controller to compensate for hysteresis whist a PID feedback controller was used to account for other effects, in [137] a inverse nonlinear differential equation hysteresis model was used as the feedforward controller and a PI controller was used as a feedback controller, and in [16] a inverse extended Coleman-Hodgdon model was used as the feedforward controller and a feedback controller designed via loop shaping techniques was employed.

Another control scheme involving feedback and inverse model feedforward is shown in **Figure 10** [97,138]. In this scheme, the feedback controller is a high gain controller used for nonlinear effect suppression such that the closed loop system has linear dynamics. Then a feedforward controller, which is the inverse linear dynamics of the closed loop system, is applied to make the output y follow the desired output y_d. This control scheme can be seen as being evolved from the open-loop control scheme shown in **Figure** 7. The rationale behind this scheme is that the accurate modeling and inversion of the nonlinear effects such as hysteresis are complicated whist the inversion of a linear plant is relatively easy to compute [139]. However the low gain margin problem with the feedback loop exists in this scheme [19]. Also, similar to the open-loop control schemes, this scheme is sensitive to the disturbances acting outside the closed-loop system.

4.4. Disturbance Observer Based Schemes

Recently, a technique call disturbance observer is used for the compensation of hysteresis and other nonlinear/uncertain effects in PEAs. In this scheme, a PEA is modeled as a linear dynamic system $G(s)$, so when subject to the same input voltage u_{pea}, the output of the PEA y and that of the corresponding $G(s)$ are different due to hysteresis and other nonlinear/uncertain effects. The disturbance observer utilizes u_{pea} and y to estimate an additional input voltage u_d such that if $u_{pea}+u_d$ is applied to $G(s)$, the output of $G(s)$ is the same as y. u_d is referred to as an input disturbance to $G(s)$ that represents the hysteresis and other nonlinear/uncertain effects. As such, to compensate for hysteresis and other nonlinear/uncertain effects in the PEA, one only needs to subtract u_d from u_{pea} before u_{pea} being applied to the PEA. A control scheme with such a disturbance observer is shown in **Figure 11**. This scheme has been shown to be effective in both step [140] and high frequency tracking (200 Hz sinusoidal) [141] operations.

5. Emerging Issues in PEA Modeling and Control

Based on the reviews in the previous sections, some emerging issues regarding the modeling and control of PEAs and PEA-driven positioning stages are discussed in this section.

5.1. Modeling Issues

5.1.1. Limitations of Some Existing Hysteresis Models of PEAs

While the modeling of creep and vibration dynamics of PEAs have been relatively well address, the existing models of hysteresis, especially the rate-independent hysteresis models reviewed in Section 3.1.2, still possess issues, as follows. Though being highly effective in representing both the major hysteresis loops and the minor hysteresis loops, it is noted in the literature [40,53-56,58] that the Preisach hysteresis model cannot at the same time represent the initial ascending curve when the PEA is subject to nonnegative voltage inputs. This is due to the fact that the Preisach hysteresis operator has only two saturation values and without a separate initial section that connects the origin to the hysteresis loop (**Figure 2(a)**). The PI hysteresis models and the MRC hysteresis model do not suffer from this problem due to the existence of initial sections in their respective hysteresis operator besides the hysteresis loop (**Figures 3(a)** and **4(a)(2))**. However, to limit the computational effort, the number of hysteresis operators used in PI or MRC hysteresis models in practice is usually very small (typically around 10) compared to the Preisach hysteresis model (infinity). As a result, the PI and MRC models tend to be less accurate than the Preisach hysteresis model in representing the hysteresis loops, especially in the cases of the small minor loops. So, given a tractable amount of computational effort, there is a contradiction between the capability of representing all hysteresis phenomena, including one-sided hysteresis (the PI and MRC hysteresis models), and the accuracy (Preisach hysteresis models) in the existing rate-independent hysteresis models. This remains an issue to be solved in future research.

5.1.2. Applications of Different Actuator Structures

The applications of PEA-driven positioning systems with structures different from the conventional ones (those can be treated as a mass fixed to one end of the PEA) also pose modeling problems. A typical example is the PEA-driven stick-slip actuator shown in **Figure 12**, in which an end-effector is supported and guided by a movable platform that is driven by a PEA, and it works as follows. During the course of slow expansion of the PEA, the end-effector moves along with the platform. If the PEA suddenly contracts, the end-effector slides on the platform because the force due to inertia becomes larger than the friction between the end-effector and the movable platform. As a result, the end-effector moves a step, ΔS, with respect to its original position. Such steps can be accumulated to achieve a theoretically unlimited displacement (actually limited by the size of the moveable platform). The modeling of the dynamics of the end-effector displacement has not been well addressed in the literature. Chang and Li [142] developed a model for the PEA-driven stick-slip actuator without considering the dynamics of the PEA while the friction involved was modeled as Coulomb friction, which is over simplified. By taking into account presliding friction, models were developed and reported in more recent studies on stick-slip actuators, e.g. [143]. However, issues including the influence of PEA nonlinearities and the end-effector mass on the performance of the PEA-driven stick-slip actuator remain to be addressed.

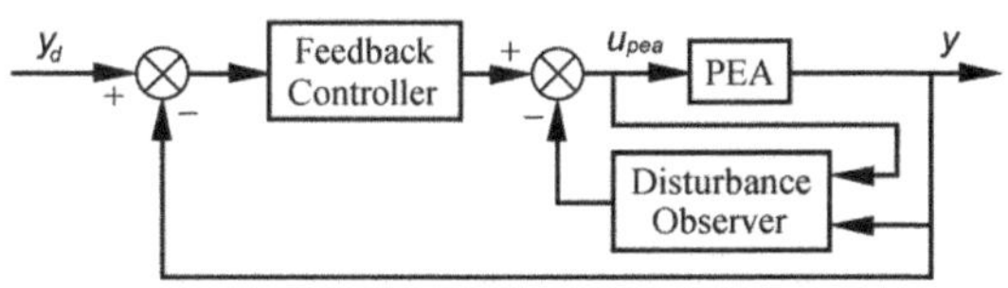

Figure 11. Augmenting the feedback control loop with a disturbance observer.

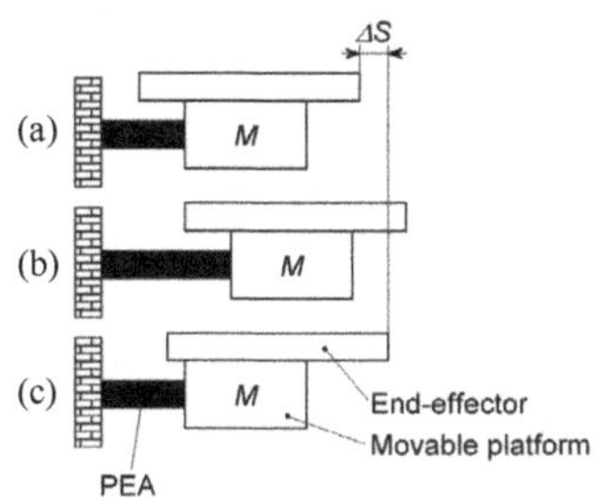

Figure 12. Actuation sequence of a PEA-driven stick-slip actuator; (a) Start; (b) Slow expansion and (c) Fast contraction.

5.2. Control Issues

5.2.1. Controller Design

As can be seen from Section 4.2, sliding mode control has been one of the most promising techniques for PEA control applications due to their capability to largely suppress hysteresis and creep in practice. However, there are two remaining problems. One problem of implementing the sliding mode control techniques on PEAs is chattering. This is a kind of high frequency vibration in the plant output displacement induced by the discontinuous control signal generated by a sliding mode controller, and it can induce wear, noises, and even resonance and instability in the plant. Chattering is especially severe in plants with fast responses, such as PEAs, because they cannot filter out the high frequency component in the switching control signal. Another problem is that the bounds of the matched uncertainties need to be known for designing the discontinuous control signal, but such bounds are usually difficult to determine in practice while conservatively setting the bounds to large values only leads to more severe chattering [110]. Existing research efforts have been focused on solving the chattering problem. For example, a method referred to as the sliding mode with boundary layer control was developed by approximating the discontinuous control signal with a continuous one, but with the cost of introducing nonzero steady state error, and chattering is still significant if the bounds of the matched uncertainties are large [110]. Higher order sliding surfaces are also used for chattering suppression, such as the integral sliding mode technique in [92,144] and the PID sliding surface in [115]. However, the second problem remains to be solved.

5.2.2. State Estimation

Many advanced control techniques reviewed in Section 4, such as sliding mode control [114], require state feedback. Since usually not all of the states of a PEA model are measurable in application, state observers must be used. However, the problem of estimating the states of PEA models has not been well addressed in the literature, as discussed below.

Currently, many kinds of state observers are applicable to PEA control and they can be classified into two categories: non-model-based and model-based. Non-model-based filters/differentiators, such as the low-pass filter plus ideal differentiator, the α-β filter [145], the high-gain differentiator [146], the integral-chain differentiator [146], and the sliding mode differentiator based on the super-twisting algorithm [147] (given that the states needed are the derivatives of the measured plant output), usually generate large phase lags if the desired level of noise suppression is enforced (for the first four aforementioned methods) or excessive chattering in noisy systems (for the last method). Compared to non-modelbased filters/differentiators, model-based observers, e.g., the extended Kalman filter [148], the unscented Kalman filter [149,150], and the high-gain observer [114], can generate more accurate estimations if the imperfection of the model can be ignored. However, with the presence of uncertainties such as the effects of hysteresis, creep, and external load on a PEA, however, the performances of the model-based observers degrade. For improvement, in the cases that the system uncertainties can be treated as a lumped unknown input to the system model, a kind of model-based observers should be used to estimate the system states even with the presence of the unknown input, which is referred to as unknown input observers (UIOs) in the literature. Many UIOs have been reported in the past three decades, including the full-order UIO [151], the reduced-order UIO [152], the UIO designed based on projection operator [153], and the sliding-mode-based observers (SMOs) [110,153-155].

Applications of these UIOs require that the observer matching condition be satisfied [156,157], which states that the rank of the product of the output matrix and the unknown input matrix in the state space model of the system must be equal to that of the unknown input matrix [110]. However, existing PEA models, such as the one in [92], do not meet this condition. Attempts to relax the observer matching condition have been reported [156,157], but the resultant UIOs were very complicated. As such, a UIO with a simpler structure and the capability of relaxing the observer matching condition still needs to be developed in future research for the use in PEA tracking control.

5.3. Modeling and Control of Multi-Degree-of-Freedom (Multi-DOF) PEA-Driven Positioning Systems

All of the above discussions have been focused on single-DOF PEA-driven positioning systems. However, there are modeling and control issues uniquely related to multi-DOF PEA-driven positioning systems and they are briefly discussed in this section.

The multi-DOF PEA-driven positioning systems reported in the literature can be divided in to two categories: series mechanisms and parallel mechanisms. Among which, multi-DOF PEA-driven series mechanisms [16] and the xyz-type parallel mechanisms (translations along *x*-, *y*- and *z*-axis) such as those in [16,158-160] can usually be treated as a series of individual 1-DOF PEA-driven mechanisms without significant coupling between the moving axes, as such their modeling and control can be well address with the techniques applied to the single-DOF systems.

The remaining multi-DOF PEA-driven parallel mechanisms whose moving axes are significantly coupled can be further divided into fully-actuated systems, *i.e.* the number of actuation is the same as the number of DOF [17,114,161-167], and over-actuated systems, *i.e.* the number of actuation is the more than the number of DOF [20,112].

For the fully-actuated systems, their models are usually established based on the geometric structure of the mechanism and mechanics. Since most of such mechanisms use flexure hinges instead of conventional hinges to avoid friction and backlash [17,114,158,159-166], the modeling of the flexure hinges becomes an important part in modeling such mechanisms [114,160,166]. Two kinds of workflows can be adopted for modeling the multi-DOF PEA-driven positioning systems. For the first kind of workflows, a kinematic model is firstly constructed based on the geometric structure of the mechanism, and then either a phenomenological [114] or physical [160] dynamic model can be established based on the kinematic model. The second kind of workflows involves direct identification of the dynamic model of the mechanism without establishing the kinematic model [16].

Many control schemes that have been applied to control single-DOF PEA-driven positioning systems can also be extended to control fully-actuated PEA-driven positioning systems, e.g. PID control [112,168], H_2 control [168], and sliding mode control [112] since such methods can be easily extended to fully-actuated multi-input-multi-output cases.

Compared to the fully-actuated systems, the over-actuated systems have certain advantages in the performance point of view, e.g. singularity elimination, dexterity improvement, and better load carrying ability [169]. However, the over-actuated structure poses difficulties in both modeling and control. Specifically, the PEAs in the mechanism need to cooperate well or excessive internal forces will occur and may damage the mechanism, whilst the involvement of PEA nonlinearities makes such cooperation difficult to model and control. As a result, very few researches on the modeling and control of over-actuated PEA-driven parallel mechanism have been reported compared to its popularity in macro parallel mechanisms. The only example is [112], in which the modeling and control of a $x\theta_y\theta_z$ 3-DOF positioning system driven by 4 PEAs are concerned. Specifically, in [112] each PEA in the mechanism is modeled and controlled individually and uses only the output displacements of the corresponding PEA as feedback, while the inverse kinematic model of the mechanism is used to generate the desired displacement for each PEA according to the desired position of the end-effector. The cross-axis couplings are considered as disturbances. As such, the control problem is transformed into several 1-DOF control problems, which are much simpler. However, in this scheme the actual position of the end-effector is not closed-loop-controlled, and the requirement of being able to measure the displacement of each PEA in the mechanism limits the generality of this scheme. As such, more general schemes for modeling and controlling over-actuated multi-DOF PEA-driven positioning systems that can relax the aforementioned requirement are still to be developed in the future.

6. Conclusions

Though possessing fine resolution, high actuating forces, and fast responses, the nonlinear effects make modeling and control of PEAs for the use in micro- and nanopositioning challenging. Researches for solving such problems are abundant, and the resulting typical methods developed are reviewed in this paper, from which some major conclusions can be drawn: 1) the decoupled-structure PEA models have the advantages of being more accurate and flexible as compared to the undecoupled ones; 2) in the decoupled-structure PEA models, the sub-models concerning vibration dynamics and creep have been relatively well addressed in the literature; 3) however, among the existing hysteresis sub-models, for a tractable among of computation efforts, a contradiction between the capability of representing all hysteresis phenomena and maintaining model accuracy remains to be solved; 4) among the control schemes reviewed, the sliding mode control is among the most promising ones as applied to PEA positioning since when designed based on the decoupled-structure PEA models, it can completely reject the hysteresis and creep effects in theory; 5) however, there are still issues associated with sliding mode control (chattering) and the corresponding state estimators (the incompatibility between the observer matching condition and the existing PEA models); finally, 6) the modeling and control of some PEA-driven positioning systems with unconventional structures, e.g. PEA-driven stick-slip actuators and over-actuated multi-DOF PEA-driven positioning systems, are largely untouched in the literature, thus remain to be addressed in future research.

REFERENCES

[1] D. Croft, G. Shed and S. Devasia, "Creep, Hysteresis, and Vibration Compensation for Piezoactuators: Atomic Force Microscopy Application," *Journal of Dynamic Systems, Measurement, and Control*, Vol. 123, No. 1, 2001, pp. 35-43.

[2] Q. Zou, K. K. Leang, E. Sadoun, M. J. Reed and S. Devasia, "Control Issues in High-Speed AFM for Biological Applications: Collagen Imaging Example," *Asian Journal of Control*, Vol. 6, No. 2, 2004, pp. 164-178.

[3] R. Kassies, K. O. Van der Werf, A. Lenferink, C. N. Hunter, J. D. Olsen, V. Subramaniam and C. Otto, "Combined AFM and Confocal Fluorescence Microscope for Applications in Bio-Nanotechnology," *Journal of Microscopy*, Vol. 217, No. 1, 2005, pp. 109-116.

[4] H. Song, G. Vdovin, R. Fraanje, G. Schitter and M. Verhaegen, "Extracting Hysteresis from Nonlinear Measurement of Wavefront-Sensorless Adaptive Optics System," *Optics Letters*, Vol. 34, No. 1, 2009, pp. 61-63.

[5] W. Yang, S.-Y. Lee and B.-J. You, "A Piezoelectric Actuator with a Motion-Decoupling Amplifier for Optical Disk Drives," *Smart Materials and Structures*, Vol. 19, No. 6, 2010, pp. 065027.1-065027.10.

[6] G. Stöppler and S. Douglas, "Adaptronic Gantry Machine Tool with Piezoelectric Actuator for Active Error Compensation of Structural Oscillations at the Tool Centre Point," *Mechatronics*, Vol. 18, No. 8, 2008, pp. 426-433.

[7] S. R. Viswamurthy, A. K. Rao and R. Ganguli, "Dynamic Hysteresis of Piezoceramic Stack Actuators Used in Helicopter Vibration Control: Experiments and Simulations," *Smart Materials and Structures*, Vol. 16, No. 4, 2007, pp. 1109-1119.

[8] M. S. Senousy, F. X. Li, D. Mumford, M. Gadala and R. K. N. D. Rajapakse, "Thermo-Electro-Mechanical Performance of Piezoelectric Stack Actuators for Fuel Injector Applications," *Journal of Intelligent Material Systems and Structures*, Vol. 20, No. 4, 2009, pp. 387-399.

[9] J.-J. Wei, Z.-C. Qiu, J.-D. Han and Y.-C. Wang, "Experimental Comparison Research on Active Vibration Control for Flexible Piezoelectric Manipulator Using Fuzzy Controller," *Journal of Intelligent and Robotic Systems*, Vol. 59, No. 1, 2010, pp. 31-56.

[10] D. Shu, J. Maser, M. Holt, R. Winarski, C. Preissner, A. Smolyanitskiy, B. Lai, S. Vogt and G. B. Stephenson, "Optomechanical Design of a Hard X-Ray Nanoprobe Instrument with Nanometer-Scale Active Vibration Control," *AIP Conference Proceedings*, Vol. 879, 2007, pp. 1321-1324.

[11] H. Zhou and B. Henson, "Analysis of a Diamond-Shaped Mechanical Amplifier for a Piezo Actuator," *The International Journal of Advanced Manufacturing Technology*, Vol. 32, No. 1-2, 2007, pp. 1-7.

[12] M. Muraokaa and S. Sanada, "Displacement Amplifier for Piezoelectric Actuator Based on Honeycomb Link Mechanism," *Sensors and Actuators A*: *Physical*, Vol. 157, No. 1, 2010, pp. 84-90.

[13] Y. Zhang, G. Liu and J. Hesselbach, "On Development of a Rotary-Linear Actuator Using Piezoelectric Translators," *IEEE/ASME Transactions on Mechatronics*, Vol.

11, No. 5, 2006, pp. 647-650.

[14] J. Li, R. Sedaghati, J. Dargahi and D. Waechter, "Design and Development of a New Piezoelectric Linear Inchworm Actuator," *Mechatronics*, Vol. 15, No. 6, 2005, pp. 651-681.

[15] R. Merry, R. Van de Molengraft and M. Steinbuch, "Modeling of a Walking Piezo Actuator," *Sensors and Actuators A*: *Physical*, Vol. 162, No. 1, 2010, pp. 51-60.

[16] R. Merry, M. Uyanik, R. Van de Molengraft, R. Koops, M. Van Veghel and M. Steinbuch, "Identification, Control and Hysteresis Compensation of a 3 DOF Metrological AFM," *Asian Journal of Control*, Vol. 11, No. 2, 2009, pp. 130-143.

[17] T.-F. Lu, D. C. Handley, Y. K. Yong and C. Eales, "A Three-DOF Compliant Micromotion Stage with Flexure Hinges," *Industrial Robot*: *An International Journal*, Vol. 31, No. 4, 2004, pp. 355-361.

[18] S. W. Lee, K.-G. Ahn and J. Ni, "Development of a Piezoelectric Multi-Axis Stage Based on Stick-and-Clamping Actuation Technology," *Smart Materials and Structures*, Vol. 16, No. 6, 2007, pp. 2354-2367.

[19] S. Devasia, E. Eleftheriou and S. O. R. Moheimani, "A Survey of Control Issues in Nanopositioning," *IEEE Transactions on Control Systems Technology*, Vol. 15, No. 5, 2007. pp. 802-823.

[20] P. Instrumente, "The World of Nanopositioning and Micropositioning 2005/2006," Physik Instrumente, Karlsruhe, 2005.

[21] I. Mayergoyz and G.Bertotti (Eds.), "The Science of Hysteresis," Vol. 3, Elsevier, St. Louis, 2005.

[22] H. J. M. T. S. Adriaens, W. L. De Koning and R. Banning, "Modeling Piezoelectric Actuators," *IEEE/ASME Transactions on Mechatronics*, Vol. 5, No. 4, 2000, pp. 331-341.

[23] J. A. Main and E. Garcia, "Piezoelectric Stack Actuators and Control System Design: Strategies and Pitfalls," *Journal of Guidance, Control, and Dynamics*, Vol. 20, No. 3, 1997, pp. 479-485.

[24] T. Fett and G. Thun, "Determination of Room-Temperature Tensile Creep of PZT," *Journal of Materials Science Letters*, Vol. 17, No. 22, 1998, pp. 1929-1931.

[25] W. T. Ang, F. A. Garmón, P. Khosla and C. N. Riviere, "Modeling Rate-Dependent Hysteresis in Piezoelectric Actuator," 2003 *IEEE/RSJ International Conference on Intelligent Robots and Systems*, Las Vegas, 27 October-1 November 2003, pp. 1975-1980.

[26] W. T. Ang, P. K. Khosla and C. N. Riviere, "Feedforward Controller with Inverse Rate-Dependent Model for Piezoelectric Actuators in Trajectory-Tracking Applications," *IEEE/ASME Transactions on Mechatronics*, Vol. 12, No. 2, 2007, pp. 134-142.

[27] R. S. Robinson, "Interactive Computer Correction of Piezoelectric Creep in Scanning Tunneling Microscopy Images," *Journal of Computer-Assisted Microscopy*, Vol. 2, No. 1, 1996, pp. 53-58.

[28] X. B. Chen, Q. Zhang, D. Kang and W. Zhang, "On the Dynamics of Piezoactuated Positioning Systems," *Review of Scientific Instruments*, Vol. 79, No. 11, 2008, pp. 116101.1-116101.3.

[29] W. Chen and C. S. Lynch, "A Micro-Electro-Mechanical Model for Polarization Switching of Ferroelectric Materials," *Acta Materialia*, Vol. 46, No. 15, 1998, pp. 5305-5311.

[30] W. Seemann, A. Arockiarajan and B. Delibas, "Modeling and Simulation of Piezoceramic Materials Using Micromechanical Approach," *European Congress on Computaional Methods in Applied Sciences and Engineering* 2004 (*ECCOMAS* 2004), Jyväskylä, 24-28 July 2004.

[31] A. Arockiarajan, B. Delibas, A. Menzel and W. Seemann, "Studies on Nonlinear Electromechanical Behavior of Piezoelectric Materials Using Finite Element Modeling," *IWPMA* 2005, *2nd International Workshop on Piezoelectric Materials and Applications in Actuators*, Heinz Nixdorf Institute, Paderborn, 22-25 May 2005.

[32] B. Delibas, A. Arockiarajan and W. Seemann, "Intergranular Effects on Domain Switchings in Polycrystalline Piezoceramics," *IWPMA* 2005, *2nd International Workshop on Piezoelectric Materials and Applications in Actuators*, Heinz Nixdorf Institute, Paderborn, 22-25 May 2005.

[33] M. Kamlah and Q. Jiang, "A Constitutive Model for Ferroelectric PZT Ceramics under Uniaxial Loading," *Smart Materials and Structures*, Vol. 8, No. 4, 1999, pp. 441-459.

[34] M. Kamlah and C. Tsakmakis, "Phenomenological Modeling of the Non-Linear Electro-Mechanical Coupling in Ferroelectrics," *International Journal of Solids and Structures*, Vol. 36, No. 5, 1999, pp. 669-695.

[35] M. Kamlah and U. Böhle, "Finite Element Analysis of Piezoceramic Components Taking into Account Ferroelectric Hysteresis Behavior," *International Journal of Solids and Structures*, Vol. 38, No. 4, 2001, pp. 605-633.

[36] C. M. Landis, J. Wang and J. Sheng, "Microelectromechanical Determination of the Possible Remanent Strain and Polarization States in Polycrystalline Ferroelectrics and the Implications for Phenomenological Constitutive Theories," *Journal of Intelligent Material Systems and Structures*, Vol. 15, No. 7, 2004, pp. 513-525.

[37] J. Schröder and H. Romanowski, "A Simple Coordinate Invariant Thermodynamic Consistent Model for Nonlinear Electro-Mechanical Coupled Ferroelectrica," *European Congress on Computaional Methods in Applied Sciences and Engineering* 2004 (*ECCOMAS* 2004), Jyväskylä, 24-28 July 2004.

[38] J. Schröder and H. Romanowski, "A Thermodynamically Consistent Mesoscopic Model for Transversely Isotropic Ferroelectric Ceramics in a Coordinate-Invariant Setting,"

Achive of Applied Mechanics, Vol. 74, No. 11-12, 2005, pp. 863-877.

[39] S. Klinkel, "A Thermodynamic Consistent 1D Model for Ferroelastic and Ferroelectric Hysteresis Effects in Piezoceramics," *Communications in Numerical Methods in Engineering*, Vol. 22, No. 7, 2006, pp. 727-739.

[40] T. Hegewald, B. Kaltenbacher, M. Kaltenbacher and R. Lerch, "Efficient Modeling of Ferroelectric Behavior for the Analysis of Piezoceramic Actuators," *Journal of Intelligent Material Systems and Structures*, Vol. 19, No. 10, 2008, pp. 1117-1129.

[41] IEEE, "ANSI/ IEEE Std. 176-1987: IEEE Standard on Piezoelectricity," The Institute of Electrical and Electronics Engineers, New York, 1988.

[42] P. G. Harper, "Kinematic Theory of Piezoelectric Hysteresis," *Journal of Applied Physics*, Vol. 52, No. 11, 1981, pp. 6851-6855.

[43] D. C. Jiles and D. L. Atherton, "Theory of the Magnetisation Process in Ferromagnets and Its Application to the Magnetomechanical Effect," *Journal of Physics D: Applied Physics*, Vol. 17, No. 6, 1984, pp. 1265-1281.

[44] R. C. Smith and C. L. Hom, "Domain Wall Model for Ferroelectric Hysteresis," Technical Report CRSC-TR99-07, Center for Research in Scientific Computation, New York, 1999.

[45] T. Hegewald, M. Kaltenbacher and R. Lerch, "Characterization and Modeling of Piezoelectric Stack Actuators," *IWPMA* 2005, *2nd International Workshop on Piezoelectric Materials and Applications in Actuators*, Heinz Nixdorf Institute, Paderborn, 22-25 May 2005, pp. 99-106.

[46] E. Preisach, "On the Magnetic Aftereffect," *Zeitschrift für Physik A Hadrons and Nuclei*, Vol. 94, No. 5-6, 1935, pp. 277-302.

[47] I. D. Mayergoyz, "Mathematical Models of Hysteresis," *Physical Review Letters*, Vol. 56, No. 15, 1986, pp. 1518-1521.

[48] I. D. Mayergoyz and G. Friedman, "Generalized Preisach model of Hysteresis," *IEEE Transactions on Magnetics*, Vol. 24, No. 1, 1988, pp. 212-217.

[49] P. Hejda and T. Zelinka, "Generalized Preisach Model of Hysteresis—Theory and Experiment," *Czechoslovak Journal of Physics*, Vol. 40, No. 1, 1990, pp. 57-68.

[50] M.-J. Jang, C.-L. Chen and J.-R. Lee, "Modeling and Control of a Piezoelectric Actuator Driven System with Asymmetric Hysteresis," *Journal of the Franklin Institute*, Vol. 346, No. 1, 2009, pp. 17-32.

[51] P. N. Sreeram, G. Salvady and N. G. Naganathan, "Hysteresis Prediction for a Piezoceramic Material System," *The ASME Winter Annual Meeting*, New Orleans, 28 November-3 December 3 1993, pp. 35-42.

[52] D. C. Hughes and J. T. Wen, "Preisach Modeling and Compensation for Smart Material Hysteresis," *Proceedings of SPIE*, Vol. 2427, 1995, pp. 50-64.

[53] P. Ge and M. Jouaneh, "Generalized Preisach Model for Hysteresis Nonlinearity of Piezoceramic Actuators," *Precision Engineering*, Vol. 20, No. 2, 1997, pp. 99-111.

[54] H. Hu and R. B. Mrad, "On the Classical Preisach Model for Hysteresis in Piezoceramic Actuators," *Mechatronics*, Vol. 13, No. 2, 2002, pp. 85-94.

[55] G. Robert, D. Damjanovic and N. Setter, "Preisach Modeling of Piezoelectric Nonlinearity in Ferroelectric Ceramics," *Journal of Applied Physics*, Vol. 89, No. 9, 2001, pp. 5067-5074.

[56] G. Song, J. Zhao, X. Zhou and J. A. De Abreu-Garcia, "Tracking Control of a Piezoceramic Actuator with Hysteresis Compensation Using Inverse Preisach Model," *IEEE/ASME Transactions on Mechatronics*, Vol. 10, No. 2, 2005, pp. 198-209.

[57] X. Zhao and Y. Tan, "Neural Network based Identification of Preisach-Type Hysteresis in Piezoelectric Actuator Using Hysteretic Operator," *Sensors and Actuators A: Physical*, Vol. 126, No. 2, 2006, pp. 306-311.

[58] X. Zhao and Y. Tan, "Modeling Hysteresis in Piezo Actuator based on Neural Networks," *Advances in Computation and Intelligence*, *Lecture Notes in Computer Science*, Vol. 5370, 2008, pp. 290-296.

[59] K. Kuhnen and H. Janocha, "Adaptive Inverse Control of Piezoelectric Actuators with Hysteresis Operators," *Proceedings of European Control Conference* (*ECC*), Karsruhe, 31 August-3 September 1999, paper F 0291.

[60] P. Krejci and K. Kuhnen, "Inverse Control of Systems with Hysteresis and Creep," *IEE Proceedings—Control Theory and Applications*, Vol. 148, No. 3, 2001, pp. 185-192.

[61] K. Kuhnen and H. Janocha, "Complex Hysteresis Modeling of a Broad Class of Hysteretic Nonlinearities," *Proceedings of the 8th international conference on New Actuators*, Bremen, 10-12 June 2002, pp. 688-691.

[62] K. Kuhnen and F. Previdi, "Modeling, Identification and Compensation of Complex Hysteretic Nonlinearities: A Modified Prandtl-Ishlinskii Spproach," *European Journal of Control*, Vol. 9, No. 4, 2003, pp. 407-421.

[63] K. Kuhnen, "Modeling, Identification and Compensation of Complex Hysteretic Nonlinearities and Log(*t*)-Type Creep Dynamics," *Control and Intelligent System*, Vol. 33, No. 2, 2005, pp. 134-147.

[64] M. Goldfarb and N. Celanovic, "Modeling Piezoelectric Stack Actuators for Control of Micromanipulation," *IEEE Control Systems Magazine*, Vol. 17, No. 3, 1997, pp. 69-79.

[65] S.-H. Lee, T. J. Royston and G. Friedman, "Modeling and Compensation of Hysteresis in Piezoceramic Transducers for Vibration Control," *Journal of Intelligent Material Systems and Structures*, Vol. 11, No. 10, 2000, pp. 781-790.

[66] T.-J. Yeh, S.-W. Lu and T.-Y. Wu, "Modeling and Identi-

fication of Hysteresis in Piezoelectric Actuators," *Journal of Dynamic Systems, Measurement, and Control*, Vol. 128, No. 2, 2006, pp. 189-196.

[67] T.-J. Yeh, R.-F. Hung and S.-W. Lu, "An Integrated Physical Model that Characterizes Creep and Hysteresis in Piezoelectric Actuators," *Simulation Modelling Practice and Theory*, Vol. 16, No. 1, 2008, pp. 93-110.

[68] R. B. Mrad and H. Hu, "A Model for Voltage to Displacement Dynamics in Piezoceramic Actuators Subject to Dynamic Voltage Excitations," *IEEE/ASME Transactions on Mechatronics*, Vol. 7, No. 4, 2002, pp. 479-489.

[69] D. Song and C. J. Li, "Modeling of Piezo Actuator's Nonlinear and Frequency Dependent Dynamics," *Mechatronics*, Vol. 9, No. 4, 1999, pp. 391-410.

[70] Y. Yu, Z. Xiaob, N. G. Naganathan and R. V. Dukkipati, "Dynamic Preisach Modeling of Hysteresis for the Piezoceramic Actuator System," *Mechanism and Machine Theory*, Vol. 37, No. 1, 2002, pp. 75-89.

[71] X. Tan, R. Venkataraman and P. S. Krishnaprasad, "Control of Hysteresis: Theory and Experimental Results," Pentagon Technical Report A687934, 2001.

[72] H. Hu and R. Ben Mrad, "A Discrete-Time Compensation Algorithm for Hysteresis in Piezoceramic Actuators," *Mechanical Systems and Signal Processing*, Vol. 18, No. 1, 2004, pp. 169-185.

[73] S. Bashash and N. Jalili, "A Polynomial-Based Linear Mapping Strategy for Feedforward Compensation of Hysteresis in Piezoelectric Actuators," *Journal of Dynamic Systems, Measurement, and Control*, Vol. 130, No. 3, 2008, pp. 031008.1-031008.10.

[74] R. Venkataraman and P. S. Krishnaprasad, "Novel Algorithm for the Inversion of the Preisach Operator Smart Structures and Materials," *Proceedings of SPIE* 3984, *Smart Structures and Materials* 2000: *Mathematics and Control in Smart Structures*, 2000, pp. 404-414.

[75] R. Venkataraman and P. S. Krishnaprasad, "Approximate Inversion of Hysteresis: Theory and Numerical Results," *Proceedings of the* 39*th IEEE Conference on Decision and Control*, Sydney, Australia, 12-15 December 2000, pp. 4448-4454.

[76] J. Y. Peng and X. B. Chen, "H_2-Optimal Digital Control of Piezoelectric Actuators," *Proceedings of the* 8*th World Congress on Intelligent Control and Automation, WCICA*-2010, Jinan, 7-9 July 2010, pp. 3684-3690.

[77] U.-X. Tan, W. T. Latt, C. Y. Shee, C. N. Riviere and W. T. Ang, "Feedforward Controller of Ill-Conditioned Hysteresis Using Singularity-Free Prandtl-Ishlinskii Model," *IEEE/ASME Transactions on Mechatronics*, Vol. 14, No. 5, 2009, pp. 598-605.

[78] B. D. Coleman and M. L. Hodgdon, "A Constitutive Relation for Rate-Independent Hysteresis in Ferromagnetically Soft Materials," *International Journal of Engineering Science*, Vol. 24, No. 6, 1986, pp. 897-919.

[79] R. Bouc, "Forced Vibrations of a Mechanical System with Hysteresis," *Proceedings of the* 4*th Conference on Non-linear Oscillations*, Prague, 5-9 September 1967, pp. 315.

[80] Y.-K. Wen, "Method of Random Vibration of Hysteretic Systems," *Journal of the Engineering Mechanics Division*, Vol. 102, No. 2, 1976, pp. 249-263.

[81] M. Ismail, F. Ikhouane and J. Rodellar, "The Hysteresis Bouc-Wen Model, a Survey," *Archives of Computational Methods in Engineering*, Vol. 16, No. 2, 2009, pp. 161-188.

[82] O. Gomis-Bellmunt, F. Ikhouane and D. Montesinos-Miracle, "Control of Bouc-Wen Hysteretic Dystems: Application to a Piezoelectric Actuator," *The* 13*th Power Electronics and Motion Control Conference*, 2008. *EPE-PEMC* 2008, Poznan, 1-3 September 2008, pp. 1670-1675.

[83] M. Rakotondrabe, "Bouc-Wen Modeling and Inverse Multiplicative Structure to Compensate Hysteresis Nonlinearity in Piezoelectric Actuators," *IEEE Transactions on Automation Science and Engineering*, Vol. 8, No. 2, 2011, pp. 428-431.

[84] D. H. Wang, W. Zhu and Q. Yang, "Linearization of Stack Piezoelectric Ceramic Actuators Based on Bouc-Wen Model," *Journal of Intelligent Material Systems and Structures*, Vol. 22, No. 5, 2011, pp. 401-413.

[85] L. Deng and Y. Tan, "Modeling Hysteresis in Piezoelectric Actuators Using NARMAX Models," *Sensors and Actuators A: Physical*, Vol. 149, No. 1, 2009, pp. 106-112.

[86] A. Badel, J. Qiu and T. Nakano, "A New Simple Asymmetric Hysteresis Operator and Its Application to Inverse Control of Piezoelectric Actuators," *IEEE Transactions on Ultrasonics, Ferroelectrics and Frequency Control*, Vol. 55, No. 5, 2008, pp. 1086-1094.

[87] H. Jung, J. Y. Shim and D. Gweon, "New Open-Loop Actuating Method of Piezoelectric Actuators for Removing Hysteresis and Creep," *Review of Scientific Instruments*, Vol. 71, No. 9, 2000, pp. 3436-3440.

[88] K. Kuhnen and H. Janocha, "Compensation of Creep and Hysteresis Effects of Piezoelectric Actuators with Inverse Systems," 6*th International Conference on New Actuators, Actuator*'98, Bremen, 17-19 June 1998, pp. 309-312.

[89] H. Janocha and K. Kuhnen, "Real-Time Compensation of Hysteresis and Creep in Piezoelectric Actuators," *Sensors Actuators A*: *Physical*, Vol. 79, No. 2, 2000, pp. 83-89.

[90] J. S. Dewey, K. Leang and S. Devasia, "Experimental and Theoretical Results in Output-Trajectory Redesign for Flexible Structures," *Journal of Dynamic Systems, Measurement, and Control*, Vol. 120, No. 4, 1998, pp. 456-461.

[91] K. S. Abidi and A. Şabanoviç, "Sliding Mode Control for High-Precision Motion of a Piezostage," *IEEE Transactions on Industrial Electronics*, Vol. 54, No. 1, 2007, pp. 629-637.

[92] H. C. Liaw, B. Shirinzadeh and J. Smith, "Enhanced Sliding Mode Motion Tracking Control of Piezoelectric Actuators," *Sensors and Actuators A*: *Physical*, Vol. 138, No. 1, 2007, pp. 194-202.

[93] G. M. Clayton, S. Tien, A. J. Fleming, S. O. R. Moheimani and S. Devasia, "Inverse-Feedforward of Charge-Controlled Piezopositioners," *Mechatronics*, Vol. 18, No. 5-6, 2008, pp. 273-281.

[94] J. W. Li, X. B. Chen and W. J. Zhang, "A New Approach to Modeling System Dynamics—In the Case of a Piezoelectric Actuator with a Host System," *IEEE/ASME Transactions on Mechatronics*, Vol. 15, No. 3, 2010, pp. 371-380.

[95] J. W. Li, X. B. Chen and W. J. Zhang, "Axiomatic-Design-Theory-Based Approach to Modeling Linear High Order System Dynamics," *IEEE/ASME Transactions on Mechatronics*, Vol. 16, No. 2, 2011, pp. 341-350.

[96] K. Kuhnen, H. Janocha, D. Thull and A. Kugi, "A New Drive Concept for High-Speed Positioning of Piezoelectric Actuators," *Proceedings of the 10th International Conference on New Actuators*, Bremen, 14-16 June 2006, pp. 82-85.

[97] K. K. Leang and S. Devasia, "Feedback-Linearized Inverse Feedforward for Creep, Hysteresis, and Vibration Compensation in AFM Piezoactuators," *IEEE Transactions on Control Systems Technology*, Vol. 15, No. 5, 2007, pp. 927-935.

[98] X. Dang and Y. Tan, "An Inner Product-Based Dynamic Neural Network Hysteresis Model for Piezoceramic Actuators," *Sensors and Actuators A: Physical*, Vol. 121, No. 2, 2005, pp. 535-542.

[99] C. Ru, L. Chen, B. Shao, W. Rong and L. Sun, "A Hysteresis Compensation Method of Piezoelectric Actuator: Model, Identification and Control," *Control Engineering Practice*, Vol. 17, No. 9, 2009, pp. 1107-1114.

[100] L. Sun, C. Ru, W. Rong, L. Chen and M. Kong, "Tracking Control of Piezoelectric Actuator Based on a New Mathematical Model," *Journal of Micromechanics and Microengineering*, Vol. 14, No. 11, 2004, pp. 1439-1444.

[101] C. Ru and L. Sun, "Hysteresis and Creep Compensation for Piezoelectric Actuator in Open-Loop Operation," *Sensors and Actuators A*: *Physical*, Vol. 122, No. 1, 2005, pp. 124-130.

[102] H. H. Najafabadi, S. M. Rezaei, S. S. Ghidary, M. Zareinejad, K. Razi and R. Seifabadi, "Hysteresis Compensation of Piezoelectric Actuators under Dynamic Load Condition," *IEEE/RSJ International Conference on Intelligent Robots and Systems*, 2007. *IROS* 2007, San Diego, 29 October-2 November 2007, pp. 1166-1171.

[103] E. Kouno, "A Fast Response Piezoelectric Actuator for Servo Correction of Systematic Errors in Precision Machining," *CIRP Annals-Manufacturing Technology*, Vol. 33, No. 1, 1982, pp. 369-372.

[104] J. F. Cuttino, A. C. Miller and D. E. Schinstock, "Performance Optimization of a Fast Tool Servo for Single-Point Diamond Turning Machines," *IEEE/ASME Transactions on Mechatronics*, Vol. 4, No. 2, 1999, pp. 169-179.

[105] J. Lin, H. Chiang and C. C. Lin, "Tuning PID Control Gains for Micro Piezo-Stage in using Grey Relational Analysis," *2008 International Conference on Machine Learning and Cybernetics*, Kunming, 12-15 July 2008, pp. 3863-3868.

[106] H.-J. Shieh, Y.-J. Chiu and Y.-T. Chen, "Optimal PID Control System of a Piezoelectric Microospitioner," 2008 *IEEE/SICE International Symposium on System Integration*, Nagoya, 4 December 2008, pp. 1-5.

[107] D. Y. Abramovitch, S. Hoen and R. Workman, "Semi-Automatic Tuning of PID Gains for Atomic Force Microscopes," *Asian Journal of Control*, Vol. 11, No. 2, 2009, pp. 188-195.

[108] K. K. Tan, T. H. Lee and H. X. Zhou, "Micro-Positioning of Linear-Piezoelectric Motors Based on a Learning Nonlinear PID Controller," *IEEE/ASME Transactions on Mechatronics*, Vol. 6, No. 4, 2001, pp. 428-436.

[109] W. S. Oates and R. C. Smith, "Nonlinear Control Design for a Piezoelectric-Driven Nanopositioning Stage," Pentagon Technical Report A591444, 2005.

[110] C. Edwards and S. K. Spurgeon, "Sliding Mode Control: Theory and Applications," Taylor & Francis, Abingdon, 1998.

[111] P.-K. Huang, P.-H. Shieh, F.-J. Lin and H.-J. Shieh, "Sliding-Mode Control for a Two-Dimensional Piezo-Positioning Stage," *Control Theory* & *Applications*, *IET*, Vol. 1, No. 4, 2007, pp. 1104-1113.

[112] J.-C. Shen, W.-Y. Jywel, C.-H. Liu, Y.-T. Jian and J. Yang, "Sliding-Mode Control of a Three-Degrees-of-Freedom Nanopositioner," *Asian Journal of Control*, Vol. 10, No. 3, 2008, pp. 267-276.

[113] J.-C. Shen, J.-C. Shen, H.-K. Chiang and Y.-L. Shu, "Precision Tracking Control of a Piezoelectric-Actuated System," *Precision Engineering*, Vol. 32, No. 2, 2008, pp. 71-78.

[114] Q. Xu and Y. Li, "Dynamics Modeling and Sliding Mode Control of an XY Micropositioning Stage," *The 9th International Symposium on Robot Control* (*SYROCO*'09), Gifu, 9-12 September 2009, pp. 781-786.

[115] Y. Li and Q. Xu, "Adaptive Sliding Mode Control with Perturbation Estimation and PID Sliding Surface for Motion Tracking of a Piezo-Driven Micromanipulator," *IEEE Transactions on Control Systems Technology*, Vol. 18, No. 4, 2010, pp. 798-810.

[116] H. C. Liaw, B. Shirinzadeh and J. Smith, "Sliding-Mode

Enhanced Adaptive Motion Tracking Control of Piezoelectric Actuation Systems for Micro/Nano Manipulation," *IEEE Transactions on Control Systems Technology*, Vol. 16, No. 4, 2008, pp. 826-833.

[117] X. Chen and T. Hisayama, "Adaptive Sliding-Mode Position Control for Piezo-Actuated Stage," *IEEE Transactions on Industrial Electronics*, Vol. 55, No. 11, 2008, pp. 3927-3934.

[118] H. C. Liaw and B. Shirinzadeh, "Enhanced Sliding-Mode Constrained Motion Tracking Control of Piezo-Actuated Flexure-Based Mechanisms," *IEEE/ASME International Conference on Advanced Intelligent Mechatronics*, 2009. *AIM* 2009, Singapore, 14-17 July 2009 2009, pp. 1879-1884.

[119] G. Schitter, P. Menold, H. F. Knapp, F. Allgower and A. Stemmer, "High Performance Feedback for Fast Scanning Atomic Force Microscopes," *Review of Scientific Instruments*, Vol. 72, No. 8, 2001, pp. 3320-3327.

[120] M. S. Tsai and J. S. Chen, "Robust Tracking Control of a Piezoactuator Using a New Approximate Hysteresis Model," *Journal of Dynamic Systems, Measurement, and Control*, Vol. 125, No. 1, 2003, pp. 96-102.

[121] G. Schitter, A. Stemmer, and F. Allgöwer, "Robust Two-Degree-of-Freedom Control of an Atomic Force Microscope," *Asian Journal of Control*, Vol. 6, No. 2, 2004, pp. 156-163.

[122] A. Sebastian and S. M. Salapaka, "Design Methodologies for Robust Nano-Positioning," *IEEE Transactions on Control Systems Technology*, Vol. 13, No. 6, 2005, pp. 868-876.

[123] Y. Okazaki, "A Micro-Positioning Tool Post Using a Piezoelectric Actuator for Diamond Turning Machines," *Precision Engineering*, Vol. 12, No. 3, 1990, pp. 151-156.

[124] D. Croft, S. Stilson and S. Devasia, "Optimal Tracking of Piezo-Based Nanopositioners," *Nanotechnology*, Vol. 10, No. 2, 1999, pp. 201-208.

[125] C. J. Li, H. S. M. Beigi, S. Li and J. Liang, "Nonlinear Piezo-Actuator Control by Learning Self Tuning Regulator," *Journal of Dynamic Systems, Measurement, and Control*, Vol. 115, No. 4, 1993, pp. 720-723.

[126] H.-J. Shieh, F.-J. Lin, P.-K. Huang and L.-T. Teng, "Adaptive Tracking Control Solely Using Displacement Feedback for a Piezo-Positioning Mechanism," *IEE Proceedings——Control Theory and Applications*, Vol. 151, No. 5, 2004, pp. 653-660.

[127] X. Tan and J. S. Baras, "Adaptive Identification and Control of Hysteresis in Smart Materials," *IEEE Transactions on Automatic Control*, Vol. 50, No. 6, 2005, pp. 827-839.

[128] H. C. Liaw and B. Shirinzadeh, "Enhanced Adaptive Motion Tracking Control of Piezo-Actuated Flexure-Based Four-Bar Mechanisms for Micro/Nano Manipulation," *Sensors and Actuators A: Physical*, Vol. 147, No. 1, 2008, pp. 254-262.

[129] J.-H. Xu, "Neural Network Control of a Piezo Tool Positioner," *Canadian Conference on Electrical and Computer Engineering*, Vancouver, 14-17 September 1993, pp. 333-336.

[130] F.-J. Lin, H.-J. Shieh and P.-K. Huang, "Adaptive Wavelet Neural Network Control with Hysteresis Estimation for Piezo-Positioning Mechanism," *IEEE Transactions on Neural Networks*, Vol. 17, No. 2, 2006, pp. 432-444.

[131] F.-J. Lin, H.-J. Shieh, P.-K. Huang and L.-T. Teng, "Adaptive Control with Hysteresis Estimation and Compensation Using RFNN for Piezo-Actuator," *IEEE Transactions on Ultrasonics, Ferroelectrics and Frequency Control*, Vol. 53, No. 9, 2006, pp. 1649-1661.

[132] H. Numasato and M. Tomizuka, "Settling Control and Performance of a Dual-Actuator System for Hard Disk Drives," *IEEE/ASME Transactions on Mechatronics*, Vol. 8, No. 4, 2003, pp. 431-438.

[133] K. Leang and S. Devasia, "Hysteresis, Creep, and Vibration Compensation for Piezoactuators: Feedback and Feedforward Control," *Proceedings of the 2nd IFAC Conference on Mechatronic Systems*, Berkley, 9-11 December 2002, pp. 283-289.

[134] L. Y. Pao, J. A. Butterworth and D. Y. Abramovitch, "Combined Feedforward/Feedback Control of Atomic Force Microscopes," *Proceedings of the 2007 American Control Conference*, New York City, 9-13 July 2007, pp. 3509-3515.

[135] D. Croft and S. Devasia, "Vibration Compensation for High Speed Scanning Tunneling Microscopy," *Review of Scientific Instruments*, Vol. 70, No. 12, 1999, pp. 4600-4605.

[136] H. Hu, H. M. S. Georgiou and R. Ben-Mrad, "Enhancement of Tracking Ability in Piezoceramic Actuators Subject to Dynamic Excitation Conditions," *IEEE/ASME Transactions on Mechatronics*, Vol. 10, No. 2, 2005, pp. 230-239.

[137] C.-J. Lin and S.-R. Yang, "Modeling of a Piezo-Actuated Positioning Stage Based on a Hysteresis Observer," *Asian Journal of Control*, Vol. 7, No. 1, 2005, pp. 73-80.

[138] S. S. Aphale, S. Devasia and S. O. Reza Moheimani, "High-Bandwidth Control of a Piezoelectric Nanopositioning Stage in the Presence of Plant Uncertainties," *Nanotechnology*, Vol. 19, No. 12, 2008, pp. 125503.1-125503.9.

[139] D. Croft and S. Devasia, "Vibration Compensation for High Speed Scanning Tunneling Microscopy," *Review of Scientific Instruments*, Vol. 70, No. 12, 1999, pp. 4600-4605.

[140] S. Khan, M. Elitas, E. D. Kunt and A. Sabanovic, "Discrete Sliding Mode Control of Piezo Actuator in Nano-Scale Range," *IEEE International Conference on Industrial Technology*, 2006. *ICIT* 2006, Mumbai, 15-17 December 2006, pp. 1454-1459.

[141] J. Yi, S. Chang and Y. Shen, "Disturbance-Observer-Based Hysteresis Compensation for Piezoelectric Actuators," *IEEE/ASME Transactions on Mechatronics*, Vol. 14, No. 4, 2009, pp. 4196-4201.

[142] S. Chang and S. Li, "A High Resolution Long Travel Friction-Drive Micropositioner with Programmable Step Size," *Review of Scientific Instruments*, Vol. 70, No. 6, 1999, pp. 2276-2782.

[143] A. Bergander and J. Breguet, "Performance Improvements for Stick-Slip Positioners," *Proceedings of* 2003 *International Symposium on Micromechatronics and Human Science*, Nagoya, 19-22 October 2003, pp. 59-66.

[144] F. Castanos and L. Fridman, "Analysis and Design of Integral Sliding Manifolds for Systems with Unmatched Perturbations," *IEEE Transactions on Automatic Control*, Vol. 51, No. 5, 2006, pp. 853-858.

[145] J. H. Painter, D. Kerstetter and S. Jowers, "Reconciling Steady-State Kalman and Alpha-Beta Filter Design," *IEEE Transactions on Aerospace and Electronic Systems*, Vol. 26, No. 6, 1990, 1990, pp. 986-991.

[146] X. Wang, "High-Order Integral-Chain Differentiator and Application to Acceleration Feedback," Submitted to *Computer Science* for Possible Publication, 2011.

[147] A. Levant, "Robust Exact Differentiation via Sliding Mode Technique," *Automatica*, Vol. 34, No. 3, 1998, pp. 379-384.

[148] Q. Xu and Y. Li, "Sliding Mode Control of a Piezo-Driven Micropositioning System Using Extended Kalman Filter," *IEEE International Conference on Automation and Logistics* (*ICAL*), 2010, Hong Kong and Macau, 16-20 August 2010, pp. 427-432.

[149] J. L. Minase, T.-F. Lu and F. Wornle, "State Estimation of Nonlinear Piezoelectric Stack Actuator Hysteresis Model," *Proceedings of SPIE*, Vol. 6414, 2006, pp. 641403.1-641403.10.

[150] J. Minase, T.-F. Lu, B. Cazzolato and S. Grainger, "Adaptive Identification of Hysteresis and Creep in Piezoelectric Stack Actuators," *The International Journal of Advanced Manufacturing Technology*, Vol. 46, No. 9-12, 2009, pp. 913-921.

[151] F. Yang and R. W. Wilde, "Observers for Linear Systems with Unknown Inputs," *IEEE Transactions on Automatic Control*, Vol. 33, No. 7, 1988, pp. 667-681.

[152] M. Hou and P. C. Muller, "Design of Observers for Linear Systems with Unknown Inputs," *IEEE Transactions on Automatic Control*, Vol. 37, No. 6, 1992, pp. 871-875.

[153] S. Hui and S. Żak, "Observer Design for Systems with Unknown Inputs," *International Journal of Applied Mathematics and Computer Science*, Vol. 15, No. 4, 2005, pp. 431-546.

[154] S. H. Zak and S. Hui, S. "Output Feedback Variable Structure Controllers and State Estimators for Uncertain/Nonlinear Dynamic Systems," *IEE Proceedings D: Control Theory and Applications*, Vol. 140, No. 1, 1993, pp. 41-50.

[155] J.-L. Chang and T.-C. Wu, "Robust Disturbance Attenuation with Unknown Input Observer and Sliding Mode Controller," *Electrical Engineering*, Vol. 90, No. 7, 2008, pp. 493-502.

[156] T. Floquet, C. Edwards and S. K. Spurgeon, "On Sliding Mode Observers for Systems with Unknown Inputs," *International Journal of Adaptive Control and Signal Processing*, Vol. 21, No. 8-9, 2007, pp. 638-656.

[157] K. Kalsi, J. Lian, S. Hui and S. H. Zak, "Sliding-Mode Observers for Systems with Unknown Inputs: A High-Gain Approach," *Automatica*, Vol. 46, No. 2, 2010, pp. 347-353.

[158] X. Liu, J. Jeong and J. Kim, "A Three Translational DoFS Parallel Cube-Manipulator," *Robotica*, Vol. 21, No. 6, 2003, pp. 645-653.

[159] Q. Xu, and Y. Li, "A Novel Design of a 3-PRC Translational Compliant Parallel Micromanipulator for Nanomanipulation," *Robotica*, Vol. 24, No. 4, 2006, pp. 527-528.

[160] Y. Yue, F. Gao, X. Zhao and Q. Jeffrey Ge, "Relationship among Input-Force, Payload, Stiffness and Displacement of a 3-DOF Perpendicular Parallel Micro-Manipulator," *Mechanism and Machine Theory*, Vol. 45, No. 5, 2010, pp. 756-771.

[161] S. Awtar and A. Slocum, "Constraint-Based Design of Parallel Kinematic XY Flexure Mechanisms," *Journal of Mechanical Design*, Vol. 129, No. 8, 2007, pp. 816-830.

[162] Q. Yao, J. Dong and P. Ferreira, "A Novel Parallel-Kinematics Mechanisms for Integrated, Multi-Axis Nanopositioning: Part 1. Kinematics and Design for Fabrication," *Precision Engineering*, Vol. 32, No. 1, 2008, pp. 7-19.

[163] J. Dong, Q. Yao and P. Ferreira, "A Novel Parallel-Kinematics Mechanism for Integrated, Multi-Axis Nanopositioning: Part 2: Dynamics, Control and Performance Analysis," *Precision Engineering*, Vol. 32, No. 1, 2008, pp. 20-33.

[164] J. Dong, S. Salapaka and P. Ferreira, "Robust Control of a Parallel-Kinematic Nanopositioner," *Journal of Dynamic Systems, Measurement, and Control*, Vol. 130, No. 4, 2008, pp. 041007.1-041007.15.

[165] Y. Tian, B. Shirinzadeh and D. Zhang, "A Flexure-Based Five-Bar Mechanism for Micro/Nano Manipulation," *Sensors and Actuators A: Physical*, Vol. 153, No. 1, 2009, pp. 96-104.

[166] Y. Tian, B. Shirinzadeh and D. Zhang, "Design and Dynamics of a 3-DOF Flexure-Based Parallel Mechanism for Micro/Nano Manipulation," *Microelectronic Engineering*, Vol. 87, No. 2, 2010, pp. 230-241.

[167] K. Hu, J. Kim, J. Schmiedeler and C. Menq, "Design, Implementation, and Control of a Six-Axis Compliant

Stage," *Review of Scientific Instruments*, Vol. 79, No. 2, 2008, pp. 025108.2-025105.11.

[168] T. W. Seo, D. S. Kang and J. Kim, "Synthesis and Comparison of Fine Actuator Controllers for a 3-DOF Micro Parallel Positioning Platform," *Proceedings of the 10th International Conference on Control, Automation, Robotics and Vision*, Hanoi, 17-20 December 2008, pp. 1606-1611.

[169] M. Grotjahn, B. Heimann and H. Abdellatif, "Identification of Friction and Rigid-Body Dynamics of Parallel Kinematic Structures for Model-Based Control," *Multibody System Dynamics*, Vol. 11, No. 3, 2004, pp. 273-294.

13

Determination of Residual Stresses Induced by the Autofrettage Treatment by the X-Rays Diffraction Method

Naziha Zerari[1], Tarik Saidouni[2], Abdelouaheb Benretem[1]
[1]Laboratory of the Electromechanicaly Engineering Badji Mokhtar-Annaba University, Annaba, Algeria
[2]Laboratory of the Mechanical Systems, Path, Engineering of the Mechanical Systems, Option Structure and Production Ministry of the National Defense, School Military Polytechnic EMP, Alger, Algeria

ABSTRACT

Some meaningful advances have been made these last years to value precise and reliable way the residual stresses experimentally created by the autofrettage. The autofrettage process is used widely to introduce residual stresses into thick walled tubes; traditionally residual stresses have been measured using the Sachs method destructive or non-destructive methods. In this paper we describe the application of the X-rays diffraction; this technique permits to justify the presence of the compressive tangential residual stresses, and to value their distribution after two different autofrettage internal pressures loading. The results show that there is a large difference in the residual stresses find in the different autofrettege pressure. One can see the influence of the autofrettage's pressure quantity on residual stresses created in the thickness of the test tubes.

Keywords: Autofrettage; Elasto-Plastic; Residual Stresses; X-Rays Diffraction

1. Introduction

The mechanical piece fatigue behaviour or a structure have a strong historic with the solicitations that it undergoes during manufacture processes and during the service hours, under some working conditions. The residual stresses led by these loadings have a big influence on the life of fatigue [1,2]. The residual stresses in the mechanical pieces can result from a lot of sources, residual stress can raise or lower the mean stress experienced over a fatigue cycle. In general, we can classify them like being the harmful stresses that accelerate the starting phases of fatigue growth creeks. In other case, some additional treatments are added after the manufacture process in the specific object to produce some repressive residual stresses beneficial to the most critical structure places. It is now established that the majority of the fatigue failings can be avoided by compressive efforts [3].

This means that considerable advantage can be gained by engineering a compressive inplane stress in the near surface region, for example, by peening, autofrettage into thick walled tubes [4,5].

2. Hydraulic Autofrttage Principle

The basic principle of the hydraulic autofrettage process consists to recharge a thick cylinder by an internal pressure superior to the service pressure. If the internal pressure increases in a thick cylinder, the distortion imposed to the material cylinder is first merely elastic until the elastic limit is reached. Beyond this pressure zone (a), occurs an irreversible plastic distortion zone (b). When the internal pressure is freed after the autofrettage process (discharge), it appears two different layers along the thickness cylinder (a) and (b). The internal, deformed layers plastically, prevent the return elastically in the state of balance the deformed external layers. The thickness of the cylinder undergoes residual stresses in the external layers and compression on the internal layers [6]. These are going to be added algebraically to the stresses to generate by the ulterior application of the service pressure.

3. Determination of the Residual Stresses

3.1. Introduction

A Several methods for the determination of the residual

stresses exist currently; we can classify them in destructive or non-destructive methods. Let's mention among the non destructive methods, the X-rays diffraction and the neutrons diffraction. The destructive methods, for example, the holes method, that requires a drilling to measure the evolution of the residual stresses according to the depth of the holes pierced [7-9]. Let's recall that the residual stresses are ordinarily divided in three categories:

- Residual stresses I type: macrostresses, *i.e.* Constant over a large number of grains.
- Residual stresses II type: constant over a single grain.
- Residual stresses III type: constant over the atomic.

What follows is concerned only with residual stresses of the first type (I) [10].

3.2. Bragg's Law and Diffraction

Bragg diffraction occurs when electromagnetic radiation or subatomic particle waves with wavelength comparable to atomic spacings are incident upon a crystalline sample, are scattered in a specular fashion by the atoms in the system, and undergo constructive interference in accordance to Bragg's law. For a crystalline solid, the waves are scattered from lattice planes separated by the interplanar distance *d*. Where the scattered waves interfere constructively, they remain in phase since the path length of each wave is equal to an integer multiple of the wavelength. The path difference between two waves undergoing constructive interference is given by $2d\sin\theta$, where θ is the scattering angle. This leads to Bragg's law, which describes the condition for constructive interference from successive crystallographic planes (*h*, *k*, and *l*) of the crystalline lattice, as shown in **Figure 1**. Bragg expressed this in an equation now known as Bragg's Law:

$$n\lambda = 2\times d_{hkl}\times\sin\theta \quad (1)$$

where n is an integer, λ is the wavelength of incident wave, d is the spacing between the planes in the atomic lattice, and θ is the angle between the incident ray and the scattering planes, 2θ the diffraction angle.

A small change in the lattice parameter Δd will result in a change $\Delta\theta$ in the Bragg angle so that the lattice elastic strain is given by:

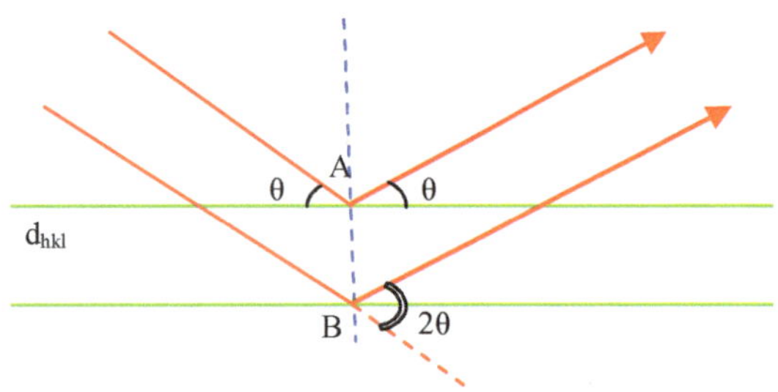

Figure 1. Bragg's law.

$$\varepsilon_{\phi\psi} = (D_d - d_0)/d_0 = -\cot g\theta_0\cdot\Delta\theta \quad (2)$$

The diffraction angle sample 2θ is near to 90°, for which one gets a good compromise between the spatial resolution and the measure precision. The ϕ angle defines the measure direction on the surface of the sample while the angle ψ defines the internal angle there between the normal to the sample and the normal to the diffracting plan. *L* and *P* are respectively the reference marks of the laboratory (reference mark of the measure as shown in **Figure 2**) and of the sample. From the mechanics laws of the continuous surroundings (Hooke laws generalized) and while supposing the homogeneous elastic and isotropy material.

3.3. Measurement of Residual Stresses Using X-Ray $\sin^2\psi$ Method

With the homogeneous material hypothesis, macroscopically isotropy and the sufficiently small grains propertied on a plane surface, we can make the stresses calculation [8]. You can use the X-rays diffraction to determine the residual Strains (macrostrains) presented on the surface, in the crystalline material [10,11].

The $\sin^2\psi$ method it consists to measure the distance variation between the plans atomic d of a same family.

Usually in X-rays diffraction, one considers the superficial measure and with stresses plan state of the materials area with a stresses plan state to the material surface. $\sigma_{rz} = \sigma_{\theta z} = \sigma_{zz} = 0$. This is a base equation for the $\sin^2\psi$ method as follows:

$$\varepsilon_{\phi\psi} = \frac{1}{2}S_1\cdot\sigma_\phi\sin^2\psi + S_2\cdot\text{trace}[\sigma] \quad (3)$$

where the main stress directions are unknown, the distributions of lattice strain should be recorde in to two variable directions ψ and ϕ. As ψ = 0° - 90°, Step is 3° and ϕ = 0° - 180° with 1 and 5 step.

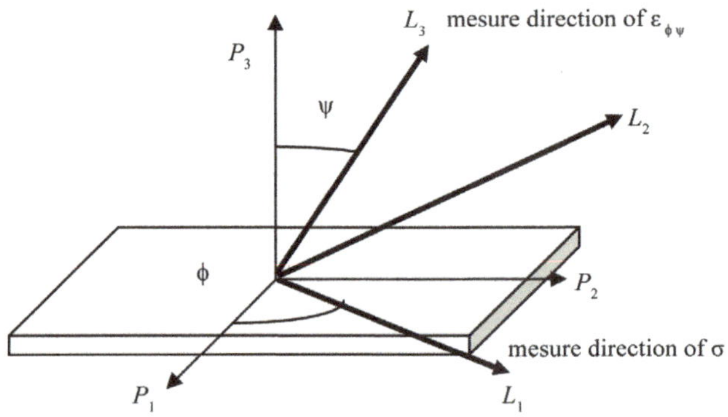

Figure 2. Coordinates systems and notations usually used in strains measure by diffraction. *L* and *P* are respectively the reference marks of the laboratory (reference mark of the measure) and of the sample.

$$\frac{1}{2}S_1 = \frac{1+\nu}{E} \quad (4)$$

$$S_2 = -\frac{\nu}{E} \quad (5)$$

$$\sigma_\phi = \left(\sigma_{rr}.\cos^2\phi + \sigma_{\theta\theta}\sin^2\phi\right) \quad (6)$$

$$\text{Trace}[\sigma] = (\sigma_{rr} + \sigma_{\theta\theta}) \quad (7)$$

3.4. Materials and Sample

The chemical composition of the sample is aluminum alloy (AU4G) given in the **Table 1**. A schematic of the simple is shown in **Figure 3**.

These mechanical properties sum up as follows:

- The limit elastic σ_0 = 100 N/mm.
- The Young's modulus E = 74000 N/mm^2.
- The tangent modulus E_p = 26,000 N/mm^2.
- Poisson's ratio $\nu = 0.33$.
- Internal diameter a =30 mm.
- External diameter b = 50 mm.
- Length of sample L = 300 mm.
- The wall thickness $K = b/a$ =1.66.

Equipment for field measurement of residual stresses by X-ray diffraction was performed in the Laboratory Materials, Minerals and Composites (LMMC), University of Boumerdes ALGERIE. **Table 2** gives the Operating condition.

3.5. Samples Preparation

For a faithful determination of the residual stresses led by the autofrettage process, it is necessary to eliminate all residual stresses that exist before in the sample tube. By a recook in 530°C during one half hour, the existing stresses led by the manufacture processes have been eliminated. We consider a thick cylinde, constituted of a material considered elasto-plastic to the springiness linear isotope and having for criteria of malleability the one of Trescas or Von-Mis. Leaving from the initial state, this cylinder is submitted to a pressure interior P that one makes grow from scratch while the outside pressure is negligible.

Table 1. Chemical composition table of the tube test.

Al	Cu	Mg	Mn
92.3%	4.7%	2.1%	0.7%

Table 2. Operating condition of X-rays test.

Operating condition	Value
Characteristic of the diffract meter	Monochromatic radiation Cr-Kα
Power supply	40 mA, 40 kV
Penetration depth	$\approx$ 27 µm
Extended values 2θ	0° - 90°
Extended from the values ϕ	ϕ = 0° - 90° step 0.5°
Extended from the values ψ	ψ = 0° - 90°, step 3°
Wavelength λ	1.54 nm

4. Results and Discussions

4.1. Residual Stresses Analysis Produced by Autofrettage Pressure 580 Bars

After The thick tube recharging by an internal pressure the position measure a peak that translates the macroscopic state stresses in the material is accompanied by a distance measure d interreticulaire. These sizes permit to characterize the macro-strains material state. The diffraction measure results after the autofrettage pressure treatment equal to 580 bar are regrouped in the **Table 3**. The d_0 value of the AU4G material in the Free State (without strains d_0 = 4.04 Å. From the Equation (2) one can calculate $\varepsilon_{\phi\psi}$, if one carries the value of the strains $\Delta d/d$ accord the value of $\sin^2\psi$ **Figure 4**. After a simplification, we can write Equation (3) as follows:

$$\varepsilon_{\phi\psi} = A\cdot\sin^2\psi + B \quad (8)$$

$$A = \frac{\Delta\varepsilon_{\phi\psi}}{\Delta\sin^2\psi} = \frac{1+\nu}{E}\cdot\left(\sigma_{rr}\cdot\cos^2\phi + \sigma_{\theta\theta}\sin^2\phi\right) \quad (9)$$

$$B = S_2\cdot[\sigma_{rr} + \sigma_{\theta\theta}] \quad (10)$$

From the **Figure 4** we can get constant B:

$$B = -0.7451 \Rightarrow A = \frac{\Delta\varepsilon_{\phi\psi}}{\Delta\sin^2\psi} = \frac{1}{2}S_1\cdot\sigma_\phi$$

To determine σ_{rr} and $\sigma_{\theta\theta}$ to solve the set of two

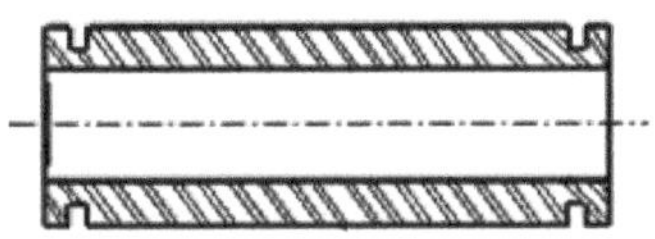

Figure 3. Test-tube.

Table 3. Result rough of measurements of the test of X-ray autofrettage with 580 bars.

2θ [°]	d [Å]
26.7908	3.89490
38.3231	3.2231
44.5462	2.8502
55.7794	2.7480
64.8949	1.4786
78.0485	1.2124
82.3749	1.0298

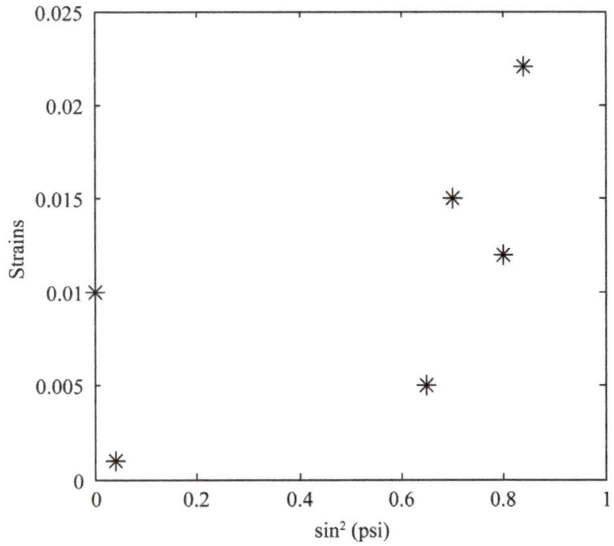

Figure 4. Evolution of the deformation according to $\sin^2\psi$ ($P = 580$ bar).

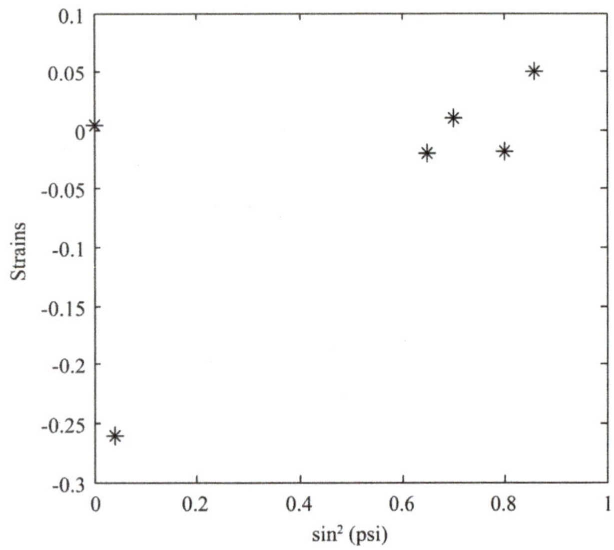

Figure 5. Evolution of the strains according to $\sin^2\psi$ ($P = 400$ bar).

Equations (9) and (10) are necessary. The results obtained are translated into a curve, shown in **Figure 6**.

We can conclude that after an autofretagge process at 580 bars we note that:

- The tangential residual stresses $\sigma_{\theta\theta}$, are important in the state of compression that in traction.
- The radial residual stresses σ_{rr} is always of constant tractions, with a very weak constant value by contribution to the tangential residual stresses.

4.2. Residual Stresses Analysis Produced by Autofrettage Pressure 400 Bars

Table 4 gives the angular positions of diffraction 2θ and the distances d inters reticular X-diffraction of the autofrettage treatment at pressure equal 400 bars.

In this case, we determine the residual stresses and going to follow the same stages that one mentioned before. We present the strains curve in function of $\sin^2\psi$ in **Figure 5**, when we can find B equal at −0.7106.

Figure 7 gives the radial and tangential residual stresses distribution at autofrettage pressure of 400 bars. By this figure we can conclude that:

- The tangential residual stresses $\sigma_{\theta\theta}$ is more important in compressions that in traction.
- The radial residual stresses σ_{rr} are always of traction, with the same feeble value by contribution to the tangent stresses.
- **Figure 8** shown the analysis results of tow thick-walled tube are subjected to different high internal pressure which expands it and leaves residual compressive stresses on the surface of the bore. One can see the influence of the autofrettage pressure on the tangential residual stresses quantity created by this treatment. It has been also observed that, if an autofrettage pressure value is exceeded, the residual stresses increase.

5. Conclusions

The X-rays diffraction technique permits to justify the

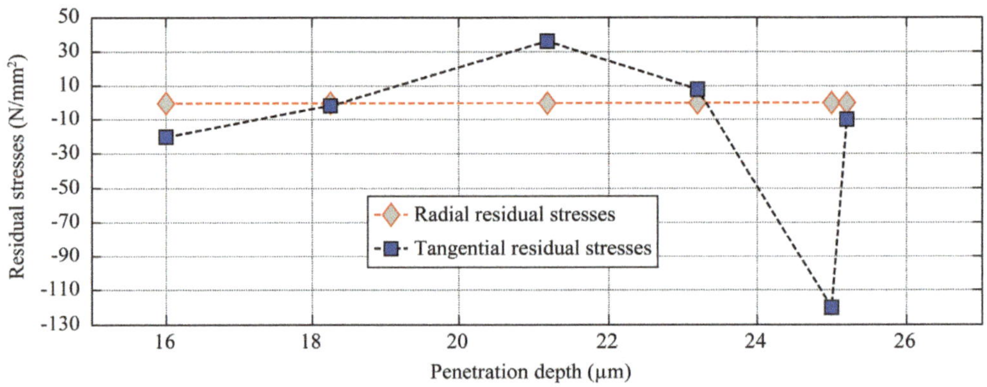

Figure 6. Distribution of the residual stresses at pressure load ($P = 580$ bar).

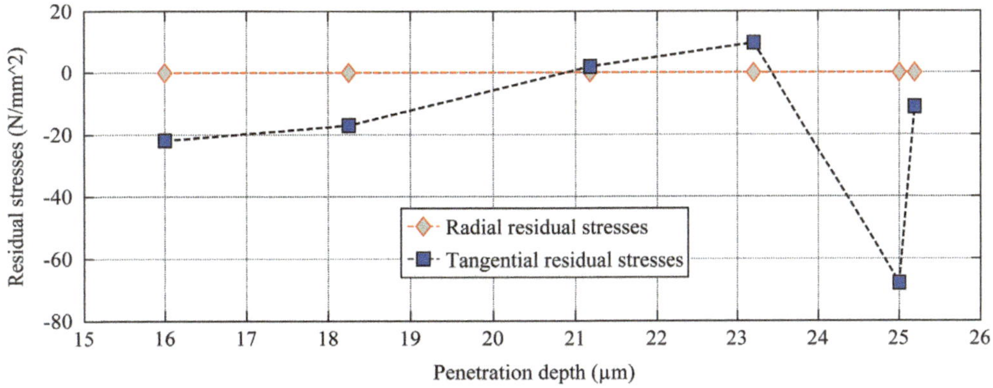

Figure 7. Distribution of the residual stresses at pressure load (P = 400 bar).

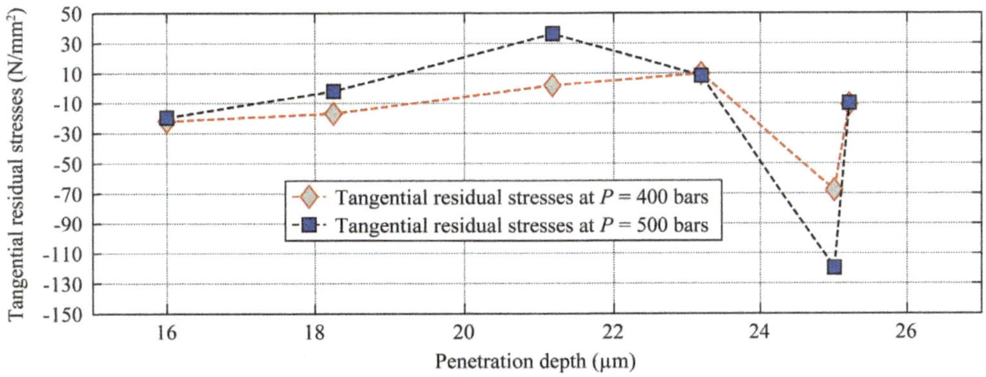

Figure 8. Distribution of the residual stresses compared for two different pressures.

Table 4. Result rough of measurements of the test of X-rays autofrettage with 400 bars.

2θ [°]	d [Å]
26.7908	3.8517
38.3231	3.4477
44.5462	3.0033
55.7794	2.5573
64.8949	1.9841
78.0485	1.5224
82.3749	1.1692

presence of the residual stresses, and to value their distribution.

The results show that the radial and tangential residual stresses distribution find by the X-rays diffraction after an autofrettage test, one can see the influence of the autofrettage pressure on residual stresses created.

REFERENCES

[1] B. Avitzur, "Determination of Residual Stress Distributions in Autofrettaged Tubing," *International Journal of Pressure Vessels and Piping*, Vol. 31, No. 3, 1988, pp. 147-157.

[2] G. A. Webster and A. N. Ezeilo, "Residual Stress Distributions and Their Influence on Fatigue Lifetimes," *International Journal of Fatigue*, Vol. 23, Suppl. 1, 2001, pp. S375-S383.

[3] S. M. Aceves, J. Martinez-Frias and F. Espinosa-Loza, "Certification Testing and Demonstration of Insulated Pressure Vessel for Vehicular Hydrogen Storage," *International Journal of Hydrogen Energy*, Vol. 31, No. 15, 2006, pp. 2274-2283.

[4] J. M. Alegre, P. Bravo and M. Preciado, "Fatigue Behaviour of an Autofrettaged High-Pressure Vessel for the Food Industry," *Engineering Failure Analysis*, Vol. 14, No. 2, 2007, pp. 396-407.

[5] H. O. Fuchs, "Techniques of Surface Stressing to Avoid Fatigue," pp. 197-229.

[6] J. Faisandier and T. Coll, "Mécanisme Hydrauliques et Pneumatiques," 8ème Edition, Ecole d'Ingénieur des Arts et Métiers et de Ecole National Supérieur de l'Aéronautique, Paris, 1999.

[7] D. George and D. J. Smith, "The Application of the Deep Hole Technique for Measuring Residual Stresses in an Autofrettaged Tube," *Proceedings of PVP* 2000, Seattle, 23-27 July 2000.

[8] M. Certiti, "Apport de la Diffraction des Neutron a

l'Analyse des Contraintes Interne," PhD Thèse, Science Chimiques, Université de Paris sud, Centre d'Orsay, 2004.

[9] M. R. Hill, "Determination of Residual Stress Based on the Estimation of Eigenstrain," PhD Thesis, Stanford University, Stanford, 1996.

[10] P. J. Withers and H. K. D. H. Bhadeshia, "Residual Stress Part 1—Measurement Techniques," *Materials Science and Technology*, Vol. 17, No. 4, 2001, pp. 355-365.

[11] G. A. Webster and A. N. Ezeilo, "Residual Stress Distributions and Their Influence on Fatigue Lifetimes," *International Journal of Fatigue*, Vol. 23, Suppl. 1, 2001, pp. S375-S383.

Comparative Study of the Impact of Corundum Particle Contaminants Size on Wear and Fatigue Life of Grease Lubricated Ball Bearings

D. Koulocheris, A. Stathis, Th. Costopoulos, G. Gyparakis
School of Mechanical Engineering, National Technical University of Athens, Athens, Greece

ABSTRACT

Ball bearings' rating life is reduced with the presence of hard particle contaminants in the lubricant. This life reduction is taken into account when calculating the modified rating life by using the contamination factor. The contamination factor is based on a general characterization of the lubrication conditions but the impact of contaminant's variables such as size, hardness and concentration level is not determined in detail. In this work, greases contaminated with hard corundum (alumina, Al_2O_3) particles of different sizes are tested aiming at finding a pattern in the relationship between particle sizes' and wear's progress. A laboratory rig is utilized for these tests and vibration analysis tools regarding bearings' condition and estimated residual life are being assessed. After the tests, optical inspections using a stereoscope verify the vibration analyses results. The experimental results show that wear is progressing faster when smaller particles are used. The crystalline structure of the corundum and the different degree of brittleness of the contaminants seem to be the reasons for this behavior, whereas severe abrasion and deformation of the raceways have been detected.

Keywords: Debris; Grease; Contaminants; Hardness; Wear; Vibration

1. Introduction and Theory Overview

Rolling element bearings are critical components of many modern engines and machinery. When bearings have been chosen correctly, handled, installed and maintained properly, their service life can be well predicted from their basic rating life L_{10} as it is calculated by their manufacturer. Usually the operating conditions are different from the ideal conditions on which the L_{10} life is based. In order to take into account these different operating conditions, the international standard ISO 281:2007 has moved towards a system approach of the bearing and has introduced the modified rating life Lnm which is based on the L_{10} life, the reliability modification factor a_1 and the modification factor α_{ISO} which is a function of lubricating conditions, lubricant contamination, operating temperature and fitting stresses.

Most of the largest bearing manufacturers state that lubricant contamination is the major cause of bearing failure before they reach their rated life. Lubricant cleanliness is taken into account when calculating the rating life. The classification of lubricant cleanliness is based on the ISO 4406:1999 cleanliness code which is limited in filtrated oil lubrication systems. For grease lubricated bearings, which have a large market share, an equivalent grease cleanliness classification has not been established in detail.

Other aspects regarding the nature of particle contaminants such as size, hardness, brittleness and concentration levels have been investigated by researchers, but they have not been quantified.

In this paper we experimentally evaluate vibration signal analysis methods for bearings under different lubricant contamination conditions, which cause accelerated wear. Self-aligned, double row, ball bearings lubricated with hard steel particle pre-contaminated grease of two different particle sizes were tested. Vibration analysis is performed in the time and in the frequency domain in order to evaluate the accuracy of each indicator under these operating conditions and lubrication. At the end of the experiments the vibration analyses findings are compared with optical inspections of the bearings.

1.1. Basic Rating Life of Rolling Element Bearings

During their operating life, bearings are carrying loads which cause variable stresses to the contact surfaces. These Hertzian contact stresses are the main cause of fatigue, determining the useful operating life. This service life of rolling element bearings can be estimated based on the Weibull distribution and its application on bearings. Based on Weibull's theory Palmgren and Lundberg [1] proposed that for a given probability n, the service life of bearing can be calculated from the equation below:

$$L_n = \left(\frac{Q_c}{Q}\right)^p \quad (1)$$

where: L_n = Basic rating life (with a probability n).

Q_c = Basic dynamic load.

Q = Actual load.

p = Constant.

Constant p is a function of the three constants *c*, *e*, *h*, regarding maximum shear stress exponents, Weibull slope and the type of the bearing. Several functions which correlate these constants have been proposed by researchers [2,3] based on different assumptions of the importance and the influence of each one. Regardless of the adopted theory, constant p has been experimentally approximated to the values 3 for ball bearings and 10/3 for roller bearings. Usually, the basic rating life L_{10} is used, expressing the expected life with 90% reliability. In addition, the basic dynamic radial C_r or axial C_a load is used in conjunction with the dynamic equivalent radial P_r or axial P_a load.

In practice, rolling element bearings show longer service lives than their basic rated L_{10} life when they operate under normal conditions. On the other hand, service lives are shorter when the operating conditions and especially temperature and lubricating conditions are worse. The need for calculating the expected service life under realistic, non-normal conditions led to the introduction of the modified rating life L_{nm} which takes into account these operating conditions and also provides results for probabilities different from 90%. According to ISO 281:2007 [4], the modified rating life L_{nm} of the bearings can be calculated from the equation below:

$$L_{nm} = \alpha_1 \times \alpha_{ISO} \times L_{10} \quad (2)$$

where: α_1 = Life modification factor for reliability.

α_{ISO} = Life modification factor, based on a systems approach of life calculation.

The life modification factor α_{ISO} is a function of the bearings operating conditions as it is shown in Equation (3).

$$\alpha_{ISO} = f\left(\frac{e_c, C_u}{P}, \kappa\right) \quad (3)$$

where: e_c = Contamination factor.

C_u = Fatigue load limit.

P = Dynamic equivalent load.

κ = Viscosity ratio of the lubricant.

The contamination factor e_c can be estimated based on some rough, simple guidelines regarding the lubrication conditions. The values of the factor have been found experimentally and they are given in relevant tables. Generally, life reduction caused by solid particles in the lubricant film is dependent on:

1) Type, size, hardness and quantity of the particles;

2) Lubricant film thickness and viscosity;

3) Bearing size and clearances.

The researches on bearings' wear due to contaminants point to several possible mechanisms of wear [5,6]. In terms of contaminant sizes, Maru *et al.* [7] and Nikas *et al.* [8] state that the critical size is in the order of the lubrication film thickness as particles entrapped in the contact zone are causing local oil starvation, increased heating and material melting, while others report that when particles larger than the oil film thickness enter the contact zone, they cause stress peaks and permanent indentations in the raceway as they are over rolled. These stress peaks lead to a reduced life of the rolling bearing due to fatigue.

Usually the contaminants are larger than the typical lubricant film but rather small compared with the contact bodies. As they enter the contact zone it is difficult to move the bearing elements apart, so they are deformed. Ductile contaminants are rolled into pellets and brittle contaminants are crushed into fragments, as shown in **Figure 1**. The damage to the contact surfaces is controlled by the size of these deformed particles which in turn depends on the hardness of the ductile particles and the toughness of the brittle particles [9].

When brittle particles enter the contact zone they fracture down to a critical size. This critical size depends on the material toughness and this size is larger on materials with higher fracture toughness [10,11]. Brittle particles of low toughness cause small dents before their fracture to small fragments and these small dents have minimal interaction with the subsurface stresses which cause fatigue. Surface initiated cracking is more possible due to higher surface stresses on these cases.

On the other hand, particles of high toughness are causing larger dents and grooves when trapped in the contact zone before they fracture [12]. The size of these dents depends on the contact gap, counter surface velocities, fracture toughness, fragment size and contact surface hardness [6].

Denting is causing high stresses and the continuous over rolling is leading to a reduction in the fatigue life. The bearing life is found to be decreased by a factor of ten when brittle particles with high toughness have been

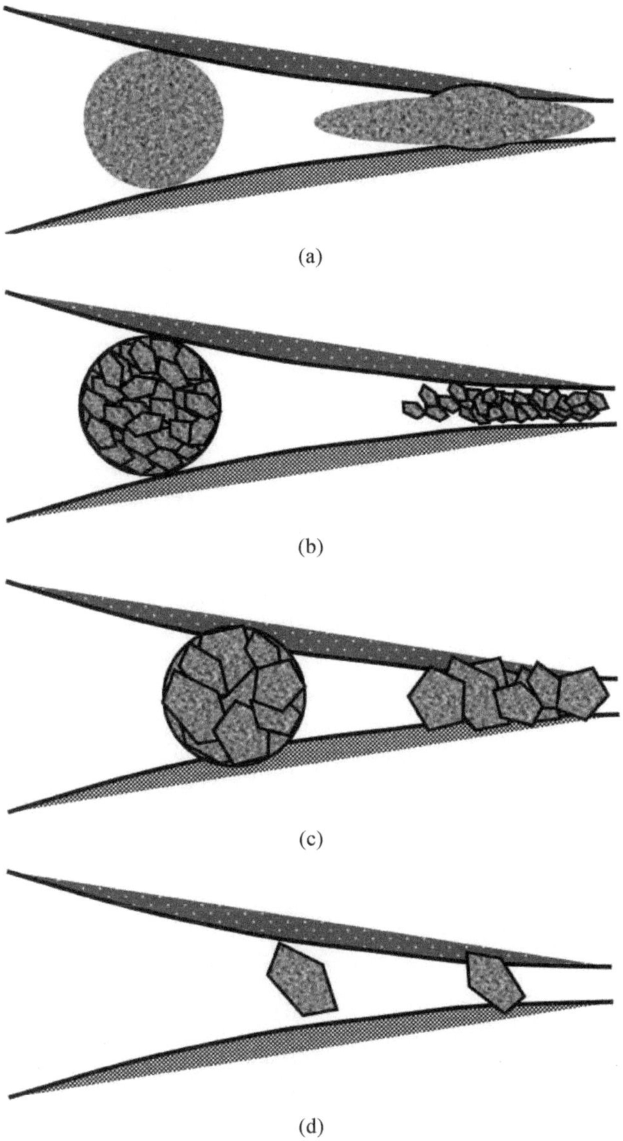

Figure 1. Particle behavior as it enters the contact zone in relation with its characteristics [6]. (a) A ductile particle deforms in the inlet zone and becomes a platelet as it is compressed. (b) A low-toughness brittle particle fractures early in the inlet zone and produces small fragments that may imbed the contact surfaces. (c) A high-toughness brittle particle fractures late in the inlet zone and produces large fragments that imbed the contact surfaces. (d) A small, tough particle behaves rigidly.

used [11]. Dents, grooves and surface deformation are causing abnormal operation of the bearing. When these faults are present the vibration signals from the bearing are increasing and with the proper analysis the size and the location of these faults can be identified.

1.2. Vibration Signal Analysis on Rolling Element Bearings

Rolling element bearings produce mechanical vibrations and noise as they rotate. When faults are present, these vibrations are increased as the motion of the rolling elements is more disturbed. These faults can be minor cracks or even spalls due to fatigue, denting, scuffing, scoring, wear, or any other kind of deformation of the contact surfaces. The vibration signal can be further processed in the time and in the frequency domain.

In time domain, data are processed with statistical and arithmetical methods. The most common values that are measured or calculated are [13,14]:

1) The minimum and the maximum amplitude (min-max).

2) The peak count.

3) Mean μ and standard deviation σ.

4) Root Mean Square (RMS).

5) The crest factor (peak to rms).

6) Statistical moments of higher order such as skewness S and kurtosis K.

Data which have been recorded in time domain are can easily be represented in the frequency domain with the use of a Fourier transformation. The most common Fourier transform method is the Discrete Time Fourier Transform (DTFT) which is calculated efficiently with the use of a Fast Fourier Transform (FFT) algorithm. In frequency domain various types of analyses can be performed such as spectral analysis, cepstrum analysis or bispectral analysis [15,16].

Certain faults and malfunctions exhibit signal peaks in characteristic frequencies. Faults in rolling element bearings produce signals in frequencies which depend on the geometrical characteristics of the bearing and the rotation frequency f.

Most studies regard experimental setups in laboratory conditions with minimum distortion and low noise to affect the measured signals. In most real life situations things are more complicated. Vibrations and noise are introduced into the signal due to angular and axial misalignment, shaft imbalance, improper support and fixings, electrical noise, transducer position, load variation. These parameters are limiting the use of time domain analysis as the magnitude of the measured signal is dominated mostly by noise rather than the measured characteristic. In these situations frequency domain analysis seems to present more useful results.

2. Experiments

2.1. Experimental Setup

The setup simulates a common industrial arrangement where a shaft is radially loaded on its free end. The schematic of the bearing test rig on which the experiments are carried out is shown in **Figure 2**, [17,18]. Three different types of bearings are used, having the same bore diameter. This selection has been made in order to use in each

Figure 2. Photo of the Test Rig.

housing bearings with different characteristic frequencies and as a result to have a clearer indication about the source of the measured vibration signal. All bearings have conical internal rings with a bore diameter of 35 mm and they are mounted on the shaft with the use of proper adapter sleeves.

Two Kistler type 8792A25T three-axial accelerometers are mounted with magnets on the top of the two plumper type housings. Visualization and processing of the signals is made through HBM CATMAN software installed on a Laptop PC. The naming of the axes is following the naming of the accelerometer's axes which is: **x** for the axis of the shaft, **z** for the axis of the load direction and **y** for the axis vertical to the plane xz. Also, a K-type thermometer was used in order to measure the shaft, bearings and housings temperatures and the ambient temperature.

2.2. Pre-contaminated Grease

The pre-contaminated grease was made by mixing solid contaminant particles with clean grease. The grease used was SKF LGMT3 mineral oil based, lithium soap thickened grease with base oil viscosity of 120 - 130 mm^2/s at 40°C and 12 mm^2/s at 100°C. The solid particles which were used as contaminants were corundum particles with a hardness of about 2000 HV. Two samples with different particle size were used as shown on **Table 1**.

In order to achieve similar volumetric concentration of particles in the contaminated grease, a quantity of 0.75 cm^3 of particles was mixed with 50 grams of clean grease. Most of the quantity used remains in the housing and only a small amount of grease is recirculating. The actual amount of particles which enter the contact zone is only a fraction of the used quantity.

2.3. Procedure of Experimentation

Two tests were conducted using the above configuration. At the beginning of the experiments the rear housing and the bearing was cleaned from the preservation oil and then the bearing was filled with clean grease. The same procedure has been followed for the front housing using one of the grease mixtures described above.

A load of 2800 N was applied to the flanged housing resulting in a load of 3500 N at the front bearing and a load of 700 N at the rear bearing. The motor and shaft rotation speed was set to 2400 rpm (40 Hz). The signal from six channels was recorded, corresponding to the three axes of each of the two accelerometers. The sampling rate was set to 2400 samples per second (sampling frequency) for each channel.

The duration of the experiments was 14 hours for each test, representing about 2 millions of revolutions. After the completion of each test, the front housing was opened and cleaned from the remaining grease. The bearing was dismantled and its parts were cleaned from the grease. After cleaning, the parts of each bearing were subjected to optical inspection with a LEICA stereoscope.

During the tests, the vibration signals from the accelerometers were monitored in real time and recorded All data were recorded and saved in. BIN type file format which can be further processed with Matlab software.

3. Data and Calculations

According to ISO 281:2007, the L_{nm} life of the tested bearings can be calculated from the Equations (1) and (2). For the applied load the L_{10} life of each bearing is calculated based on the manufacturer's basic dynamic load. The results are given in **Table 2**.

For the modified basic rating life L_{nm} the life modification factor for reliability α_1 equals to 1 for a 90% reliability. The life modification factor for system approach

Table 1. Material and size table of the used contaminants.

test	Material	Mesh size	Particle size (μm)	Average size (μm)
1	Corundum	180 - 220	62 - 73	68
2	Corundum	80 - 120	105 - 177	141

Table 2. Loads and Basic Rating Life calculations for the all testing rig's bearings.

	SKF 1207 EKTN9 "front"	SKF 2207 EKTN9 "rear"	SKF YSA 207 - 2FK "load"
Basic dynamic load	19,000 N	30,700 N	25,500 N
Maximum allowed static load	6000 N	8800 N	15,300 N
Actual load	3500 N	700 N	2800 N
L_{10} in revolutions	160×10^6	$84,400 \times 10^6$	760×10^6
L_{10} in hours (at 2400 rpm)	1110	585,800	5250

a_{ISO} depends on the contamination factor e_c, the fatigue load limit C_u, the actual load P and the viscosity ratio κ.

The contamination factor e_c can be estimated from the given guidelines. The bearing's pitch diameter d is about 54 mm and the viscosity ratio κ of the grease used at the test temperature of the bearings (65°C) is about 3.5. For grease lubrication under high cleanliness, e_c can be as high as 0.8 for the dimensions of the tested bearing (Pitch diameter d ≈ 54 mm) and for the viscosity ratio κ of 3.5. In severe contamination conditions, e_c is in the range of 0.05 and in very severe contamination it tends to zero. The bearings manufacturer provides also an advanced method of calculating the rating life according to the size of the contaminants and their hardness. The results from these calculations are shown on **Tables 3** and **4**.

The characteristic bearing potential frequencies of each bearing can be calculated with the use of Equations (9)-(12). The dimensions of the bearings are required but instead, the manufacturer provides these values according to the operation speed. For the rotational speed of 2400 rpm (40 Hz) and for a contact angle of 0° the characteristic frequencies are given in **Table 5**.

Table 3. Modified Rating Life calculations for various lubricant conditions for the tested bearing.

Lubricant condition	e_c	a_{ISO}	L_{10m}
High cleanliness	0.8	3.8	4240
Normal cleanliness	0.6	2.6	2850
Slight to typical contamination	0.18	0.77	856
Severe contamination	0.05	0.36	400
Very severe contamination	0	0.1	111

Table 4. Modified Rating Life calculations according to SKF.

test	Particle hardness	Average size (μm)	e_c	a_{ISO}	L_{10m}
1	2000 HV	68	0.046	0.35	387
2	2000 HV	141	0.015	0.23	256

Table 5. Frequencies of potential damage in the tested ball bearings.

	SKF 1207 EKTN9 "front"	SKF 2207 EKTN9 "rear"	SKF YSA 207 - 2FK "load"
f	40.0 Hz	40.0 Hz	40.0 Hz
f_c	16.8 Hz	15.8 Hz	15.8 Hz
f_{bpfo}	251.0 Hz	190.0 Hz	143.0 Hz
f_{bpfi}	349.0 Hz	290.0 Hz	217.0 Hz
f_r	119.0 Hz	89.1 Hz	92.1 Hz
$2 x f_r$	238.0 Hz	178.0 Hz	184.0 Hz

4. Experimental Results

The measured signals are processed in time domain and in frequency domain. The axes of greater importance are the x axis, which is the axis of the shaft and represents the axial loads of the bearing, and the z axis which is the axis of the applied radial load. The y axis, which is the axis vertical to the applied load, shows similar trends to the others but with much lower signal amplitudes.

4.1. Vibration Analysis in the Time Domain

Due to the numerous sources of vibrations as various particles enter the loading zone and the randomness of their nature, the statistical variables of higher order such as skewness and kurtosis do not give any useful results. The importance of these variables is grater when a fault appears on a bearing running under normal conditions. In these situations, the aforementioned variables, which normally show small fluctuation, will change significantly. Such behavior has been successfully reported by other researchers [13,14] when they created an artificial single point defect.

As it is shown in the plots of acceleration peaks and rms values in **Figure 3**, in the first test where the smaller particles were used, the amplitudes begin from lower levels compared with the second test. It is also clearly visible that the amplitude of the acceleration peaks and rms value is increasing on both cases. In the first test, the increase rate is much higher between the 6th and the 10th hour of operation. During that period, the amplitude of acceleration peaks increased from 5 g to 15 g on the x axis and from 1 g to 5 g on the z axis. Acceleration rms values increased too, from 2 g to 4 g on the x axis and from 0.25 g to 0.9 g on the z axis. After the 10th hour of operation, the increase rate was lower. In the second test, acceleration peaks and rms values increase with a constant rate, much slower than the first test.

The increasing acceleration amplitudes indicate that the failures are growing in size and the performance of the bearing is reduced. Generally, the overall acceleration measurement can indicate if there is an abnormal operation and if alarm levels are set, it can indicate potential failures.

4.2. Vibration Analysis in the Frequency Domain

The vibration analysis in the time domain can show a malfunction of the bearing only when it is affecting the vibration levels. The exact source of this malfunction is unknown and only assumptions can be made. In our case, the source of the increased vibrations could not be identified only from the vibration analysis results. Vibrations could be generated by particles entering randomly the contact zone, or by permanent deformations on the raceways. In the first case, vibrations do not follow a specific

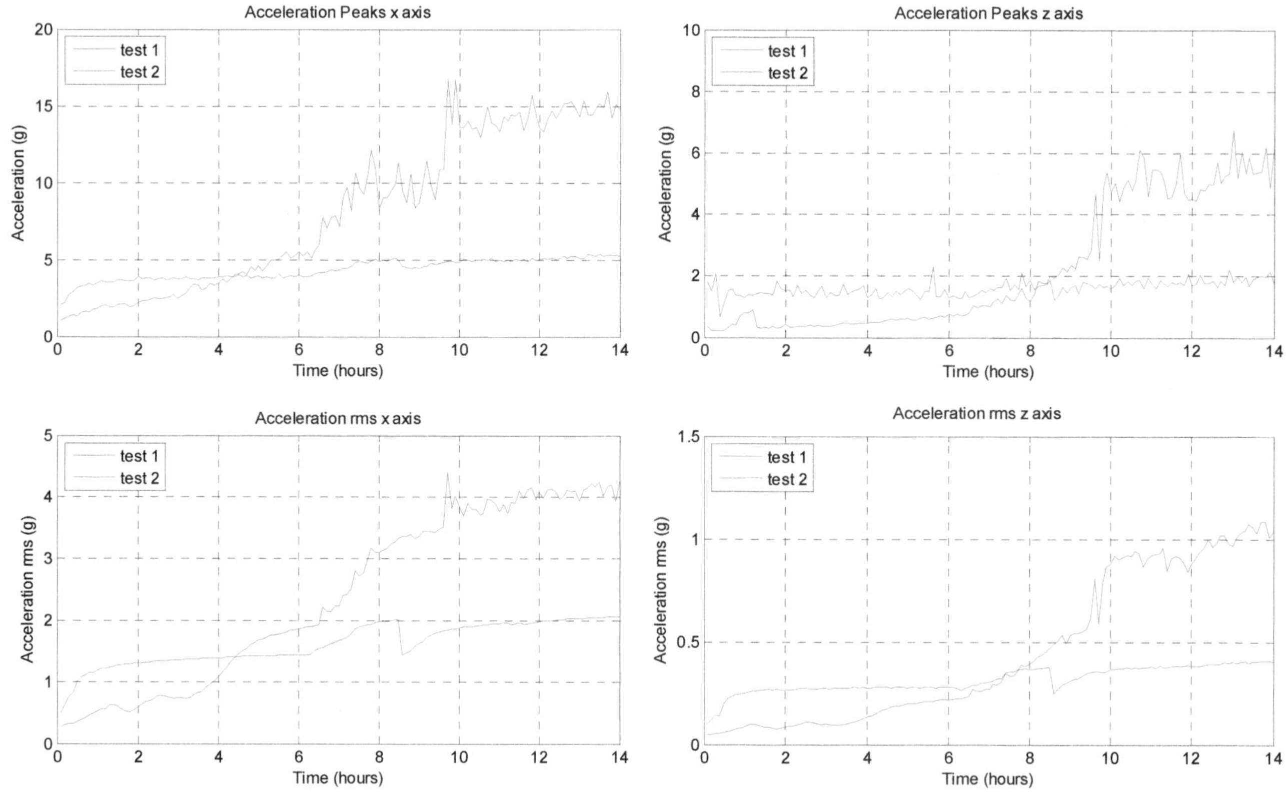

Figure 3. Acceleration Peaks and RMS values using corundum particle contaminants.

pattern, but in the second case the amplitude of vibrations on the bearings potential damage frequencies, shown on **Table 5**, is increased.

Processing the acquired data in the frequency domain, results in the identification of the true source of the vibrations and in the detection of possible faults. Frequency spectrum plots of the obtained data are created by implementing a fast Fourier transform in Matlab software environment. In order to reduce the spectral leakage and make the dominant frequencies easier to be identified, the power spectrum is used, where the outputs of the FFT are raised to the square. The frequency spectra plots of the tests are shown in **Figures 4** and **5**. Five samples of each test data series are plotted. In the first test using smaller particle contaminants noticeable higher amplitude is occurring on both x and z axes, at the frequency of 251.2 Hz, which is the bearing's potential damage frequency for the outer raceway, indicating a failure at this part. In the second test, using larger particles, peaks occur at the frequencies of 348.2 Hz and 388.1 Hz. The frequency at 348.2 Hz corresponds to the bearing's potential damage frequency for the inner raceway but the frequency at 388.1 Hz does not match any of the characteristic frequencies shown on **Table 5**. This frequency seems to occur due to sliding of the balls relatively to the motion of the inner ring, which results in a higher rate of contact between a certain point of the inner ring and the balls, causing vibrations in higher frequency.

Other bearing characteristic frequencies can be identified when plotting the results of the Fourier transform without raising them to the square. When comparing them with the dominant frequencies mentioned above it is clearly that failures have occurred at the relevant parts.

The results of the frequency domain analysis conform to the results of the time domain analysis, indicating an increase of a fault at a certain location in each test. When both techniques are used a potential fault and its impact on the overall performance of the bearing can be identified.

4.3. Visual Inspection and Verification of Results

After the completion of the tests, the bearings were dismantled and cleaned from the greases. The bearing elements and the faults on them were photographed with the use of a stereoscope and a digital camera attached to it. The findings from the visual inspections are conforming to the results of the vibration analyses.

On both bearings the rolling marks on the raceways can be seen. Only small indentations exist, heavier in the second test where larger particles used (**Figures 6** and **7**). In the first test a significant deformation is visible at the points where the balls enter and leave the load zone. As abrasion mechanisms removed material from the race-

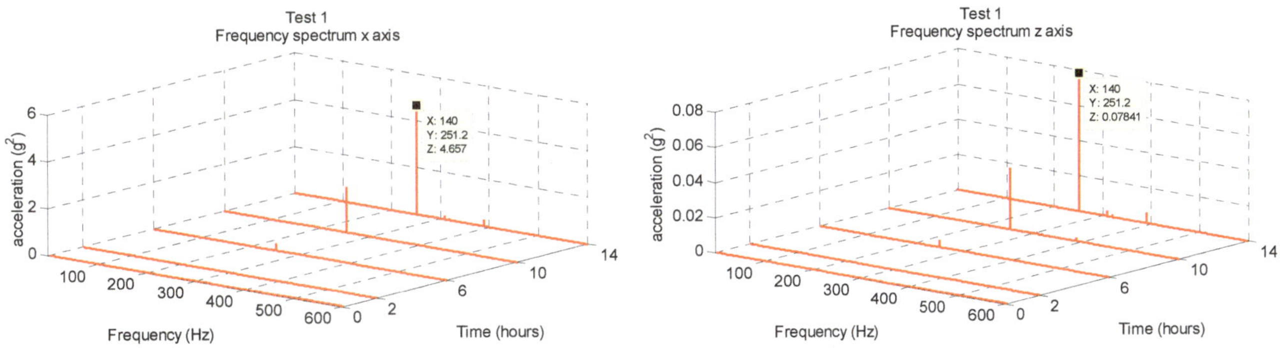

Figure 4. Frequency spectra plots of the test using corundum particles with average size of 68 μm.

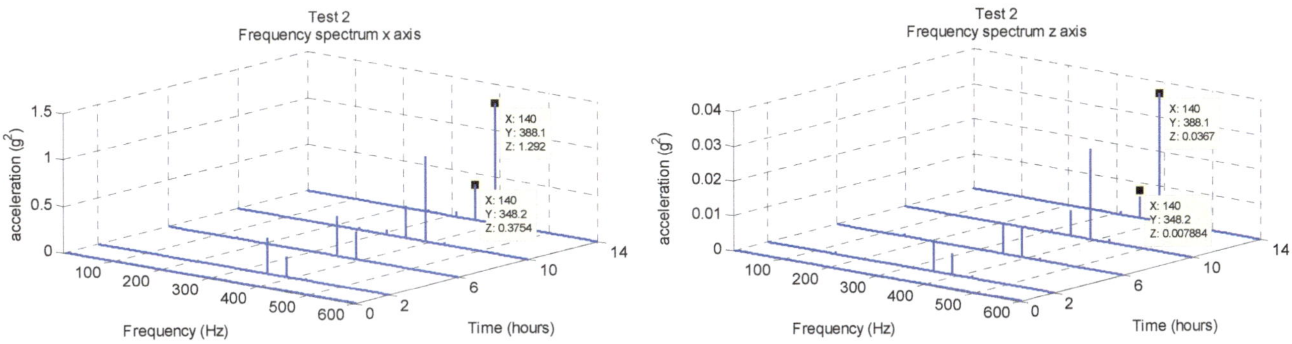

Figure 5. Frequency spectra plots of the test using corundum particles with average size of 141 μm.

Figure 6. Light abrasion of the outer ring raceway with small indentations (test 1).

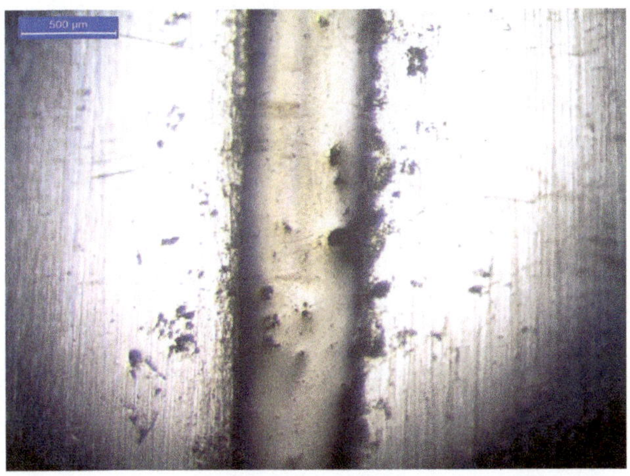

Figure 7. Abrasion of the outer ring raceway with heavier indentations and particles imbedded in the contact area (test 2).

ways, the gap between the balls and the raceways on the unloaded zone increased. The inertia of the balls caused the deformation at the entrance and the exit of the load zone as it is shown in **Figure 8**. In this case balls were skidding and vibrating causing wave like deformation on the outer raceways (**Figure 9**).

The width of the affected contact zone is about 600 - 700 μm for both tests on the load zone and about 1 mm on the unloaded zone. The width of the affected contact area of the inner ring is about 2 mm, wider than that of the outer raceways due to the different curving. On the inner ring the surface on the contact zone has been polished and on the second test there is also some small scale pitting (**Figures 10** and **11**).

On the second test a deformation on the inner raceway is also visible in **Figure 11**. This groove is probably the failure indicated by the vibration analysis results on the frequency domain.

Generally, when the larger particles were used, the bearing seems to have rougher wear marks. The deformation of the raceways is not following the same pattern

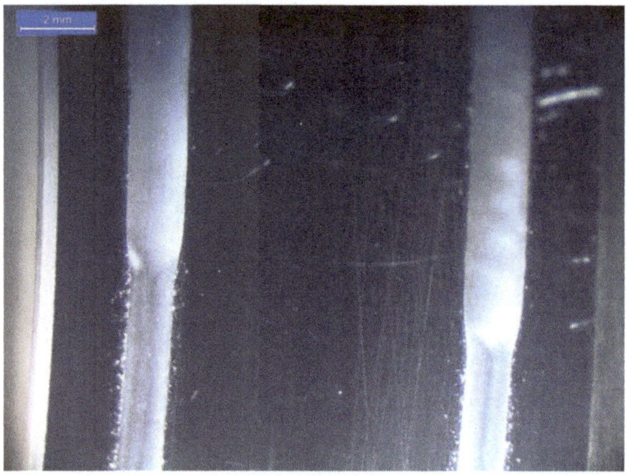

Figure 8. Heavy abrasion with a noticeable different deformation at the point of entering the contact zone (test 1).

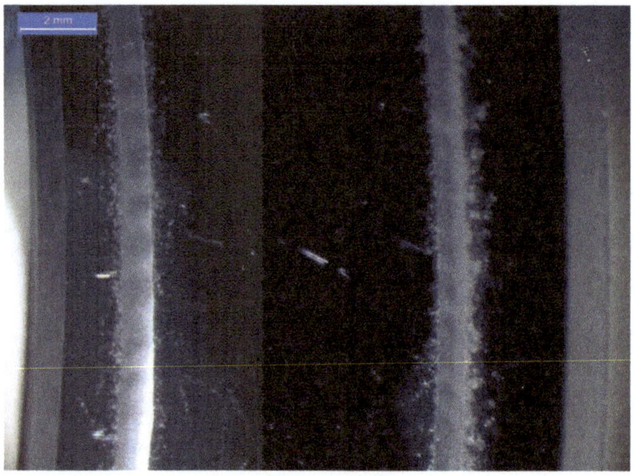

Figure 9. Abrasion of the outer ring raceway with a wave like deformation of the raceways (test 2).

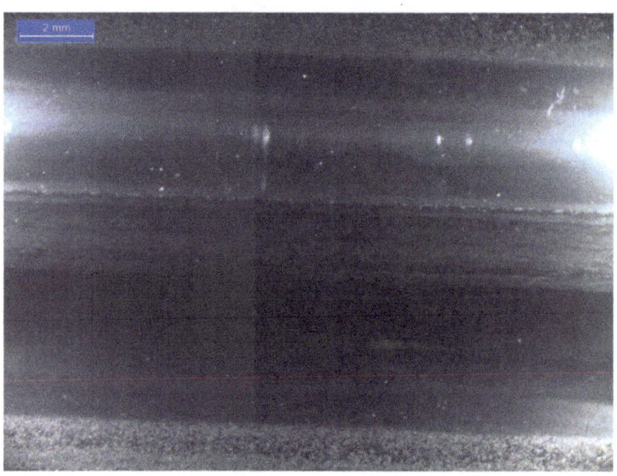

Figure 10. Light wear on the inner raceways with polishing of the contact zone (test 1).

and cannot be correlated with the particle size. In both cases the deformations were significant and the total damage seemed to be critical for the operation of the bearing.

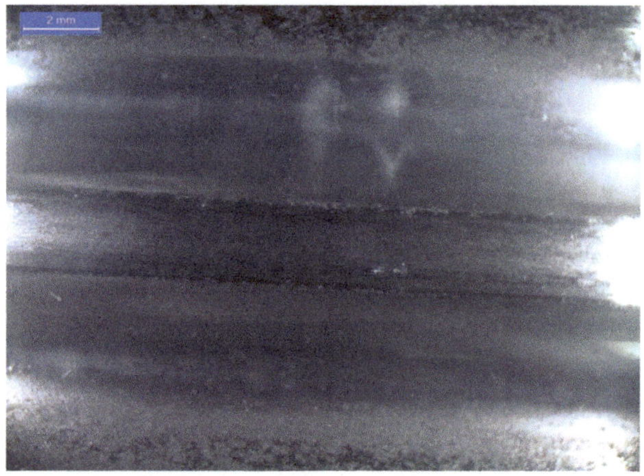

Figure 11. Visible pitting on the raceways and a single fault in one of the two inner raceways (test 2).

5. Discussion

As brittle particles enter the contact zone they are smashed down to smaller fragments until a certain critical size depending on their fracture toughness. Due to the vortexes of the lubricant, new or already fragmented particles recirculate and they can reenter the contact zone in a different position. This particle recirculation is causing continuous abrasive wear.

Particles do not have exactly the same size, fracture toughness and brittleness and their behavior cannot be predicted precisely. It is expected that larger particles should cause heavier wear than smaller particles. The findings from the contacted tests showed that in both cases the wear was similar. Material from the raceways has been removed and deformations have occurred. Larger particles caused heavier indentations and material removed from the raceways has been melted and adhered to the side bands of the raceways due to excessive heat.

Generally, the vibration levels are increasing as wear and deformation of the raceways propagate. When using vibration analysis in the frequency domain it is possible to identify the main source of these vibrations. In the first test, where the smaller particles were used, the vibration amplitudes in the time domain and in the frequency domain reached higher levels compared to the second test. This finding was not expected but it was the result of a larger deformation on the entrance and exit of the balls in the load zone.

As the fragmentation rate of the particles depends on their fracture toughness and their initial size, it can be assumed that the two particle samples used in the experiments might not have the same fracture toughness. Some of the larger particles might have been more brittle and they smashed down to fragments of critical size easier. The distribution of fracture toughness among the particles is also important and it might not was the same either.

The vibration levels of the bearings and their overall performance on the test rig indicate that they have reached their useful operational life after less than 14 hours of operation in both cases. This service life is much lower than the calculated service life of 110 hours considering very severe contamination (e_c = 0 and α_{ISO} = 0.1).

6. Conclusions

Measuring of vibration has an increasingly important role as a tool in assisting predictive and preventive maintenance. When monitoring rolling element bearings, both vibration analyses in the time domain and in the frequency domain can give useful results. The vibration analysis in the time domain can indicate if there is an abnormal operation of a bearing and show the trend of the amplitude increase. In the frequency domain, vibration analysis can indicate whether the increased vibrations are caused from a certain bearing defect or from external sources. The increase of the vibration amplitude on the characteristic frequencies of the bearing can also indicate the progress of that specific fault. In the case of grease contaminated with tough brittle particles, vibration analysis can indicate the severity of wear and monitor its progress.

From the findings of conducted tests it is concluded that the size of the contaminant particles has a minor role in the abrasive wear. As particle size in both cases is greater than the thickness of the lubrication film, they cause significant wear, deforming the bearings' raceways by removing material. Even when they are smashed down to smaller fragments, they continue to cause wear as they recirculate within the lubricant and affect different areas. Although greater particles seem to cause heavier indentations, smaller particles are causing severe abrasive wear too. The fracture toughness of the particles is also important as it is possible that larger particles might have less fracture toughness than smaller particles due to different microcrystal structure.

It is difficult to establish an equation that describes and quantifies the wear progress involving the parameters of size and concentration levels of the tough brittle particles due to the stochastic nature of the wear mechanisms. The experimental useful life of the tested bearings was smaller than the life calculated according to ISO 281: 2007 considering the worst lubricating conditions, with the value of the contamination factor equal to zero, indicating that actual wear can reduce the bearing life even more. In order to establish new values for the contamination factor and the factor of ISO considering parameters such as particle size, brittleness and fracture toughness, a large number of further experiments are necessary.

REFERENCES

[1] A. Palmgren and G. Lundberg, "Dynamic Capacity of Rolling Bearings," *Acta Polytechnica*, Vol. 1, No. 3, 1947, p. 7.

[2] T. A. Harris, M.N. Kotzalas, "Advanced Concepts of Bearing Technology," 5th Edition, CRC Press, Boca Raton, 2007.

[3] E. V. Zaretsky, J. V. Poplawski and C. R. Miller, "Rolling Bearing Life Prediction—Past, Present and Future, NASA TM-2000-210529, NASA Glenn Research Center, 2000, p. 13.

[4] International Organization for Standardization, ISO 281: 2007, Rolling Bearings—Dynamic Load Ratings and Rating Life, 2007.

[5] M. Godet, "The Third Body Approach: A Mechanical View of Wear," *Wear*, Vol. 100, No. 1-3, 1984, pp. 437-452.

[6] G. K. Nikas, "A State-of-the-Art Review on the Effects of Particulate Contamination and Related Topics in Machine-Element Contacts," *Proceedings of the Institution of Mechanical Engineers, Part J: Journal of Engineering Tribology*, Vol. 224, No. 5, 2010, pp.453-479.

[7] M. M. Maru, R. S. Castillo and L. R. Padovese, "Study of Solid Contamination in Ball Bearings through Vibration and Wear Analyses," *Tribology International*, Vol. 40 No. 3, 2007, pp. 433-440.

[8] G. K. Nikas, R. S. Sayles and E. Ioannides, "Effects of Debris Particles in Sliding/Rolling Elastohydro Dynamic Contacts," *Proceedings of the Institution of Mechanical Engineers, Part J: Journal of Engineering Tribology*, Vol. 212, No. 5, 1998, pp. 333-342.

[9] R. S. Dwyer-Joyce, "Predicting the Abrasive Wear of Ball Bearings by Lubricant Debris," *Wear*, Vol. 233-235 1999, pp. 692-701.

[10] R. S. Dwyer-Joyce, "The Life Cycle of a Debris Particle," *Tribology and Interface Engineering Series*, Vol. 48 2005, pp. 681-690.

[11] R. S. Dwyer-Joyce, J. C. Hamer, R. S. Sayles and E. Ioannides, "Surface Damage Effects Caused by Debris in Rolling Bearing Lubricants, with an Emphasis on Friable Materials," *Proceedings of the Institution of Mechanical Engineers Symposium on Rolling element bearings*, 1990, pp. 1-8.

[12] R. S. Sayles, "Debris and Roughness in Machine Element Contacts: Some Current and Future Engineering Implications," *Proceedings of the Institution of Mechanical Engineers, Part J: Journal of Engineering Tribology*, Vol. 209, No. 3, 1995, pp. 149-172.

[13] T. Karacay and N. Akturk, "Experimental Diagnostics of Ball Bearings Using Statistical and Spectral Methods," *Tribology International*, Vol. 42, 2009, pp. 836-843.

[14] X. Niu, L. Zhu and H. Ding, "New Statistical Moments for the Detection of Defects in Rolling Element Bearings," *International Journal of Manufacturing Technology*, Vol. 26, 2005, pp. 1268-1274.

[15] C. T. Yiakopoulos and I. A. Antoniadis, "Cyclic Bispectrum Patterns of Defective Rolling Element Bearing Vibration Response," *Forsch Ingenieurwes*, Vol. 70, 2006, pp. 90-104.

[16] K. Raptis, G. Papadopoulos, Th. Costopoulos and A. Tsolakis, "Experimental Study of Load Sharing in Roller-Bearing Contact by Caustics and Photoelasticity," *American Journal of Engineering and Applied Science*, Vol. 4, No. 2, 2011, pp. 294-300. http://thescipub.com/abstract/10.3844/ajeassp.2011.294.300

[17] I. Zotos and Th. Costopoulos, "A Study on the Increase of Numerical Stability and Accuracy of the Transfer Matrix Method," *Journal of Mathematics and Statistics*, Vol. 4, No. 4, 2008, pp. 208-212. http://thescipub.com/abstract/10.3844/jmssp.2008.208.212

[18] I. Zotos and Th. Costopoulos, "On the Use of Rolling Element Bearings Models in Precision Maintenance," *American Journal of Engineering and Applied Science*, Vol. 2, No. 2, 2009, pp. 344-352.

Analytical Solutions of Dynamic Crack Models of Bridging Fiber Pull-Out in Unidirectional Composite Materials

Yuntao Wang[1], Yunhong Cheng[2], Nianchun Lü[3], Jin Cheng[4]
[1]College of Mechanical Engineering, Liaoning Technical University, Fuxin, China
[2]Department of Civil Engineering, Northeastern University, Shenyang, China
[3]School of Material Science and Engineering, Shenyang Ligong University, Shenyang, China
[4]Department of Astronautics and Mechanics, Harbin Institute of Technology, Harbin, China

ABSTRACT

An elastic analysis of an internal central crack with bridging fibers parallel to the free surface in an infinite orthotropic anisotropic elastic plane was analyzed, and the crack extension should occur in the format of self-similarity. When the fiber strength is over its maximum tensile stress, the fiber breaks. By means of complex variable functions, the problem considered can be easily translated into Reimann-Hilbert mixed boundary value problem. Utilizing the built dynamic model of bridging fiber pull-out in unidirectional composite materials, analytical solutions of the displacements, stresses and stress intensity factors under the action of increasing loads Pt^5/x^5, Px^5/t^4 are obtained, respectively. After those analytical solutions were used by superposition theorem, the solutions to arbitrary complex problems were acquired.

Keywords: Composite Materials; Bridging Fibers; Analytical Solutions; Crack; Variable Loads

1. Introduction

It is well known that the matrix cracking as well as fracture process of the bridging fibers is one of the significant mechanisms of the cracking expansion in fiberreinforced composite materials, such as unidirectional fiber-reinforced brittle matrix composites [1,2], and threedimensional fiber-reinforced composites with an orthogonal fiber structure [3]. Literature [4] proposed an approach for the assessment of the distribution of the traction force for a crack with bridging fibers in an infinite, orthotropic elastic plane under a uniform remove tension stress. Most researchers, such as Woo, Lee and Tsai [5-7] etc, almost investigated static problems of composite materials; moreover, they obtained only numerical solutions. Literature [8] set up a model of bridging fiber pull-out, but it also acquired the numerical solutions under the static conditions. It is indispensable to consider the mechanical analysis of matrix cracking with bridging fibers, so as to evaluate the distribution of the axis traction force in each fiber. However, the fractures of composite materials often arise in dynamic conditions, so accordingly it is extremely important to research their fracture dynamics problems. In an orthotrpic medium, elastodynamics crack problems were studied and closed solutions were also gained, but bridging fiber pull-out problems weren't dealt with in literatures [9,10]. Bridging fiber pull-out is very complicated and cockamamie in dynamic fracture process of composite materials, so a lot of difficulty must be overcome in studying dynamic crack expansion problems on bridging fiber pull-out of composite materials. When composite materials occur in a crack, bridging fiber pull-out often exists ahead of the crack tips, and this is a frequent phenomenon.

Because the fiber failure is governed by maximum tensile stress, which appears at the crack plane, the fiber breaks and hence the crack propagation should occur in a self-similar fashion. The fiber breaks along a transverse line and therefore present a notch [8,11-12]. When a crack runs at higher velocity, bridging fiber pull-out still exists in the dynamic case of composite materials, which are more important than those in the statics.

The problem under consideration is that of a crack, moving in one plane, presumed to nucleate from an infinitesimally small micro-crack with maximum velocity from the start. This modality of symmetrical crack, running with constant velocity V in both the positive and negative directions of the x-axis, has been researched by

Broberg [13] and Craggs [14]. Both considered motions in materials postulated to be homogeneous and isotropic, as regards stress-strain relationships and fracturing characters. If the fiber failure is governed by maximum tensile stress, which appears at the crack plane, the fiber breaks and hence the crack expansion should occur in the modality of self-similarity [13,14]. The fiber breaks along a transverse line and therefore presents a notch [7, 11-12]. When a crack runs at higher speed, bridging fibers still exist in the dynamic situation of composite materials, which are more significant than those in the statics. Since bridging fibers can result in stabilizing effect of crack extension problem along the original notch plane, the dynamic fracture influence on bridging fibers of composite materials will be shown in detail; at the same time, stresses and displacements as well as stress intensity factors are deduced properly.

In this paper, the dynamic expansion problem on an internal central crack with bridging fibers of composite materials is analyzed by the ways of Keldysh-Sedov mixed boundary value problem, and analytical solutions for unidirectional reinforced material with fibers parallel to the free surface are shown. First, the solution of a sole dislocation in an elastically half-plane is derived from the uses of complex variable analysis. The crack is then displayed in terms of a consecutive distribution of dislocation. This solution which has relation to a bridging fiber force induces a system of self-similar functions with dislocation density as unknown quantities. Then self-similar functions are resolved analytically by means of Keldysh-Sedov's method.

2. A Dynamic Model of Bridging Fiber Pull-out of Composite Materials

When fiber-reinforced composite materials occur a crack, phenomenon of bridging fiber pull-out will often occur ahead of the crack tips and off x-axis on occasion, and analyzing crack problems in this situation is more difficult. Composite materials are often referred to as orthotropic aniostropic body in virtue of the direction of their fibers, while bridging fibers play an important role in their strength, consequently queries on bridging fiber pull-out are one of the most complex advancing tasks in mechanics of composite materials. When a crack moves with high speed, bridging fiber pull-out phenomenon of composite materials is still likely to exist. Because the problems of bridging fibers are more complicated and cockamamie, there is a lot of difficulty in mathematical calculations. In order to resolve dynamic fracture queries of bridging fiber pull-out of unidirectional composite materials, it is indispensable to establish a suitable symmetrical dynamic model of bridging fiber pull-out, hence fracture dynamics problems of bridging fiber pull-out of composite materials are effectively solved.

2.1. Characterization of Dynamic Fracture Problems Concerning Bridging Fiber Pull-Out

The problem of an internal central crack with bridging fiber pull-out of composite materials is analyzed under the dynamic conditions by means of Reimann-Hilbert mixed boundary value, and that analytical solutions for unidirectionally reinforced composite materials with bridging fibers parallel to the free surface are presented. This solution in conjunction with a bridging fiber force gives rise to a system of self-similar functions with dislocation density as unknown units. The self-similar functions are solved analytically using Reimann-Hilbert method. In order to settle efficaciously fracture problems on bridging fibers of composite materials, it is inevitable to establish dynamic models of bridging fibers. Since bridging fibers can conduce a stabilizing effect on crack extension problems along the original notch plane, the dynamic fracture effect of bridging fiber pull-out in composite materials will be shown, at the same time, stresses and displacements as well as stress intensity factors are deduced appropriately. In order to resolve efficiently fracture problems of bridging fiber pull-out of composite materials, proper dynamic models of bridging fiber pull-out must be built. Only this approach, can a dynamic crack expansion problem of bridging fiber pull-out of composite materials obtain content solutions.

2.2. Base of A Dynamic model of Bridging Fibers

The crack is postulated to nucleate an infinitesimally small micro-crack situated along the x-axis in the form of self-similarity with the high speed extension, and to run symmetrically in the positive and negative x directions with the constant crack tip velocity V in the matrix. Bridging fiber pull-out of composite materials discussed is modeled as a two-dimensional region, having a sole row of parallel, same, equally spaced fibers, separated by matrix [15,16]. Initial breaks originate from an arbitrary number broken fibers leading to the fiber breaks along a transverse line and therefore present a notch. Furthermore, an arbitrary number of self-similar (off-axis) fiber breaks, *i.e.* fiber pull-out, with symmetry at the origin of coordinates and along a transverse line are also considered. A schematic of a dynamic model of bridging fiber pull-out configuration is described in **Figure 1**. Since the configuration in **Figure 1** is symmetry both in geometry and loading, respectively, with respective to y-axis, only the right half-plane of the zone needs to be considered for analyses. The fibers and the matrix are taken to be linearly elastic. It is further presumed that the fibers have a much higher elastic modulus in the axial direction than the matrix and therefore the fibers are usually taken as supporting all of the axial loadings in composite materi-

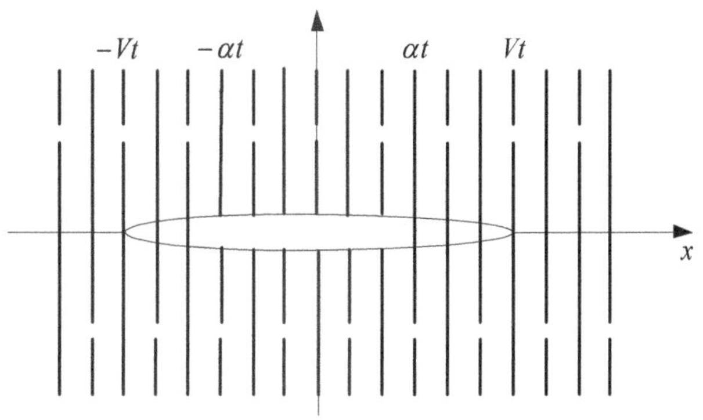

Figure 1. Schematic of a dynamic model of bridging fiber pull-out configuration.

als. Load is transferred between adjacent fibers through the matrix by a straightforward shear mechanism [15,16]. The shear stresses are independent of transverse displacements and the equilibrium equation in the fiber direction reduces to an equation in the longitudinal displacements alone, as is a typical of shear-lag theory [11,12]. The solution approaches and modeling procedure put forward by [11] will be utilized, which discussed static problems, the fiber fractures in turns occur along two sole planes, *i.e.* the fiber fracture was self-similar fiber (off-axis) break. By virtue of the point, break lie is same in geometry off x-axis, *i.e.* break is symmetry about the origin of the coordinate. In short, the fiber fracture was self-similar fiber (off-axis) break and presented a notch. Bridging fiber pull-out occur ahead of the crack tips and off x-axis. Bridging fibers do not break in the area of the crack tips along the crack plane, but the others fracture off the further of the crack tips, *i.e.* at the central section of crack or notch. When the crack runs, fibers continuously break with the constant velocity α according to my assumption. As shown in **Figure 1**, at $y = 0$, the realm of crack or notch in matrix is $|x| < Vt$; and the fibers broke at the extent of $|x| < \alpha t$; while the zone of bridging fibers is $\alpha t < |x| < Vt$, respectively [15,16]. At $y \neq 0$, the bridging pull-out positions locate at $|x| > Vt$ [8].

Evidently, the dynamic model of crack extension problem with bridging fiber pull-out in **Figure 1** is clarified by that in **Figure 2**. This is a model of symmetrical crack propagation, running with constant velocity V in both the positive and negative directions of x-axis, at the same time, bridging fibers fracture with constant velocity α. The area of bridging fibers has the symmetrical state with respect to y-axis [15,16]. Each bridging fiber is replaced by a pair of vertical traction forces which act at the points with the same x-coordinate on the upper and lower crack surfaces, but in an opposite direction. Each bridging fiber is postulated to be balanced with the fracture load of a fiber from the matrix. The present model has the symmetries of geometrical and mechanical conditions with respect to the x- and y-axes on account of the symmetrical

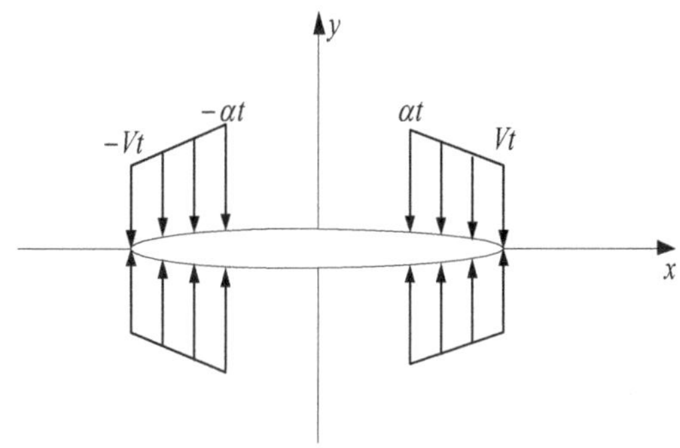

Figure 2. Dynamic model of crack-face bridging fiber zone.

crack expansion. At $y = 0$, traction forces act in the zone of $\alpha t < |x| < Vt$, which represent fibrous tensile stresses which don't act in the rest of the crack [15,16]. Bridging fibers of composite materials are usually arranged tightly, separated by matrix, therefore bridging fiber traction forces are presumed to be distributed consecutively. At $y = 0$, $\alpha t < |x| < Vt$, bridging fiber pull-out has symmetry with respective to the origin of the coordinate; the displacements of the crack face are not the same, but traction forces of bridging fibers are identical [15-17]. In short, traction forces of bridging fibers are homogenous in this section, whose magnitude is P assumed. On the other hand, when the crack extends with high speed, the magnitude of crack will increase with time t; the longer the crack spreads, the more fibers break. The above analyses are postulated that fibers in matrix are distributed uniformly, and each fiber has the identical strength, moreover, the fracture fibers and matrix simultaneity occur in the same segment of the crack expansion plane [7,15-17]. It is distinct that traction forces are larger near the points of $\pm\ \alpha t$, and they are smaller close to the points of $\pm\ Vt$. When the crack runs at high speed, its dimension has relation to the parameters x and t, and the surfaces of the crack subjected to loads must also be related to x and t. When the fracture occurs, both the fiber and the matrix are in the same plane of crack expansion [15-17]. Of course, this is an assumed model which maybe not coincident with that in practicality. We can reasonably interpret this to mean that, outside the crack or notch, conditions are steady-state because the crack had no effect it, after all we can only expect the crack to affect appreciably the part of the body which lies in a certain proximity to it.

3. Universal Expressions of Electrodynamics Equations for Orthotropic Anisotropy

In order to solve efficaciously fracture dynamics queries of bridging fibers of composite materials, solutions will be attained under the action of point forces for mode I motive crack. In terms of the theorem of generalized functions, the problems dealt with unlike boundary conditions will be facilely translated into Reimann-Hilbert mixed boundary value problem by means of self-similar functions, then correlative solutions will be obtained.

Postulate at $y = 0$ that there are any number of loaded sections and displacement sections along the x-axis, and the ends of these segments are running with unlike constant velocity. At $t = 0$, the half-plane is at rest. In these sections the loads and displacements are discretionary linear compages of the following functions [15-22]:

$$\frac{\mathrm{d}^k f_{k_1}(x)}{\mathrm{d}x^k} \cdot \frac{\mathrm{d}^s f_{s_1}(t)}{\mathrm{d}t^s} \tag{1}$$

$$\text{Where } f_i(\xi) = \begin{cases} 0 & \xi < 0 \\ \xi^i & \xi > 0 \end{cases} \tag{2}$$

Here k, k_1 and s, s_1 are discretionary integer positive numbers [15-22].

A discretional successive function of two variables x and t may be shown as a linear superposition of Equation (1), therefore resolving loads or displacements with the form of Equation (2) will possess significance in princple. Introduce the linear differential operator as well as inverse:

$$L = \frac{\partial^{m+n}}{\partial x^m \partial t^n}, \text{ inverse: } L^{-} = \frac{\partial^{-m-n}}{\partial x^{-m} \partial t^{-n}} \tag{3}$$

Here $+m +n$, $-m -n$ and 0 represent the $(m + n)$ th order derivative, the $(m + n)$ th order integral and function's self. It is facile to prove that there exist constants m and n, when L is put into Equations (1), (2), homogeneous functions of x and t of zeroth dimension (homogeneous) are gained. The coefficients m, n will be called the indices of self-similarity [15-22].

Using relative expressions of elastodynamics equation of motion for an orthotropic anisotropic body [15-22]:

For the case when function Lv is homogeneous [5,6]:

$$v^0 = Lv, \quad \sigma_y^0 = L\sigma_y \tag{4}$$

For the case when function $L\sigma_y$ is homogeneous:

$$v^0 = \partial Lv/\partial t, \quad \sigma_y^0 = \partial L\sigma_y/\partial t \tag{5}$$

The relative self-similar functions are as [15-23]:

$$v^0 = \mathrm{Re}W(\tau), \quad \sigma_y^0 = (1/t)\mathrm{Re}F(\tau) \tag{6}$$

Where: v^0 and σ_y^0 in Equations (4)-(6) are the notation in [15-17,20-23], and they are relevant variables τ and t which directly work out displacements and stresses by the course of respective calculations in Equation (7).

$$W'(\tau) = [D_1(\tau)/D(\tau)]F(\tau) \tag{7}$$

Where: $\tau = x/t$, $F(\tau)$, $W(\tau)$ are self-similar functions. The values of $D_1(\tau)/D(\tau)$ can be ascertained from Appendix 1 of literatures [15-17,20-22], indicated here are only: $D_1(\tau)/D(\tau)$ in the neighborhood of the subsonic speeds is purely imaginary for the considered values. Thus, elastodynamics problems for an orthotropic anisotropic body studied can be changed into seeking the sole unknown function problems of $F(\tau)$ and $W(\tau)$ for which must meet the boundary-value conditions. In the universal situation this is Riemann-Hilbert problem in the theory of complex functions (in the simplest cases we have Keldysh-Sedov or Dirchlet problem), this kind of problem is facilely settled by the usual approaches, for example, in the books by Muskhelishvili [24,25].

Fracture dynamics problems will be investigated for an infinite orthotropic anisotropic body. Assume at the initial moment $t = 0$ a crack occurs at the origin of coordinates and begins spreading at constant velocity V (for the subsonic velocities) along the positive direction of x-axis; and at $t < 0$, the half-plane was at rest. The surfaces of the crack are subjected to the unlike types of loads under the plane strain states.

4. Fundamental Format of the Solution of Symmetrical Dynamic Extension Query Concerning Mode I Crack

At the initial moment $t = 0$, a micro-crack is supposed to appear in an orthotropic anisotropy. Let the Cartesian coordinate axes align with the axes of elastic symmetry of the body. The problem considered is restricted to motion in the x-y-plane. The crack is moving symmetrically with constant velocity V along the positive and negative directions of x-axis respectively. The problems will be changed into the following boundary condition queries:

$$\begin{aligned} &\sigma_y(x,0,t) = f_1(x,t), \ |x| < Vt \\ &v(x,0,t) = 0, \ |x| > Vt \end{aligned} \tag{8}$$

Introducing the variable $\tau = x/t$. By means of the above correlative expressions and $t\delta(x) = \delta(x/t)$ in the theory of generalized functions [26-28], the boundary conditions can be transformed as:

$$\begin{aligned} &\mathrm{Re}F(\tau) = f_2[\tau, \delta(\tau)], \ |\tau| < V \\ &\mathrm{Re}W'(\tau) = 0, \ |\tau| > V \end{aligned} \tag{9}$$

In the light of the relationship of $F(\tau)$ and $W(\tau)$ in Equation (7) and the previous conditions, the format of sole unbeknown function $W'(\tau)$ can be confirmed:

$$W'(\tau) = f_3[\tau, \xi(\tau)] \tag{10}$$

The problems can reduce to Keldysh-Sedov problem:

$$\begin{aligned} &\mathrm{Re}\xi(\tau) = 0, \ |\tau| > V; \\ &\mathrm{Im}\xi(\tau) = 0, \ |\tau| < V \end{aligned} \tag{11}$$

Considering symmetry and the infinite point of the plane corresponding to the origin of coordinates of the physical plane as well as singularities of the stress at the crack tip [29,30], the solution in the above problems can be readily deducted by literatures [15-16,20] as:

$$\xi(\tau)=T\left[(V-\tau),(V-\tau)\right] \tag{12}$$

Using Equations (6) or (7), we will easily obtain the stress, the displacement and the stress intensity factor under the conditions of mode I crack extension problems.

5. The Solutions of Practical Problems

In order to solve symmetrical dynamics problems with bridging fiber pull-out of unidirectional composite materials, solutions will be found under the conditions of different loads for mode I expanding crack. According to the theorem of generalized functions, the unlike boundary condition problems investigated will be translated into Keldysh-Sedov mixed boundary value problem by the approaches of self-similar functions, and the corresponding solutions will be acquired.

1) Suppose at the initial moment $t = 0$ a crack occurs at the coordinate origin and begins to spread at constant velocity V in both directions along the x-axis. The edges of the crack are subjected to normal point force Pt^5/x^5, moving at a constant velocity β along the positive direction of x-axis, where $\beta < V$; at $t < 0$ the half-plane was at rest. The boundary conditions will be as follows:

$$\begin{aligned}&\sigma_y(x,0,t)=-Pt^5/x^5\cdot\delta(x-\beta t),\ |x|<Vt\\&v(x,0,t)=0,\ |x|>Vt\end{aligned} \tag{13}$$

In this case the displacement will distinctly be homogeneous functions, in which $L = 1$. Using $\tau = x/t$ and the theory of generalized functions [26-28] and Equations (4) and (6), the first of Equation (13) can be as:

$$\begin{aligned}\operatorname{Re}F(\tau)&=-Pt^5/x^5\cdot t\delta(x-\beta t)\\&=-P\tau^{-5}\delta(\tau-\beta)\quad |\tau|<V\end{aligned} \tag{14}$$

In the light of Equation (7), boundary conditions (14) will be further rewritten:

$$\begin{aligned}&\operatorname{Re}\left[W'(\tau)\frac{D(\tau)}{D_1(\tau)}\right]=-P\tau^{-5}\cdot\delta(\tau-\beta),\ |\tau|<V\\&\operatorname{Re}\left[W'(\tau)\right]=0,\ |\tau|>V\end{aligned} \tag{15}$$

Deducting from the above-mentioned formulae, the unique solution of $W'(\tau)$ must have the modality:

$$W'(\tau)=\xi(\tau)/\left[\tau^5(\tau-\beta)\right] \tag{16}$$

$\xi(\tau)$ has no singularity in the domain of $|\tau| < V$, while $D_1(\tau)/D(\tau)$ is purely imaginary for the subsonic speeds, consequently $\xi(\tau)$ must be purely real in this segment. Thus, question (15) can conduce the following problems:

$$\begin{aligned}&\operatorname{Re}\xi(\tau)=0,\ |\tau|>V;\\&\operatorname{Im}\xi(\tau)=0,\ |\tau|<V\end{aligned} \tag{17}$$

According to symmetry and the conditions of the infinite point of the plane corresponding to the origin of coordinates of the physical plane as well as singularities of the stress of the crack tip [29,30], the sole solution of the Keldysh-Sedov problem (17) under the given conditions must have the following modality:

$$\xi(\tau)=A\left(V^2-\tau^2\right)^{-1/2} \tag{18}$$

where A is an unknown constant.

Inserting Equation (18) into (16), (7), one can gain:

$$W'(\tau)=A/\left[\tau^5(\tau-\beta)\left(V^2-\tau^2\right)^{1/2}\right] \tag{19}$$

$$F(\tau)=\frac{A\cdot D(\tau)/D_1(\tau)}{\tau^5(\tau-\beta)\sqrt{V^2-\tau^2}} \tag{20}$$

Then putting Equation (20) into (14), at $\tau\to\beta$, constant A can be determined from that

$$A=-P\sqrt{V^2-\beta^2}/\left\{\pi\cdot\operatorname{Im}\left[D(\beta)/D_1(\beta)\right]\right\} \tag{21}$$

Putting Equation (20) into (6) and (4), at the surface $y = 0$, the stress σ_y, the displacement v and the stress intensity factor $K_1(t)$ are acquired, respectively:

$$\sigma_y=\operatorname{Im}\left[\frac{D(\tau)}{D_1(\tau)}\right]\cdot\frac{A}{\tau^5(\tau-\beta)\sqrt{x^2-V^2t^2}},\quad |x|<Vt \tag{22}$$

$$K_1(t)=\sqrt{\frac{\pi}{t}}\cdot\frac{A}{V^{11/2}(V-\beta)}\cdot\operatorname{Im}\left[\frac{D(V)}{D_1(V)}\right] \tag{23}$$

Replacing Equation (19) into (4), (6), after integrating with respective to τ one can obtain the displacement v:

$$\begin{aligned}v=\frac{A}{\beta}\cdot\operatorname{Re}\int_{\infty}^{x/t}\Big[&-\frac{1}{\beta^4\tau}-\frac{1}{\beta^3\tau^2}-\frac{1}{\beta^2\tau^3}-\frac{1}{\beta\tau^4}\\&-\frac{1}{\tau^5}+\frac{1}{\beta^4(\tau-\beta)}\Big]\frac{1}{\sqrt{V^2-\tau^2}}\mathrm{d}\tau\end{aligned} \tag{24}$$

Utilizing correlative integral formulas [31] to yield:

$$\int\frac{1}{\tau\sqrt{V^2-\tau^2}}\mathrm{d}\tau=\frac{1}{V}\ln\frac{V-\sqrt{V^2-\tau^2}}{\tau} \tag{25}$$

$$\int\frac{1}{\tau^2\sqrt{V^2-\tau^2}}\cdot\mathrm{d}\tau=-\frac{\sqrt{V^2-\tau^2}}{V^2\tau} \tag{26}$$

$$\int\frac{1}{\tau^3\sqrt{V^2-\tau^2}}\mathrm{d}\tau=-\frac{\sqrt{V^2-\tau^2}}{2V^2\tau^2}+\frac{1}{2V^3}\ln\frac{V-\sqrt{V^2-\tau^2}}{\tau} \tag{27}$$

$$\int\frac{\tau^{-4}}{\sqrt{V^2-\tau^2}}\mathrm{d}\tau=-\frac{\sqrt{V^2-\tau^2}}{3V^2\tau^3}-\frac{2\sqrt{V^2-\tau^2}}{3V^4\tau} \tag{28}$$

$$\int \frac{1}{\tau^5\sqrt{V^2-\tau^2}}\mathrm{d}\tau = \frac{3}{8V^5}\ln\frac{V-\sqrt{V^2-\tau^2}}{\tau} - \frac{1}{4V^2}\frac{\sqrt{V^2-\tau^2}}{\tau^4} - \frac{3}{8V^4}\frac{\sqrt{V^2-\tau^2}}{\tau^2} \tag{29}$$

$$\int \frac{1}{(\tau-\beta)\sqrt{V^2-\tau^2}}\mathrm{d}\tau = \frac{1}{\sqrt{V^2-\beta^2}}\ln\left|\frac{V^2-\beta\tau-\sqrt{(V^2-\beta^2)(V^2-\tau^2)}}{V(\beta-\tau)}\right| \tag{30}$$

The crack extends along the x-axis, therefore $W(\tau)$ can be worked out in the operation of the definite integral, we take constant $C = 0$. Then putting Equations (25)-(30) into (24), the displacement v is given as follows:

$$v = \frac{-A}{\beta V}\left(\frac{1}{\beta^4}+\frac{1}{2\beta^4V^2}+\frac{3}{8V^4}\right)\ln\left|\frac{Vt-\sqrt{V^2t^2-x^2}}{x}\right| + \frac{A}{\beta^4\sqrt{V^2-\beta^2}}\ln\left|\frac{V^2t-\beta x-\sqrt{(V^2-\beta^2)(V^2t^2-x^2)}}{V(\beta t-x)}\right| + \frac{A}{\beta^2V^2}\left(\frac{1}{\beta^2}+\frac{t}{2\beta x}+\frac{t^2}{3x^2}+\frac{2}{3V^2}+\frac{\beta t^3}{4x^3}+\frac{3\beta t}{8V^2x}\right)\frac{\sqrt{V^2t^2-x^2}}{x},\ |x|<Vt \tag{31}$$

By means of the solution of Equation (31), the bridging fiber fracture speed α can be readily attained as:

$$v = \frac{-A}{\beta V}\left(\frac{1}{\beta^4}+\frac{1}{2\beta^4V^2}+\frac{3}{8V^4}\right)\ln\left|\frac{V-\sqrt{V^2-\alpha^2}}{\alpha}\right| + \frac{A}{\beta^4\sqrt{V^2-\beta^2}}\ln\left|\frac{V^2-\beta\alpha-\sqrt{(V^2-\beta^2)(V^2-\alpha^2)}}{V(\beta-\alpha)}\right| + \frac{A}{\beta^2V^2}\left(\frac{1}{\beta^2}+\frac{1}{2\beta\alpha}+\frac{1}{3\alpha^2}+\frac{2}{3V^2}+\frac{\beta}{4\alpha^3}+\frac{3\beta}{8V^2\alpha}\right)\frac{\sqrt{V^2-\alpha^2}}{\alpha},\ |x|=\alpha t \tag{32}$$

Each fibre has same strength [15-17,20-22] according to assumption, hence the bridging fiber fracture strength must be equal. Where Δ can be ascertained by an axial tensile test of bridging fibers of composite materials with V and β regarded as known constants, respectively.

In terms of this approach, the bridging fiber fracture speed α can be only gained numerical solution, because it can not be shown in the format of explicit function.

2) With all conditions remaining the same as those in the above sample, the applied loads become variable loads Px^5/t^4. The boundary conditions of the problem will be as follows:

$$\begin{aligned} &\sigma_y(x,0,t) = -Px^5/t^4\cdot\delta(x-\beta t),\ |x|<Vt \\ &v(x,0,t)=0,\ |x|>Vt \end{aligned} \tag{33}$$

In this case the stresses are apparently homogeneous functions, in which $L = 1$, making use of the theory of generalized functions [26-28] and Equations (5) and (6), the first of the boundary conditions (33) can be rewritten in the following modality:

$$\begin{aligned} \operatorname{Re}F(\tau) &= -(-4)Px^5/t^5\cdot t\delta(x-\beta t) \\ &= 4P\tau^5\delta(\tau-\beta) \qquad |\tau|<V \end{aligned} \tag{34}$$

Owing to the derivative of Dirac's function equaling zero at $x \neq \beta t$, the above expression will be deducted.

In terms of Equation (7), boundary conditions (33) will be further rewritten:

$$\begin{aligned} &\operatorname{Re}\left[W'(\tau)\frac{D(\tau)}{D_1(\tau)}\right] = 4P\tau^5\cdot\delta(\tau-\beta),\ |\tau|<V \\ &\operatorname{Re}\left[W'(\tau)\right]=0,\ |\tau|>V \end{aligned} \tag{35}$$

Known from the above, the sole solution of $W'(\tau)$ is:

$$W'(\tau) = \xi(\tau)\cdot\tau^5/(\tau-\beta) \tag{36}$$

$\xi(\tau)$ has no singularity in the domain of $|\tau| < V$, while $D_1(\tau)/D(\tau)$ is purely imaginary for the subsonic speeds, so $\xi(\tau)$ must be purely real in this area. Thus, question (36) will be the following boundary value problems:

$$\begin{aligned} &\operatorname{Re}\xi(\tau)=0,\ |\tau|>V; \\ &\operatorname{Im}\xi(\tau)=0,\ |\tau|<V \end{aligned} \tag{37}$$

In the light of the symmetrical conditions and singularities of the stress as well as the infinite point of the plane corresponding to the coordinate origin of the physical plane, the sole solution of Keldysh-Sedov problem (37) takes the format:

$$\xi(\tau) = A\left(V^2-\tau^2\right)^{-3/2} \tag{38}$$

where A is an unknown constant.

Putting Equation (38) into (37), (7), one can attain:

$$W'(\tau) = A\tau^5\Big/\left[(\tau-\beta)\left(V^2-\tau^2\right)^{3/2}\right] \tag{39}$$

$$F(\tau) = \frac{A\cdot\tau^5\cdot D(\tau)/D_1(\tau)}{(\tau-\beta)\left(V^2-\tau^2\right)^{3/2}} \tag{40}$$

Putting Equation (40) into (35), we have the result:

$$\mathrm{Re}\left[\frac{D(\tau)}{D_1(\tau)}\cdot\frac{A\tau^5}{(\tau-\beta)\left(V^2-\tau^2\right)^{3/2}}\right] = \mathrm{Re}\left[\frac{4P\tau^5}{\pi(\tau-\beta)}i\right],\quad |\tau|<V \tag{41}$$

At $\tau\to\beta$, constant A can be determined from that:

$$A=\frac{4P\left(V^2-\beta^2\right)^{3/2}}{\pi\cdot\mathrm{Im}\left[D(\beta)/D_1(\beta)\right]} \tag{42}$$

In an orthotropic isotropic body, the disturbance range of elastic wave can be shown by the circular area of radius c_1t and c_2t. Here c_1 and c_2 are the velocities of longitudinal and transverse waves ($c_1 > c_2$) of elastic body, respectively. In an orthotropic anisotropic body, the disturbance range of elastic wave is not the circular area and can not exceed threshold value $C_d=\left(C_{11}/\rho\right)^{1/2}$ of elastic body, where C_{11} is an elastic constant of materials. At $|x|>C_dt$, with $\mathrm{Im}\left[D_1(\tau)/D(\tau)\right]=0$, thus displacements and stresses are zero with the initiate cases; and this shows that disturbance of elastic wave can not exceed C_dt.

Then inserting Equation (40) into (6) and (5), at the surface $y = 0$, the stress σ_y and dynamic stress intensity factor $K_1(t)$ are gained, respectively:

$$\sigma_y=A\cdot\mathrm{Re}\int_{\infty}^{x/t}-\frac{\tau^4\cdot\mathrm{Im}\left[D(\tau)/D_1(\tau)\right]}{(\tau-\beta)\left(\tau^2-V^2\right)^{3/2}}\mathrm{d}\tau,\ |x|<Vt \tag{43}$$

$$K_1(t)=\sqrt{\pi t}\cdot\frac{AV^{5/2}}{(V-\beta)}\cdot\mathrm{Im}\left[\frac{D(V)}{D_1(V)}\right] \tag{44}$$

The limit of Equation (44) belongs to the form $0\cdot\infty$, which should be only changed into the type of ∞/∞, then its result can be worked out by the approach of L'Hospital theorem [31,32]. Dynamic stress intensity factor $K_1(t)$ slowly augments from zero and even reaches or exceeds fracture toughness of this material, because unique variable t lies in the numerator, and the rest are also referred to as real constants.

Equation (39) can rewrite as follows

$$W'(\tau)=\frac{A\tau^5}{(\tau-\beta)\left(V^2-\tau^2\right)^{3/2}} = \left[\tau^4+\beta\tau^3+\beta^2\tau^2+\beta^3\tau+\beta^4+\frac{\beta^5}{\tau-\beta}\right]\frac{A}{\left(V^2-\tau^2\right)^{3/2}} \tag{45}$$

Integrating Equation (45), one will attain $W(\tau)$. But it has six items, separate denotation is more convenient, Integral formulas are used in literature [31], now assume:

$$W_1(\tau)=\int\frac{A\tau^4}{\left(V^2-\tau^2\right)^{3/2}}\mathrm{d}\tau = \frac{A\tau}{2}\sqrt{V^2-\tau^2}+\frac{AV^2\tau}{\sqrt{V^2-\tau^2}}-\frac{3AV^2}{2}\arcsin\frac{\tau}{V} \tag{46}$$

$$W_2(\tau)=\int W_2'(\tau)\mathrm{d}\tau=\beta\int\frac{A\tau^3}{\left(V^2-\tau^2\right)^{3/2}}\mathrm{d}\tau = A\beta\sqrt{V^2-\tau^2}+\frac{A\beta V^2}{\sqrt{V^2-\tau^2}} \tag{47}$$

$$W_3(\tau)=\int W_3'(\tau)\mathrm{d}\tau=\beta^2\int\frac{A\tau^2}{\left(V^2-\tau^2\right)^{3/2}}\mathrm{d}\tau = A\beta^2\left[\frac{\tau}{\sqrt{V^2-\tau^2}}-\arcsin\frac{\tau}{V}\right] \tag{48}$$

$$W_4(\tau)=\beta^3\int\frac{A\tau}{\left(V^2-\tau^2\right)^{3/2}}\mathrm{d}\tau=\frac{A\beta^3}{\sqrt{V^2-\tau^2}} \tag{49}$$

$$W_5(\tau)=\beta^4\int\frac{A}{\left(V^2-\tau^2\right)^{3/2}}\mathrm{d}\tau=\frac{A\beta^4\tau}{V^2\sqrt{V^2-\tau^2}} \tag{50}$$

Denominator in Equation (45) contains this term $(\tau-\beta)\left(V^2-\tau^2\right)^{3/2}$, calculation will not be preformed in the light of integral formulae, therefore integral format must be translated into integral which can be fulfilled.

For the sake of convenience, we assume: $X=V^2-\tau^2$

By variable replacement: $\tau_1=\tau-\beta$, this term X can be rewritten as follows:

$$X=V^2-\tau^2=V^2-\beta^2-2\beta\tau_1-\tau_1^2 \tag{51}$$

Known from it, the following relationship in literature [31] is: $a_1=V^2-\tau^2$, $b_1=-2\beta$, $c=-1$, $D=4a_1c-b_1^2=-4V^2$.

Integrating the sixth term of Equation (45) in terms of relevant formulae in literature [31], we will gain $W_6(\tau)$:

$$\begin{aligned}W_6(\tau)&=\beta^5\int\frac{A\mathrm{d}\tau}{(\tau-\beta)X^{3/2}}=A\beta^5\int\frac{\mathrm{d}\tau_1}{\tau_1X^{3/2}}\\&=\frac{A\beta^5}{a_1}\left[\frac{1}{\sqrt{X}}+\int\frac{\mathrm{d}\tau_1}{\tau_1\sqrt{X}}-\frac{b_1}{2}\int\frac{\mathrm{d}\tau_1}{X^{3/2}}\right]\\&=\frac{A\beta^5}{a_1}\left[\frac{D-b_1^2-2b_1\beta}{D\sqrt{X}}+\frac{2b_1\tau}{D\sqrt{X}}\right.\\&\quad\left.-\frac{1}{\sqrt{a_1}}\ln\left|\frac{\sqrt{X}+\sqrt{a_1}}{\tau-\beta}+\frac{b_1}{2\sqrt{a_1}}\right|\right]+C\end{aligned} \tag{52}$$

Known from Equation (45): $W(\tau)=W_1(\tau)+W_2(\tau)+W_3(\tau)+W_4(\tau)+W_5(\tau)+W_6(\tau)$.

The crack runs along x-axis, hence $W(\tau)$ comprising Equations (46)-(50) and (52) can be performed in the definite integral operation, one takes constant C = 0. Known from Equation (45):
$W(\tau)=W_1(\tau)+W_2(\tau)+W_3(\tau)+W_4(\tau)+W_5(\tau)+W_6(\tau)$.
The crack runs along the x-axis, so $W(\tau)$ comprising Equations (46)-(50) and (53) can be performed in the definite integral operation, one takes constant $C = 0$.

Making use of relative integral formulas [31] to yield:

$$\int \frac{1}{\sqrt{V^2-\tau^2}} \mathrm{d}\tau = \arcsin\frac{\tau}{V} \tag{53}$$

$$\int \frac{\sqrt{V^2-\tau^2}}{\tau} \mathrm{d}\tau = \sqrt{V^2-\tau^2} + V\ln\frac{V-\sqrt{V^2-\tau^2}}{\tau} \tag{54}$$

$$\int \frac{\sqrt{V^2-\tau^2}}{\tau^2} \mathrm{d}\tau = -\frac{\sqrt{V^2-\tau^2}}{\tau} - \arcsin\frac{\tau}{V} \tag{55}$$

Putting Equation (46) into (6), (5), the divisional displacement v_1 will be obtained as:

$$v_1 = -\frac{Ax}{2}\cdot \mathrm{Re}\int_{\infty}^{x/t}\left[\frac{1}{\tau}\sqrt{V^2-\tau^2} + \frac{2V^2}{\tau\sqrt{V^2-\tau^2}} - \frac{3V^2}{\tau^2}\arcsin\frac{\tau}{V}\right]\mathrm{d}\tau \tag{56}$$

Now putting Equations (54), (26) into (56), there results the divisional displacement v_1:

$$v_1 = -\frac{Ax}{2t}\cdot\sqrt{V^2t^2-x^2} - \frac{3AV^2t}{2}\cdot\arcsin\frac{x}{Vt},\ |x|<Vt \tag{57}$$

Inserting Equations (47) into (6), (5), by means of Equations (55), (26), the displacement v_2 are obtained:

$$v_2 = A\beta\left[2\sqrt{V^2t^2-x^2} + x\arcsin\frac{x}{Vt}\right],\ |x|<Vt \tag{58}$$

Replacing Equations (48) into (6), (5), by application of Equation (25), there results the sub-displacement v_3:

$$v_3 = -A\beta^2 t\cdot\arcsin\frac{x}{Vt},\ |x|<Vt \tag{59}$$

Substituting Equation (49) into (6), (5), by means of Equation (26) there results the sub-displacement v_4 as:

$$v_4 = -A\beta^3 x\cdot\mathrm{Re}\left(-\frac{\sqrt{V^2-\tau^2}}{V^2\tau}\right)\Bigg|_{\infty}^{x/t} = \frac{A\beta^3}{V^2}\sqrt{V^2t^2-x^2},\ |x|<Vt \tag{60}$$

Then putting Equation (50) into (6), (5), by means of Equation (25), there results the sub-displacement v_5:

$$v_5 = -\frac{A\beta^4 x}{V^3}\ln\left|\frac{Vt-\sqrt{V^2t^2-x^2}}{x}\right|,\ |x|<Vt \tag{61}$$

Now substituting Equation (52) into Equations (6), (5), there results the divisional displacement v_6 as:

$$v_6 = -\frac{A\beta^5 x}{a_1}\mathrm{Re}\int_{\infty}^{x/t}\frac{1}{\tau^2}\left[\frac{D-b_1^2-2b_1\beta}{D\sqrt{X}} - \frac{1}{\sqrt{a_1}}\ln\left|\frac{\sqrt{X}+\sqrt{a_1}}{\tau-\beta}+\frac{b_1}{2\sqrt{a_1}}\right| + \frac{2b_1\tau}{D\sqrt{X}}\right]\mathrm{d}\tau \tag{62}$$

Integral of the second term of Equation (62) without comprising coefficient can be written as:

$$\begin{aligned}\int -\frac{1}{\tau^2\sqrt{a_1}}\ln\left|\frac{\sqrt{X}+\sqrt{a_1}}{\tau-\beta}+\frac{b_1}{2\sqrt{a_1}}\right|\mathrm{d}\tau &= \frac{1}{\sqrt{a_1}}\int \ln\left|\frac{\sqrt{X}+\sqrt{a_1}}{\tau-\beta}+\frac{b_1}{2\sqrt{a_1}}\right|\mathrm{d}\left(\frac{1}{\tau}\right) \\ &= \frac{1}{\tau\sqrt{a_1}}\ln\left|\frac{\sqrt{X}+\sqrt{a_1}}{\tau-\beta}+\frac{b_1}{2\sqrt{a_1}}\right| - \frac{1}{\beta}\int\frac{1}{\tau}\cdot\frac{\mathrm{d}\tau}{\sqrt{X}} + \frac{1}{\beta}\int\frac{1}{\tau_1}\cdot\frac{\mathrm{d}\tau_1}{\sqrt{X}}\end{aligned} \tag{63}$$

Using integral formulas in Literature [31], one gains:

$$\int\frac{\mathrm{d}\tau_1}{\tau_1\sqrt{X}} = -\frac{1}{\sqrt{a_1}}\ln\left|\frac{\sqrt{X}+\sqrt{a_1}}{\tau_1}+\frac{b_1}{2\sqrt{a_1}}\right| \tag{64}$$

where: $a_1 = V^2-\tau^2$, $b_1 = -2\beta$, $c=-1$, $D = 4a_1c-b_1^2 = -4V^2$.

Putting Equations (64), (25) into (63), the following representation is given as:

$$\begin{aligned}\int -\frac{1}{\tau^2\sqrt{a_1}}\ln\left|\frac{\sqrt{X}+\sqrt{a_1}}{\tau-\beta}+\frac{b_1}{2\sqrt{a_1}}\right|\mathrm{d}\tau &= \left(\frac{1}{\tau}-\frac{1}{\beta}\right)\frac{1}{\sqrt{a_1}}\ln\left|\frac{\sqrt{X}+\sqrt{a_1}}{\tau-\beta}+\frac{b_1}{2\sqrt{a_1}}\right| \\ &- \frac{1}{\beta V}\ln\frac{V-\sqrt{V^2-\tau^2}}{\tau}\end{aligned} \tag{65}$$

Inserting Equations (65), (26) into (62), the divisional displacement v_7 will be obtained as follows:

$$v_6 = \frac{A\beta^5}{V^2a_1}\cdot\sqrt{V^2t^2-x^2} + \frac{A\beta^4 x}{V^3}\ln\left|\frac{Vt-\sqrt{V^2t^2-x^2}}{x}\right| - \frac{A\beta^5}{a_1^{3/2}}\left(t-\frac{x}{\beta}\right)\ln\left|\frac{\sqrt{V^2t^2-x^2}+t\sqrt{a_1}}{x-\beta t}+\frac{b_1}{2\sqrt{a_1}}\right|\ |x|<Vt \tag{66}$$

The displacement v is the sum of divisional displacement: $v = v_1+v_2+v_3+v_4+v_5+v_6$. The addition of Equations (57)-(61) and (66), the displacement v is acquired as:

$$v = \frac{-A\beta^5}{a_1^{3/2}}\left(t - \frac{x}{\beta}\right)\ln\left|\frac{\sqrt{V^2t^2 - x^2} + t\sqrt{a_1}}{x - \beta t} + \frac{b_1}{2\sqrt{a_1}}\right|$$
$$+ A\left(\beta x - \frac{3V^2 t}{2} - \beta^2 t\right)\cdot \arcsin\frac{x}{Vt} \quad (67)$$
$$+ A\left(2\beta + \frac{\beta^3}{a_1} - \frac{x}{2t}\right)\sqrt{V^2t^2 - x^2}, \quad |x| < Vt$$

where: $a_1 = V^2 - \beta^2$, $b_1 = -2\beta$, $c = -1$, $D = 4a_1c - b_1^2 = -4V^2$.

Using the similar ways as that for finding Equation (33), put $|x| = \alpha t$ into (67) while regarding V, β and t as known constants, respectively. Bridging fiber fracture speed α can be only gained numerical solution, because it can also not be shown in the form of explicit function.

6. Law of Dynamic Stress Intensity Factor

In the light of practical situations of concrete problems, mutative law of dynamic stress intensity factor should be interpreted better. The corresponding parameters are put into Equations (23), (44) to plot $K_1(t)$ as a function of time t, and their numerical solutions are facilely gained. The following constants are as [8,17,21-22,33-36]:

C_{11} = 19.24 GPa; C_{12} = 1.25 GPa; C_{11} = 17.83 GPa; P = 200 N; C_{66} = 1.00 GPa; V = 300 m/s; β = 200 m/s; ρ = 4.9 × 1000 N/m^3.

Known from Equation (23), dynamic stress intensity factor $K_1(t)$ reduces tardily and has obvious singularity on account of unique variable t in the denominator, and the rest units are regarded as real constants. Such a tend is shown in **Figure 3**. This variable current is similar to the result of Literatures [8,17,21,22,33-38]. It is known from Equation (44) that dynamic stress intensity factor $K_1(t)$ increases slowly from zero and even reaches or overruns fracture toughness of this material, because sole variable t lies in its numerator, while the rest quantities are also referred to as real constants. This result must conduce the structural instability, as shown in **Figure 4**. This trend is similar to the result in references [8,17, 21,22,33-36,38-40], consequently it is also right.

The relative numerical values between dynamic stress intensity factor $K_1(t)$ and time t are expressed in **Tables 1, 2** in terms of curves in **Figures 3, 4**, respectively.

7. Conclusions

By means of correlative expression: $f(x,y,t) = t^n f(x/t, y/t)$, where n is an integral number, the problem considered can be facilely changed into homogeneous function of x and t of zeroth dimension, *i.e.* self-similar functions. All suffice the relationship of this function, and thus the analytical solutions can be gained by Equations (4)-(7) with homogeneous function of variable τ. This measure can use not only in electrodynamics [15-20,22,33-34], but also in electrostatics [24,30] and even in other situations, referring to literatures [30, 41].

Analytical solutions of the symmetrical dynamic crack extension model for bridging pull-out of unidirectional composite materials were found by way of the theoretical usage of a complex variable function. The method developed in this paper based on the approaches of the self-similar functions makes it conceivable to obtain the con-

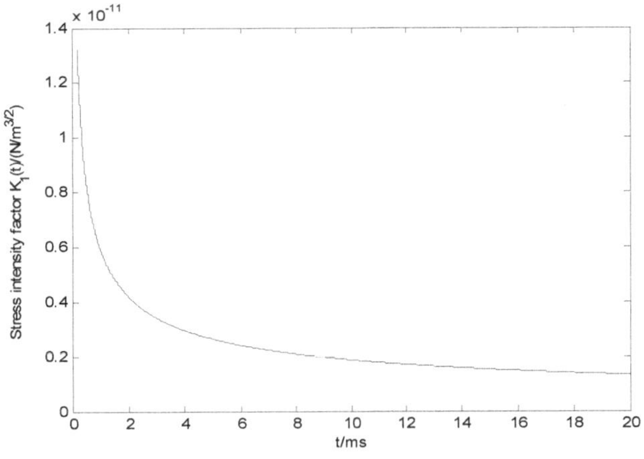

Figure 3. Dynamic stress intensity factor $K_I(t)$ versus time t.

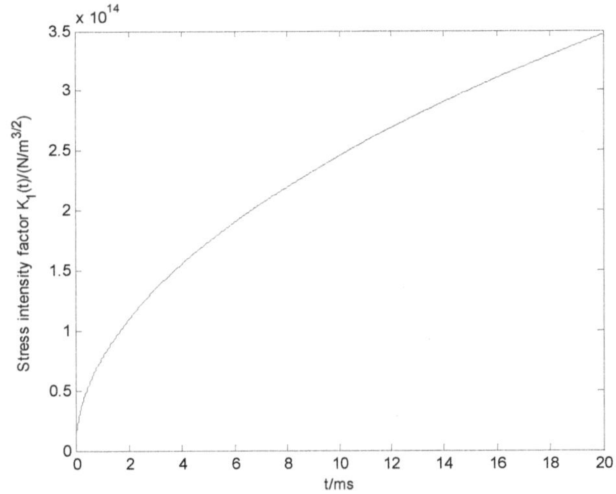

Figure 4. Dynamic stress intensity factor $K_I(t)$ versus time t.

Table 1. Relative numerical values between $K_1(t)$ versus t.

t/ms	4	8	12	16	20
$K_1(t)$	2.9552	2.0896	1.7062	1.4776	1.3216

Table 2. Relative numerical values between $K_1(t)$ versus t.

t/ms	4	8	12	16	20
$K_1(t)$	1.5511	2.1936	2.6866	3.1023	3.4684

crete solution of this model and bridging fiber fracture speed α. According to the concrete boundary conditions, self-similar function $W(\tau)$ can be easily deducted by means of corresponding to variable τ. Consequently, analytical solutions of stresses, displacements and dynamic stress intensity factors will be readily worked out. This is referred to as the analogous class of dynamic problem of the elasticity theory. However, the present solution occurs to be the simplest and intuitive of all alternative approaches appearing by so far. Indeed, we have succeeded in a mixed Keldysh-Sedov boundary value problem on a half-plane. The query is of adequate actual interest, since all the members of structures in which fractures may move are of finite dimensions and are frequently in the modality of long strips. The ways of solution are based exclusively on techniques of analytical-function theory and are simple and compendious. By making some observations regarding the solution of the mixed boundary value problem we have rather decreased the amount of the computative work needed to resolve such a crack expansion query.

8. Acknowledgements (Heading 5)

Project is supported by the open foundation of the engineering centre of hydraulic and equipment for mining of Liaoning province of China (No. CMHT-201207).

REFERENCES

[1] D. B. Marshall, B .N. Cox and A .G. Evens, "The Mechanics of Matrix Cracking in Brittle-Matrix Fiber Composites," *Acta Metallurgical*, Vol. 33, No. 11, 1985, pp. 2013-2021.

[2] B. Budiansky, J. W. Hutchinson and A. G. Evens, "Matrix Fracture in Fiber-Reinforced Ceramics," *Journal of the Mechanics and Physics of Solids*. Vol. 34, No. 2, 1986, pp. 167-189.

[3] M. Ji and H. Ishikawa, "Analysis of an Internal Central Crack with Bridging Fibers in a Finite Orthotropic Plate," *International Journal of Engineering Science*, Vol. 35, No. 4, 1997, pp. 549-560.

[4] D. B. Marshall and B. N. Cox. "Tensile Fracture of Brittle Matrix Composites: Influence of Fiber Strength," *Acta Metallurgical*, Vol. 35, No. 11, 1987, pp. 2607-2619.

[5] Z.-M. Wang, "Mechanics and Structural Mechanics of Composite Materials," Publisher of Machinery Industry, Beijing, 1991.

[6] G.-L. Shen. "Mechanics of Composite Materials," Tsinghua University Press, Beijing, 1996.

[7] C. W. Woo and Y. H. Wang, "Analysis of an Internal Crack in a Fine Anisotropic Plate," *International Journal of Fracture*, Vol. 62, No. 2, 1993, pp. 203-208.

[8] J. C. Lee, "Analysis of Fiber Bridged Crack near a Free Surface in Ceramic Matrix Composites," *Engineering Fracture Mechanics*, Vol. 37, No. 2, 1990, pp. 209-219.

[9] W. T. Tsai and I. R. Dharani, "Non Self-Similar Fiber Fracture in Unidirectional Composites," *Engineering Fracture Mechanics*, Vol. 44, No. 1, 1993, pp. 43-49.

[10] W. N. Liu, "Stress Ahead of the Tip of a Finite-Width Center-Crack in Fiber-Reinforced Composite Specimens: Subjected to Non-Linearly Distributed Bridging Stresses," *International Journal of Fracture*, Vol. 70, No. 1, 1994, pp. 31-35.

[11] K. Liao and K. Reifsnider, "A Tensile Strength Model for Unidirectional Fiber-Reinforced Brittle Matrix Composite," *International Journal of Fracture*, Vol. 106, No. 1, 2000, pp. 95-115.

[12] V. Tamuzs, S. Tarasovs and U. Vilks, "Progressive Delamination and Fibre Bridging Modeling in Double Cantilever Beam Composite Specimens," *Engineering Fracture Mechanics*, Vol. 68, No. 5, 2001, pp. 513-525.

[13] A. Piva and E. Viola, "Crack Propagation in an Orthotropic Media", *Engineering Fracture Mechanics*, Vol. 29, No. 5, 1988, pp. 535-547.

[14] J. De and B. Patra, "Elastodynimic Crack Problems in An Orthotrpic Medium through Complex Variable Approach," *Engineering Fracture Mechanics*, Vol. 41, No. 5, 1998, pp. 895-909.

[15] K. B. Broberg, "The Propagation of a Brittle Crack," *Arkve Fysik*, Vol. 18, No. 2, 1960, pp. 159-192.

[16] Y. W. Craggs, "The Growth of a Disk-Shaped Crack," *International Journal of Engineering Science*, Vol. 4, No. 2, 1966, pp. 113-124.

[17] J. G. Goree and R. S. Gross, "Analysis of a Unidirectional Composite Containing Broken Fibers and Matrix Damage," *Engineering Fracture Mechanics*, Vol. 33, No. 3, 1979, pp. 555-578.

[18] G. P. Cherepanov and E. F. Afanasov, "Some Dynamic Problems of the Theory of Elasticity—A Review," *International Journal of Engineering Science*, Vol. 12, No. 8, 1974, pp. 665-690.

[19] G. P. Charepanov, "Mechanics of Brittle Fracture," Nauka, Moscow, 1973.

[20] C. Atkinson, "The Propagation of a Brittle Crack in Anistropic Material," *International Journal of Engineering Science*, Vol. 3, No. 1, 1965, pp. 77-91.

[21] N.-C. Lü, X.-G. Li, Y.-H. Cheng and J. Cheng, "Fracture Dynamics Problem on Mode I Semi-Infinite Crack," *Archive of Applied Mechanics*, Vol. 81, No. 9, 2011, pp. 1181-1193.

[22] N. C. Lü, Y. H. Cheng. X. G. Li and J. Cheng, "Dynamic Propagation Problem of Mode I Semi-Infinite Crack

Subjected to Superimpose Loads," *Fatigue & Fracture of Engineering Materials & Structures*. Vol. 33, No. 3, 2010, pp. 141-148.

[23] N. C. Lü, Y. H. Cheng. X. G. Li and J. Cheng, "An Asymmetrical Dynamic Model for Bridging Fiber Pull-Out of Unidirectional Composite Materials," *Meccanica*, Vol. 47, No. 5, 2012, pp. 1247-1260.

[24] N. I. Muskhlishvili, "Singular Integral Equations," Nauka, Moscow, 1968.

[25] N. I. Muskhlishvili, "Some Fundamental Problems in the Mathematical Theory of Elasticity," Nauka, Moscow, 1966.

[26] F. D. Gakhov, "Boundary-Value Problems," Fitzmatigiz, Moscow, 1963.

[27] R. F. Hoskins, "Generalized Functions," Ellis Horwood, Chichester, 1979.

[28] X. S. Wang, "Singular Functions and Their Applications in Mechanics," Scientific Press, Beijing, 1993.

[29] G. C. Sih, "Mechanics of Fracture 4. Elastodynamics Crack Problems," Noordhoff, Leyden, 1977.

[30] R. P. Kanwal and D. L. Sharma, "Singularity Methods for Eastostatics," *Journal of Elasticity*, Vol. 6, No. 4, 1976, pp. 405-418.

[31] Editorial Group of Mathematics Handbook, "Mathematical Handbook," Advanced Education Press, Beijing, 2002.

[32] Teaching Office of Mathematics of Tongji University. "Advanced Mathematics," Advanced Education Press, Beijing, 1994.

[33] K. C. Wu, "Dynamic Crack Growth in Anisotropic Material," *International Journal of Fracture*, Vol. 106, No. 1, 2000, pp. 1-12.

[34] X.-G. Li, Y.-H. Cheng, N.-C. Lv, G.-D. Hao and J. Cheng, "A Dynamic Asymmetrical Crack Model of Bridging Fiber Pull-Out in Unidirectional Composite Materials," *Journal of Mechanical Science and Technology*, Vol. 25, No. 9, 2011, pp. 2297-2309.

[35] N. C. Lv, Y. H. Cheng. H. L. Si and J. Cheng, "Dynamics of Asymmetrical Crack Propagation in Composite Materials," *Theoretical and Applied Fracture Mechanics*, Vol. 47, No. 3, 2007, pp. 260-273.

[36] N. C. Lü, Y. H. Cheng and J. Cheng, "Mode I Crack Tips Propagating at Different Speeds under Differential Surface Tractions," *Theoretical and Applied Fracture Mechanics*, Vol. 46, No. 3, 2006, pp. 262-275.

[37] A. S. Kobayashi, "Dynamic Fracture Analysis by Dynamic Finite Element Method. Generation and Prediction Analyses," In: *Nonlinear and Dynarnic Fracture Mechanics*, American Society of Mechanical Engineers, New York, 1979, pp. 19-36.

[38] K. Ravi-Chandar and W. G. Knauss, "An Experimental Investigation into Dynamic Fracture: Part 1, Crack Initiation and Arrest," *International Journal of Fracture*, Vol. 25, No. 41, 1984, pp. 247-262.

[39] K. Ravi-Chandar and W. G. Knauss, "An Experimental Investigation into Dynamic Fracture: Part 2, Microstructural Aspects," *International Journal of Fracture*, Vol. 26, No. 11, 1984, pp. 65-80.

[40] K. Ravi-Chandar and W. G. Knauss, "An Experimental Investigation into Dynamic Fracture: Part 3, on Steady-State Crack Propagation and Crack Branching," *International Journal of Fracture*, Vol. 26, No. 2, 1984, pp. 141-152.

[41] L. A. Galin, "Contact Problems in Elasticity Theory," GITTL, Moscow, 1953.

Artificial Neural Networks for Controlling the Temperature of Internally Cooled Turning Tools

Frank Wardle[1], Timothy Minton[2], Saiful Bin Che Ghani[2], Paul Förstmann[3], Martin Roeder[3], Sebastian Richarz[3], Fiona Sammler[3]
[1]UPM Ltd., Swindon, England
[2]Brunel University, London, England
[3]Technische Universität, Berlin, Germany

ABSTRACT

By eliminating the need for externally applied coolant, internally cooled turning tools offer potential health, safety and cost benefits in many types of machining operation. As coolant flow is completely controlled, tool temperature measurement becomes a practical proposition and can be used to find and maintain the optimum machining conditions. This also requires an intelligent control system in the sense that it must be adaptable to different tool designs, work piece materials and machining conditions. In this paper, artificial neural networks (ANN) are assessed for their suitability to perform such a control function. Experimental data for both conventional tools used for dry machining and internally cooled tools is obtained and used to optimise the design of an ANN. A key finding is that both experimental scatter characteristic of turning and the range of machining conditions for which ANN control is required have a large effect on the optimum ANN design and the amount of data needed for its training. In this investigation, predictions of tool temperature with an optimised ANN were found to be within 5°C of measured values for operating temperatures of up to 258°C. It is therefore concluded that ANN's are a viable option for in-process control of turning processes using internally controlled tools.

Keywords: Control Systems; In-Process Control; Artificial Neural Network; Machine Tools

1. Introduction

A closed loop cooling system integrated with a novel design of tool insert, **Figure 1**, has many advantages: it eliminates a potential health hazard for machine operators; it is environmentally friendly as there is no coolant wastage; it consumes less power; swarf is easier to dispose of and it has the potential to reduce machine operating costs, [1]. However, confining coolant to internal channels built into the tool insert, presents new challenges: first of all, it will inevitably change the way in which the heat produced by the cutting operation is distributed and disposed of and secondly it will change the state of lubrication at the tool—work interface. This paper is concerned with the first of these challenges where, with the new technology, there is a risk that the local temperature at the cutting edge of the tool, may, under some machining conditions, exceed a permissible threshold and therefore reduce or at least offset any benefit in tool life.

As for internally cooled tools, coolant is completely under control, the measurement of tool temperature becomes a more meaningful and practical proposition. It is not subject to large fluctuations by small adjustments of coolant flow and direction as is the case for externally applied coolant and it therefore relates, to a much greater extent, to the machining conditions, work piece material, tool design and state of wear. Thus, routinely monitoring the temperature of internally cooled tools is one way of observing and hence avoiding the conditions that may create excessively high temperatures. Furthermore, there is now a wealth of evidence to show [2,3] that for many materials there is an optimum cutting temperature which gives maximum tool life and good surface integrity so to be able to control tool temperature has real practical advantages. However this requires tool temperature measurement to be combined with an intelligent control system that can interpret the measurements and signify some form of corrective action. For example, simply knowing temperature allows a judgment to be made about

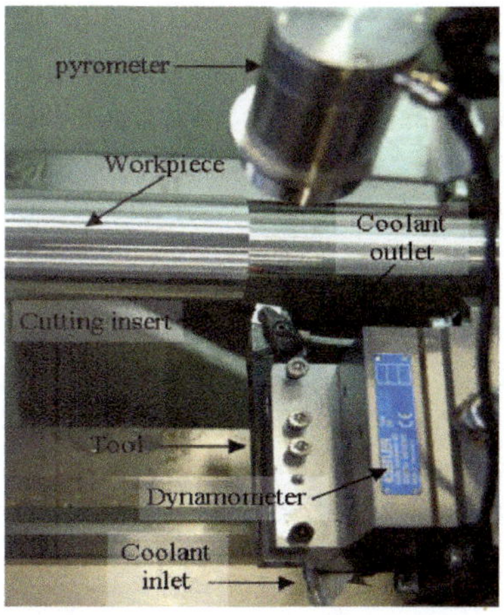

Figure 1. Experimental apparatus.

whether it is too high for the tool or work piece material, however knowing the machining conditions along with the temperature also enables the state of tool wear to be assessed when the temperature is well within material limits. Control systems based on artificial neural networks (ANN) are well suited to this form of decision making. They need to be trained with experimental data to generate a structured relationship between a number of input variables, in this case, machining conditions and an output variable-tool temperature. This relationship provides a reference for interpreting any new tool temperature measurements taken for any machining conditions within an allowable range. Furthermore, the ANN may be retrained any number of times enabling the interpretation tool temperature to be adapted to different work piece materials and tool designs.

Evaluating the benefit of using an ANN based system for controlling the temperature of internally cooled turning tools is the main objective of this paper. In this work, the tool is part of a system that incorporates a variable speed coolant delivery pump so that for a given tool design and work piece material the dominant input variables are coolant flow rate, cutting speed, cutting depth and feed/rev. Essential to the use of an ANN for controlling tool temperature in this system is how accurately it can predict tool temperature given these four input variables. But also of importance is how the prediction accuracy depends on the amount of experimental data and the time required for training.

Past work on ANN's applied to process/machine control is extensive with more than 1000 papers published since 1990. Whilst there are many types of ANN that can be used for machine tool or industrial process control those based on the Multi Layer Perceptron (MLP) have been used in more than 60% of investigations and have met with most success, [4]. They have generally employed an activation function attached to hidden neurons to improve the accuracy of prediction for the continuous and nonlinear target data characteristic of real control systems. Training has generally been performed off line utilizing back propagation or feed forward algorithms *i.e.* [5-14]. However what isn't clear from the published work is a strategy for determining the optimum ANN design for a given application. ANN designs vary widely and in many publications there is no indication as to how a particular design has been arrived at whilst in others, trial and error techniques or systematic variation of parameters have been used to determine an optimum design based on application specific, experimental data. Even in publications where systematic optimization of the ANN design has been undertaken, there have been apparently contradictory findings. For example, [15] found the best results were obtained with a low number of neurons in the hidden layer whilst [16] concluded that increasing the number of neurons improved the results. Alternatively [6] concluded that best results were achieved with an optimum number of neurons. Ref [17] presents the results of a wide ranging investigation into the effect of ANN design on prediction accuracy for data associated with wound toroidal cores. In this case, accuracy improved with the total number of neurons in the hidden layers but it didn't matter whether they were arranged in a single layer or several layers.

The lack of clear guide lines for ANN design and apparently contradictory findings of past work, at least in relation to process control, suggests ANN design and hence performance is strongly dependent on the experimental data it is required to predict. For the case of internally cooled turning tools the data may be characterized by an extremely wide dynamic range, significant experimental scatter and a non linear relationship between the output parameter—tool temperature and the machining parameters that constitute the input variables.

Turning is a universal process performed on a wide range of materials with an equally wide range of tool designs, machining speeds may vary from a few mm/s to tens of m/s, similarly depths of cut may range from a fraction of a mm to more than ten mm and tool feed/rev from tens of microns to a several mm. To determine whether ANN's can be of practical use to industry it is necessary to know how ANN design and prediction accuracy depends on data range; whether or not a single ANN design can cover a useful range of machining conditions and if so how much experimental data will be required for its training?

For a given tool and work piece material, tool temperature is primarily dependent on the machining conditions—speed, depth of cut and feed/rev. But it is also affected to a minor extent by many other factors that may collectively have a significant influence [18,19]. Tolerances on material specifications, tool geometries, the accuracy to which the work piece can be located, its initial geometry and how well it is clamped are just a few examples of variables that may influence tool temperature in an apparently random manner. How well do ANN's cope with this type of variation, is it necessary to smooth or precondition experimental data before training the ANN in order to predict the required dominant trends or can the ANN filter out these random effects?

The dependence of tool temperature on machining conditions is well known to be non linear with temperature generally approaching some limiting value as cutting speed, depth and feed/rev continue to increase in severity. ANN's with activation functions attached to hidden layer neurons are reported to cope well with non linear data but what is the best activation function to use, how does prediction accuracy depend on the degree of non linearity and how much experimental data is required for training?

In an attempt to answer the above questions and hence assess the suitability of an ANN based system for controlling the temperature of internally cooled tools, machining trials have been performed on CNC lathes fitted with conventional tools and with internally cooled tools and variable flow rate cooling systems. Conventional tools were used to generate dry machining data with which to assess the effect of ANN design on prediction accuracy; dry machining being a reference condition for internally cooled tools. The results of this study were then used to optimise ANN design for experimental data generated by internally cooled tools.

2. Dry Machining Experiments

A small desk top lathe, a Weiss WM280V-F was used for dry machining experiments. It was capable of handling work pieces of up to 250 mm diameter and 500 mm long. Spindle speed was infinitely variable from 50 to 2000 rpm and the tool feed/rev was incrementally variable between 0.07 and 0.42 mm. Motor power was limited to 2 kW peak and was the limitation to the depth of cut that could be used. The tool comprised of a tool holder with a tungsten carbide insert. The insert, DCMT060202, was diamond drilled from its underside as shown in **Figure 2**. A 0.6 mm hole positioned 1 mm from the cutting tip, was drilled to reach to within 0.8 mm of the cutting surface.

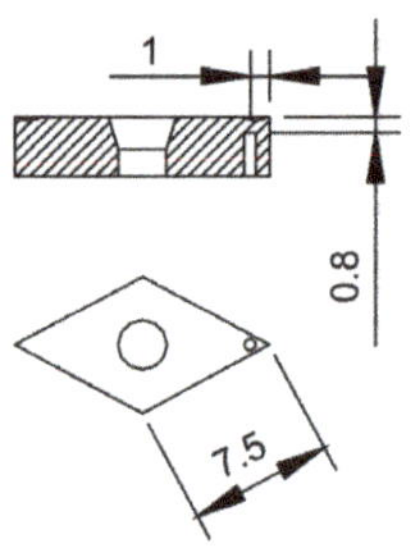

Figure 2. Diamond drilled tungsten carbide insert, type DCMT060202.

A slot was ground into the tool holder to allow a J type thermocouple, 0.5 mm diameter to access the hole in the insert and to be spring loaded against the end of the hole, 0.8 mm from the cutting surface. The lathe and tool were set up to dry machine 6082-T6 aluminium bar 75 mm in diameter and 300 mm long with a tailstock to provide rigid support to the work piece for all tests.

Experiments consisted of setting a depth of cut, the feed/rev and a surface speed and then maintaining the cut until the tool temperature reached a stable value. **Figure 3** shows a typical tool temperature—time recording. In this case the stable cutting temperature was 119.6°C. A series of 96 different cutting trials were performed covering the range of conditions: Cutting speeds, 0.2 - 3.3 m/s; Feed/rev, 0.07 - 0.42 mm; Depth of cut, 0.1 - 1.5 mm. For these conditions, recorded tool temperatures varied from 24°C to 172°C. **Figures 4-6** show the experimental data obtained for feeds of 0.07, 0.28 and 0.42 mm respectively.

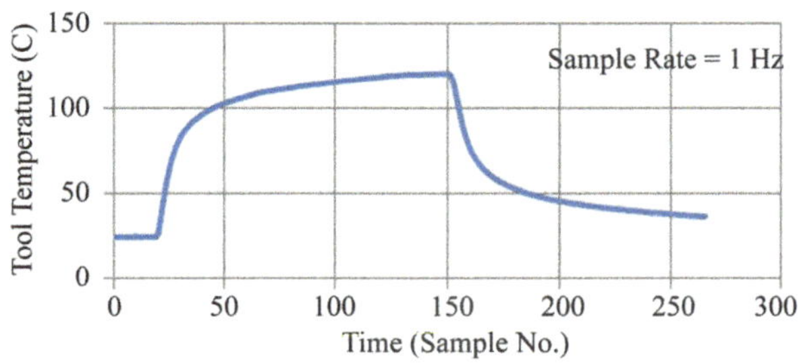

Figure 3. Typical tool temperature-time recording.

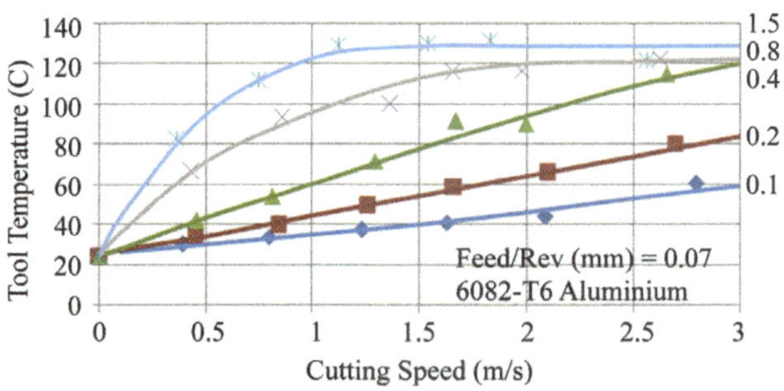

Figure 4. Experiments performed for a 0.07 mm feed/rev and 1.5, 0.8, 0.4, 0.2 and 0.1 mm depths of cut.

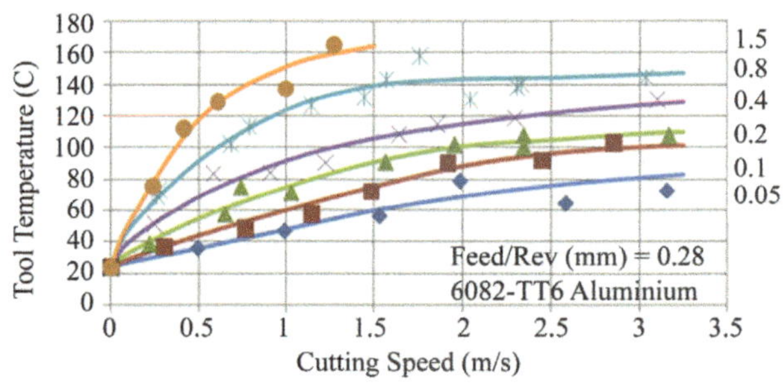

Figure 5. Experiments performed for a 0.28 mm feed/rev and 1.5, 0.8, 0.4, 0.2, 0.1 and 0.05 mm depths of cut.

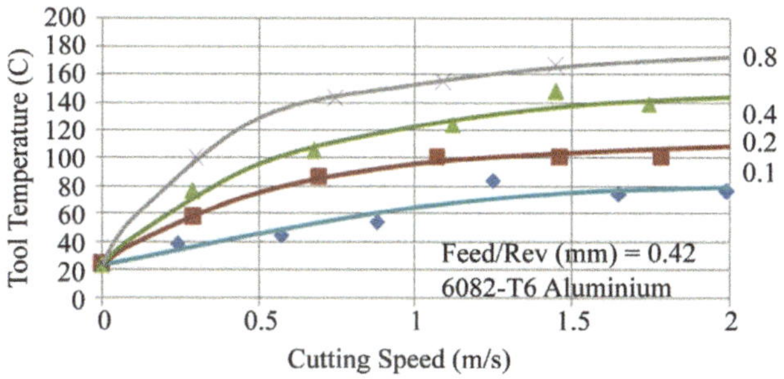

Figure 6. Experiments performed for a 0.42 mm feed/rev and 0.8, 0.4, 0.2 and 0.1 mm depths of cut.

The data exhibited significant scatter. Generally data points were within +/−5˚C of a best fit curve but isolated points, *i.e.* at 0.28 mm feed/rev, 0.8 mm depth and 1.8 and 2.1 m/s were found to be as much as 16˚C from the curve. The reason for the scatter was attributed mainly to inconsistent swarf clearance. In these experiments a chip breaker was not used and the continuous stream of swarf varied somewhat randomly in path and direction. Occasionally swarf would wrap around the work piece or tool and produce obvious fluctuations in temperature, these experiments are not included in the results. Small temperature measurement errors (estimated at less than 2˚C) also occurred as in practice the tool temperature never reached a perfectly stable value. The bulk temperature of the tool and tool holder continued to rise after the initial rapid transient increase causing the recorded tool tip temperature to continue to increase slowly. The effect of this was minimized by waiting for the rate of increase of tool temperature rate to be less than 1˚C per minute.

Experiments were performed with 4 different work pieces and two tools. Selected experiments, *i.e.* 0.28 mm feed/rev, 2.3 m/s and 0.8mm and 0.2 mm depth were repeated with the second tool on different work pieces to determine whether tool wear or variation in material properties affected temperature measurements. Within the limits of experimental scatter, agreement with the original data suggested neither tool wear nor work piece material was a significant factor in these experiments.

Figures 4-6 clearly show the relationship between tool temperature and cutting speed to be nonlinear, the degree of non linearity increasing with depth of cut. For example at a feed/rev of 0.07 mm and a depth of 1.5 mm tool temperature increases rapidly as speed is increased but reaches a maximum at 1.2 m/s. The same trend is evident at lower depths of cut but a maximum temperature is not reached within the speed range examined.

Figures 7 and **8** are extracted from **Figures 4** to **6** and show the dependence of tool temperature on depth of cut and feed rate respectively. The 3 curves in **Figure 7** are for cutting speeds of 0.5, 1.0 and 1.5 m/s whilst the 3 curves in **Figure 8** are for cut depths of 0.1, 0.2 and 0.4

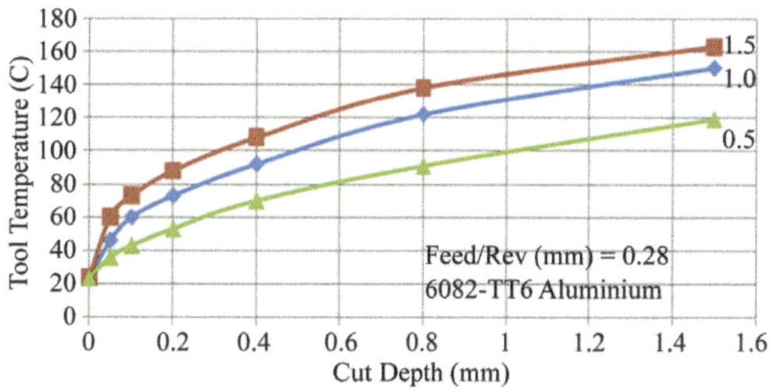

Figure 7. The effect of cut depth on tool temperature for cutting speeds of 1.5, 1.0 and 0.5 m/s.

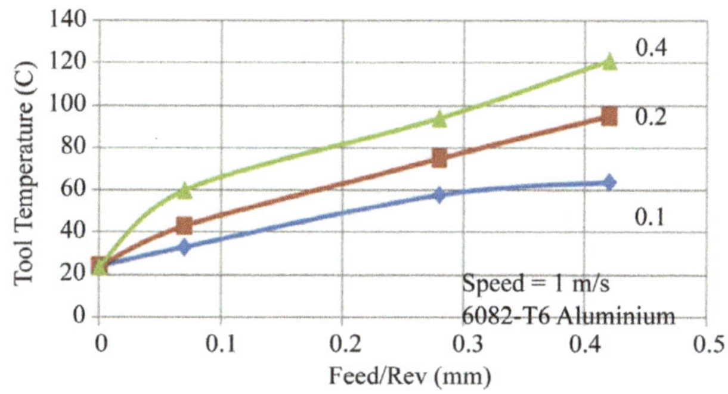

Figure 8. Effect of feed rate on tool temperature for 0.4, 0.2 and 0.1 mm depths of cut.

mm. Both figures show that the relationship between tool temperature and depth of cut or feed rate to be non linear.

3. Optimization of ANN Design

The experimental data in **Figures 4-6** was used as a basis for determining an optimum ANN design. The best fit curves to the data shows clearly defined non linear relationships between tool temperature and the machining parameters cutting speed, cut depth and feed rate. But the experimental data points show that superimposed upon the underlying trends there are small but significant random temperature measurement errors. This is broadly attributed to experimental scatter and is a characteristic of machining data. As part of a tool temperature control system, the ANN is required to provide a relationship between tool temperature and machining parameters against which real time tool temperature measurements can be compared and decisions made about the condition of the machining process and what adjustments may be required, if any. Thus, in this application the ANN is required to predict the underlying trends shown in **Figures 4-6** and filter out the random temperature measurement errors.

3.1. Application to Data Containing Experimental Scatter

The design algorithm described in [15] was used to optimize the ANN design for this application. A network with a single hidden layer and a TanH activation function was used as a starting point for the work. For the 96 experimental data points of **Figures 4-6** the effect of the number of neurons in the hidden layer upon convergence accuracy was determined by systematically varying the number of neurons and finding the minimum convergence error within 100,000 iterations. This was repeated 4 times to assess the repeatability of the convergence process. The results are shown in **Figure 9**, the curve is the average RMS convergence error and the black lines indicate maximum and minimum values. The convergence process exhibited considerable scatter and the minimum convergence error at 12.8% is considered high. Variation in the convergence process was attributed to the selection of random initial values with which to start the iteration process and the random selection of data points for testing and training. However the extent of the variation and the relatively high convergence error was attributed, at least in part, to experimental scatter within the data points used for training the network. The experimental data in **Figures 4-6** show an average error of 7% relative to best fit curves and this must limit the accuracy that the convergence process can achieve.

Figure 10 shows an example of a comparison between ANN predictions and experimental data. This is for a network with 18 neurons in the hidden layer giving a convergence error of 18%. ANN predictions do show the underlying trends in the data but careful inspection of the

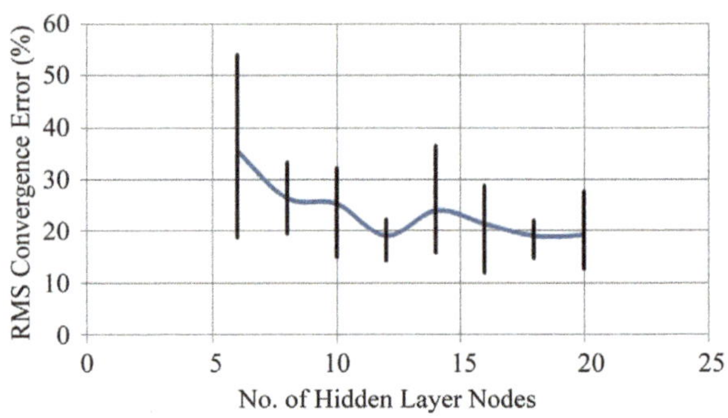

Figure 9. ANN convergence error for experimental data.

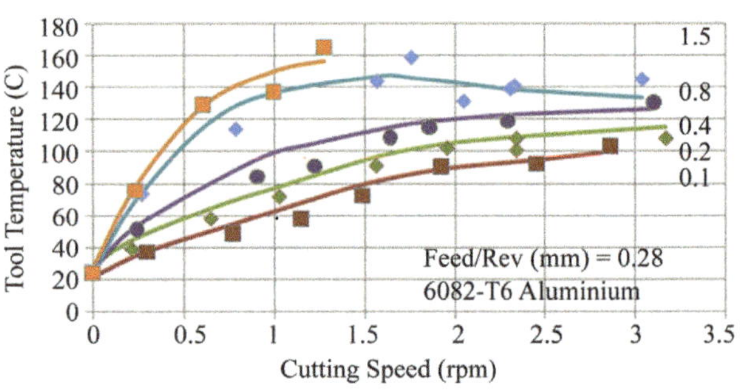

Figure 10. Comparison of ANN predictions and experimental data and 1.5, 0.8, 0.4, 0.2 and 0.1 mm depths of cut.

curves for cut depths of 0.8 and 0.4 mm reveal that the predictions depart from a best fit curve, generally overestimating tool temperature for a given cutting speed. The maximum error is 14.3°C at a speed of 1.2 m/s and a depth of 0.4 mm.

In applying ANN to machining data subject to experimental scatter it has been found that the training process gives inconsistent convergence errors so that several training exercises may be necessary to find a low convergence error. ANN does however find the under lying trends in the data but even the minimum convergence errors are relatively high and difficult to relate to the accuracy to which predictions fit the real experimental data. Furthermore the optimum number of neurons in the hidden layer is unclear. **Figure 9** shows the convergence errors to be most consistent for 12 neurons but similar convergence errors also occur for 18 neurons.

3.2. Application to Filtered Data

Removing experimental scatter from the data by using the best fit curves shown in **Figures 4-6** significantly improve the convergence accuracy of the training process.

Figure 11 shows the convergence errors and the consistency of the convergence process for the filtered experimental data to be significantly improved compared to those obtained the raw experimental data. For the TanH activation function used on both sets of data, convergence errors obtained for the filtered data were on average about half of that for the experimental data. Changing the activation function from TanH to Sigmoid gave similar convergence errors for 6 to 14 hidden neurons but a significantly reduced convergence error for 16 to 20 neurons. The lowest average convergence error was 9.3% for 20 neurons and a Sigmoidal activation function. However a network with a hidden layer of 12 neurons and a TanH activation function gave an average convergence error of 9.4% and within the limits of repeatability of the convergence process is considered to be the optimum design.

Figure 12 compares ANN predictions to the filtered data taken from **Figure 5**. The markers represent the data

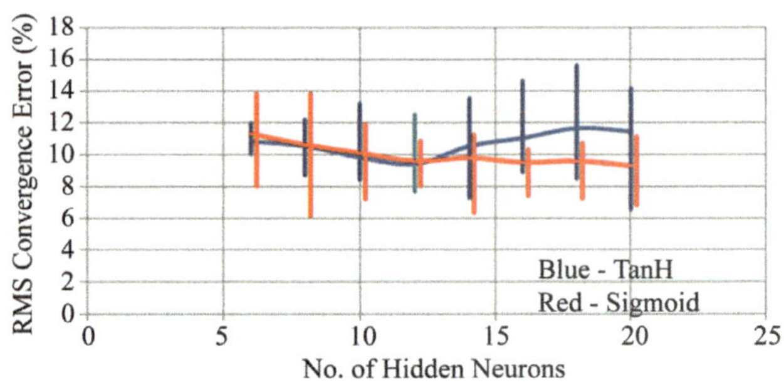

Figure 11. ANN convergence error for filtered data.

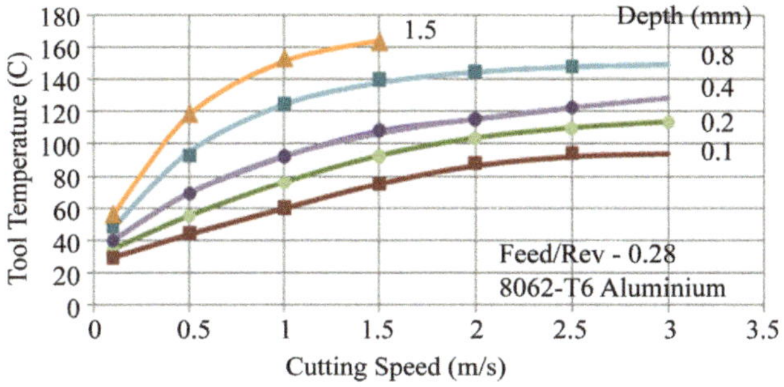

Figure 12. Comparison of ANN predictions and filtered data and 0.8, 0.4, 0.2 and 0.1 mm depths of cut.

whilst the solid lines are ANN predicted trends. The network with a hidden layer of 12 neurons, a TanH activation function and a convergence accuracy of 8.1 % was used for the predictions. **Figure 12** is equivalent to **Figure 10** but now predicted tool temperatures are within 2˚C of the filtered experimental data. The maximum error for all the data shown in **Figures 4-6** is 5˚C occurring locally for a feed/rev of 0.07 mm, a cutting depth of 0.1 mm and a speed of 0.5 m/s.

3.3. Minimum No. of Data Points for Training the ANN

Acquiring experimental data with which to train the ANN can be time consuming. Ninety six data points were obtained to generate **Figures 4-6**. Seventy eight filtered data points were used to train the network corresponding to **Figure 12**. On the assumption that reliable and consistent experimental data can be obtained by for example, repeating the same cut condition and averaging temperature measurements, then knowing how convergence error depends upon the number of data points used for training can help to minimise experimental work.

The filtered data corresponding to **Figure 12** was assumed to be representative of experimental data not subject to significant experimental scatter. The original 78 data points were reduced to 55 points by eliminating the 0.5 m/s and 2.0 m/s speed conditions and then to 36 points by also eliminating the 0.2 mm and the 0.8 mm depths of cut. All 3 sets of data were used to train the optimised ANN having a single hidden layer of 12 neurons. For each set of data the ANN was trained 5 times to determine the minimum convergence error. **Table 1** shows how the minimum convergence error depends on the number of data points.

The results show convergence error to steadily increase as the number of points used for training decreases. An example of the effect of this upon tool temperature predictions is shown in **Figure 13** where for 36 data points and a 0.28 mm/rev feed rate, ANN predictions are compared with the filtered data of **Figure 12**. Whereas for **Figure 12** ANN predictions based on 78 training data points are within 2˚C of the filtered experimental data, the error has increased to 8˚C for 36 data points. This error is for the 0.28 mm feed/rev, the maximum error over the full range of test conditions used for **Figures 4-6** was 11˚C, about the maximum acceptable in this application.

Table 1. Minimum convergence error v No. of training data points.

No. of data points	RMS convergence error (%)
78	7.8
55	9.3
36	12.2

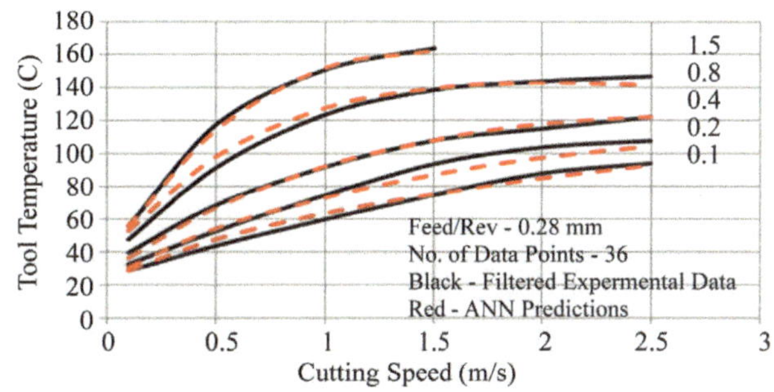

Figure 13. Example of ANN predictions for 36 training data points depths of cut—1.5, 0.8, 0.4, 0.2 and 0.1 mm.

3.4. Reduced Range of Cutting Conditions

Not every machining application needs to embrace a wide range of cutting conditions. Where only a narrow range of conditions are being used, a further reduction in the number of data points required for training the ANN is of interest.

Reducing the range of the machining conditions effectively reduces the degree of non linearity exhibited by the experimental data and it is reasonable to expect to be able to reduce the number of data points required for training the ANN without compromising convergence error. The original seventy eight filtered data points were reduced by first eliminating maximum and minimum speed conditions and then by also eliminating maximum and minimum depths of cut. The feed rates of 0.07, 0.28 and 0.42 mm/rev were retained. The effect of the number of data points used for training upon RMS convergence error for an ANN with a single hidden layer of 12 neurons and a TanH activation function is shown in **Table 2**. RMS convergence error tabulated is the minimum over 5 training processes and is shown to generally decrease with reduced number of data points and parameter range. An example of the effect of this on tool temperature predictions is shown in **Figure 14**.

For 55 data points and the cutting speed range reduced

Table 2. Convergence error v No. of data points for reduced parameter range.

No. of data points	RMS convergence error (%)	Speed range (m/s)	Depth range (mm)
78	9.4	0.1 - 2.5	0.1 - 1.5
64	7.0	0.5 - 2.5	0.1 - 1.5
55	6.8	1.0 - 2.5	0.1 - 1.5
42	6.2	1.0 - 2.0	0.1 - 1.5
32	7.1	1.0 - 2.0	0.2 - 1.5
27	5.9	1.0 - 2.0	0.2 - 1.0

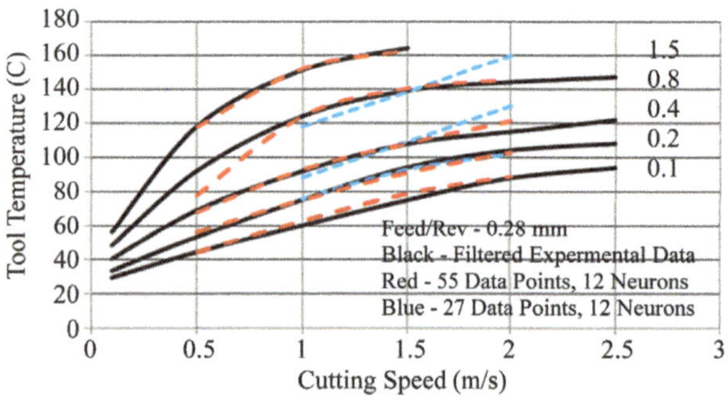

Figure 14. Effect of reduced parameter range on ANN predicted tool temperatures for 1.5, 0.8, 0.4, 0.2 and 0.1 mm depths of cut.

to 0.5 - 2.0 m/s the tool temperature predictions are accurate with one exception—the 0.5 m/s, 0.8 mm depth shows an error of 14˚C. For 27 data points and the speed range reduced to 1.0 - 2.0 m/s and depth of cut reduced to 0.2 - 0.8 mm the predictions are generally less accurate showing errors of 13˚C and 8˚C at a cutting speed of 2.0 m/s and depths of 0.8 and 0.4 mm respectively. The RMS convergence errors were similar at 6.8% for 55 and 5.9% for 27 data points. For both examples the ANN prediction accuracy in terms of absolute cutting tool temperature is of concern as in this application it is required to have a reasonable accuracy over a full range of conditions rather than good accuracy over most of the range combined with high local errors. The predictions have also failed to preserve the non linear trends exhibited by the data even though the range of the data and hence the degree of non linearity has been reduced.

In reducing the range of the experimental parameters, the characteristics of the data are also being changed and it follows that ANN design containing a single layer of 12 neurons, optimized over the full range of experimental conditions may no longer be appropriate. For the reduced parameter ranges shown in **Figure 14** the ANN design was re-optimized. For 55 data points a single hidden layer of 8 neurons produced a minimum convergence error of 5.3% whereas for 27 data points a single hidden layer of 5 neurons gave a minimum error of 3.9%. Both networks retained the TanH activation function used previously. The results are shown in **Figure 15** and in both cases the ANN predictions in terms of absolute tool temperature are greatly improved over those shown in **Figure 14**. For 27 data points prediction accuracy was within 1˚C over the full range of conditions. The improved prediction accuracies are accompanied by much faster convergence times.

3.5. Application to Internally Cooled Tools

The set up shown in **Figure 1** was used to generate experimental data for assessing the application of artificial neural networks to internally cooled tools. The machine tool was an Alpha Colchester Harrison 600 Group CNC lathe. The tool is a composite design comprising of a tool holder with integrated inlet and outlet tubes, a cooling block with integrated cooling geometry and a wear resistant insert made from tungsten carbide [ref: 20. Sun 2011]. The tool insert is a hollow shell with a fixed wall thickness which fits over the top of the cooling block, this allows the coolant to be pumped through the tool holder and pass close to the tool tip before being returned to the closed loop coolant supply. The insert was made by taking a standard SPUN type insert and using the EDM process to remove material to form the hollow. It should be noted that the inserts also have a TiN coating. The work piece was a cylindrical bar nominally 62 mm diameter and 300 mm long manufactured from AA6082-T6 aluminium. A tailstock was used to provide rigid support of the work piece for all trials. As the internal cooling circuit made it difficult to locate a thermocouple close enough to the tool tip to record a temperature representative of the cutting process an optical pyrometer, μ-Epsilon, model CTLM3- H1 CF2, was used to measure chip temperature. The pyrometer was mounted to the tool turret within the machine tool to maintain a minimum spot size of 0.45 mm at a constant working distance of 150 mm. It should be noted that any relevant parameter can be used to control tool temperature utilizing artificial neural networks. Even with a direct tool temperature measurement a transfer function will still be required to

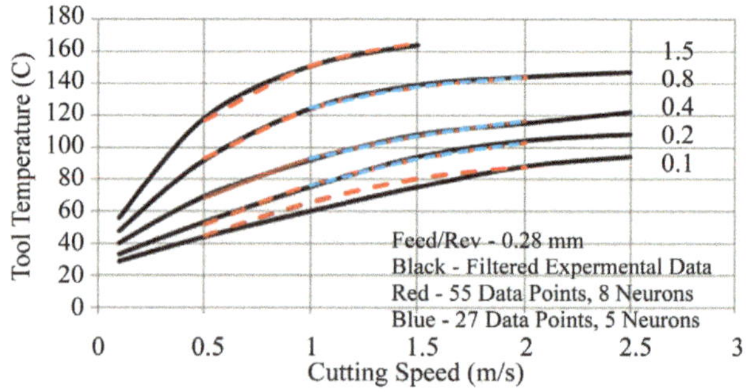

Figure 15. The effect of reduced parameter range and re-optimised ANN on tool temperature predictions for 1.5, 0.8, 0.4, 0.2 and 0.1 mm depths of cut.

relate the actual measurement to the cutting edge temperature. Thus in principle it doesn't matter what the parameter is.

Of primary interest in this assessment was to find out whether a relatively low number of experimental data points could be used to train the artificial neural network for controlling tool (chip) temperature over a relatively low range of cutting conditions. Depth of cut was limited to between 0.2 and 0.5 mm; feed/rev to between 0.1 and 0.2 mm and cutting speed to between 250 and 350 m/min. Experiments were performed with zero (dry cutting) and 0.3 l/min coolant flow rates. For each of these cooling conditions 27 data points were used to optimize and train the neural network. To minimize experimental scatter machining trials at each cutting condition were performed 3 times and the recorded temperature checked for consistency. Extraneous measurements were discarded and the trial repeated. A network with a single hidden layer of 7 neurons and a TanH activation function was found to provide convergence errors of less than 5%. The accuracy to which predicted tool (chip) temperatures fit the experimental data is shown in **Table 3**. The maximum error is less than 5˚C and is generally within 2˚C for temperatures in the range 188˚C to 258˚C.

Table 3. Experimental and predicted temperature measurements for an internally cooled tool.

Depth (mm)	Feed/rev (mm)	Speed (m/min)	Dry, measured tool temp (˚C)	Dry, predicted tool temp (˚C)	Cooled, measured tool temp (˚C)	Cooled, predicted tool temp (˚C)
0.2	0.1	250	196.534	196.534	188.82	188.82
0.35	0.1	250	226.082	2.26E+02	210.259	2.15E+02
0.5	0.1	250	242.292	2.44E+02	231.105	2.31E+02
0.2	0.15	250	198.149	1.99E+02	192.339	1.92E+02
0.35	0.15	250	221.73	2.17E+02	213.102	2.13E+02
0.5	0.15	250	248.478	2.47E+02	233.475	2.31E+02
0.2	0.2	250	195.293	1.95E+02	195.095	1.95E+02
0.35	0.2	250	223.057	2.23E+02	212.855	2.13E+02
0.5	0.2	250	251.009	2.53E+02	237.298	2.37E+02
0.2	0.1	300	199.432	1.99E+02	187.38	1.87E+02
0.35	0.1	300	226.508	2.27E+02	214.404	2.14E+02
0.5	0.1	300	252.825	2.52E+02	235.152	2.35E+02
0.2	0.15	300	196.736	1.97E+02	195.558	1.95E+02
0.35	0.15	300	224.859	2.24E+02	212.698	2.12E+02
0.5	0.15	300	249.534	2.49E+02	235.678	2.36E+02
0.2	0.2	300	202.805	2.03E+02	192.973	1.93E+02
0.35	0.2	300	221.057	2.23E+02	216.401	2.17E+02
0.5	0.2	300	252.234	2.54E+02	243.27	2.40E+02
0.2	0.1	350	200.974	2.01E+02	190.412	1.90E+02
0.35	0.1	350	226.092	2.26E+02	218.679	2.19E+02
0.5	0.1	350	248.943	2.49E+02	238.244	2.38E+02
0.2	0.15	350	199.605	2.00E+02	196.86	1.97E+02
0.35	0.15	350	232.5	2.33E+02	220.509	2.21E+02
0.5	0.15	350	254.621	2.56E+02	241.981	2.38E+02
0.2	0.2	350	203.033	2.03E+02	199.573	2.00E+02
0.35	0.2	350	225.27	2.26E+02	226.115	2.26E+02
0.5	0.2	350	258.422	2.57E+02	243.1	2.44E+02

4. Discussion

In practice tool temperature measurements for given machining conditions are prone to variation due to a number of extraneous effects. For the experiments performed in this investigation the lack of a chip breaker was a primary cause of apparently random tool temperature variations that are broadly referred to as experimental scatter. In using such data for training and optimizing ANN's it was found that minimum RMS convergence errors were large, in excess of 11%, and that the convergence process was inconsistent, exhibiting large variations in minimum convergence error. Nonetheless, the experimental data enabled an optimum design of ANN to be determined and this successfully predicted the underlying trends within the data as shown in **Figure 10**.

For the 3 input variables: cutting speed, depth and feed/rev and the parameter range investigated the optimum design of ANN had a single layer of 12 neurons and a TanH activation function. The latter was found to produce marginally lower convergence errors than a Sigmoidal activation function for less than 14 hidden neurons whereas for 16 - 20 hidden neurons the Sigmoidal function produced the lower convergence errors. Only a single hidden layer of neurons was considered as prior investigations, [17] have shown that it is the total number of hidden neurons that determine convergence error and not the number of hidden layers that they may be distributed within.

With the presence of experimental scatter and for the parameter range investigated, the 96 experimental data points were found to be sufficient for training and optimizing ANN design. In this case, there were enough data points to enable the ANN to average out experimental scatter and find the underlying trends. However, if it is required to minimize the number of data points used for training then scatter must be reduced to a negligible level to give a high degree of confidence in the data's integrity. This may be achieved by attention to experimental detail but in this investigation the experimental data was effectively filtered by using best fit curves. Using 78 filtered data points for optimizing and training the ANN also resulted in a network with 12 hidden neurons and a TanH activation function and this gave good agreement between ANN predictions and the filtered data as shown in **Figure 12**. Furthermore the convergence process was found to be more consistent.

Maintaining the parameter range and further reducing the number of data points used for training the ANN steadily increased the minimum convergence error that could be achieved. **Figures 12** and **13** show that for the same data, maximum errors between ANN predictions and filtered experimental data increased from 2°C to 8°C when the number of points used for training decreased from 78 to 36.

Reducing the parameter range together with the number of data points used for training but maintaining the optimized 12 hidden neuron network produced lower RMS convergence errors. However the errors tended to be localized so that although the RMS error was low for all data points, individual errors were high. The effect was that ANN predictions did not accurately reproduce the underling trends in the data as shown in **Figure 14**.

Reducing the parameter range together with the number of data points and re-optimizing the ANN design produced good agreement between ANN predictions and the filtered data as shown in **Figure 15**. For 27 data points tool temperature predictions were within 1°C of the filtered data, albeit for cutting speeds limited to between 1 and 2 m/s and depths to between 0.2 and 0.8 mm. The optimum ANN had only 5 neurons in a single hidden layer and produced rapid convergence. 27 data points was the minimum used in this investigation as it allowed each of the 3 input parameters to have 3 independent values—the minimum necessary to characterize non linear data.

The above findings were verified using an internally cooled tool and measuring chip temperature with an optical pyrometer. 27 experimental data points were used to train and optimize an ANN for a limited range of cutting conditions, depth: 0.2 - 0.5 mm, feed/rev: 0.1 - 0.2 mm and speed: 250 - 350 m/min. The optimum ANN had a single hidden layer of 7 neurons and predicted temperatures to within 5°C for measured temperatures in the range 188°C - 258°C.

5. Conclusions

1) ANN's can be used to predict the underlying nonlinear trends in tool temperature data obtained over a range of machining conditions, even when experimental scatter is present.

2) However ANN predictions are more accurate and fewer data points are required for training if experimental scatter is reduced or eliminated.

3) The optimum ANN design, the number of experimental data points required for training and the range of the data are interrelated. Wide ranging data requires an ANN with a relatively large number of hidden neurons and many experimental data points whereas for data obtained over a narrow range of conditions an ANN with a low number of hidden neurons and few experimental data points can provide accurate predictions.

4) The findings based on dry machining with conventional tool designs apply to internally cooled tool designs.

6. Acknowledgements

The authors gratefully acknowledge the European Commission for funding and supporting the ConTemp project of which this work is part. The authors also extend their gratitude to all ConTemp partners for their constructive contributions to the control system development.

REFERENCES

[1] E. Uhlmann and M. Roeder, "Internal Cooling of Cutting Tools," *Lambda Map Conference*, London, July 2009.

[2] F. Klocke, "Dry Cutting, Keynote Paper," *Annals of the CIRP*, Vol. 46, No. 2, 1997, pp. 519-526.

[3] V. P. Astakhov, "Effects of the Cutting Feed, Depth of Cut, and Workpiece (Bore) Diameter on the Tool Wear Rate," *International Journal of Advanced Manufacturing Technology*, Vol. 34, No. 7-8, 2007, pp. 631-640.

[4] E. Dimla Jr., P. E. Lister and N. J. Leighton, "Neural Network Solutions to the Tool Condition Monitoring Problem in Metal Cutting—A Critical Review of Methods," *International Journal of Machine Tools and Manufacture*, Vol. 37, No. 9, 1997, pp. 1291-1241.

[5] D. A. Dornfield, "Neural Network Sensor Fusion for Tool Condition Monitoring," *Annals CIRP*, Vol. 39, No. 1, 1990, pp. 101-105.

[6] T. I. Liu and E. J. Ko, "On-Line Recognition of Drill Wear via Artificial Neural Networks," *Winter Annual Meeting of the ASME, Monitoring and Control for Manufacturing Processes*, Vol. 44, Dallas, 1990, p. 101.

[7] S. Purushothaman and Y. G. Srinivasa, "A Back Propogation Algorithm Applied to Tool Wear Monitoring," *International Journal of Machine Tools and Manufacture*, Vol. 34, No. 5, 1994, pp. 625-631.

[8] Q. Liu and Y. Altintas "On-Line Monitoring of Flank Wear in Turning with Multilayered Feed-Forward Neural Networks," *International Journal of Machine Tools and Manufacture*, Vol. 39, No. 12, 1994, pp. 1945-1959.

[9] L. I. Burke and S. Rangwala, "Tool Condition Monitoring in Metal Cutting: A Neural Network Approach," *Journal of Intelligent Manufacturing*, Vol. 2, No. 5, 1991, pp. 269-280.

[10] L. I. Burke, "Competetive Learning Approaches to Tool Wear Identification," *IEEE Transactions on System, Man and Cybernetics*, Vol. 22, No. 3, 1992, pp. 559-563.

[11] A. Ruiz, D. Guinea, J. Barrios and F. Betancourt, "An Empirical Multi-Sensor Estimation of Tool Wear," *Mechanical Systems and Signal Processing*, Vol. 7, No. 2, 1993, pp. 105-119.

[12] Q. Zhou, G. S. Hong and M. Rahman, "A New Tool Wear Criterion for Tool Condition Monitoring using Neural Networks," *Engineering Applications of Artificial Intelligence*, Vol. 8, No. 5, 1995, pp. 579-588.

[13] Q. Zhou, G. S. Hong and M. Rahman, "On-Line Cutting State Recognition in Turning Using a Neural Network," *International Journal of Advanced Manufacturing Technology*, Vol. 10, No. 2, 1995, pp. 87-92.

[14] Q. Zhou, G. S. Hong and M. Rahman, "Using Neural Networks for Tool Condition Monitoring Based on Wavelet Decomposition," *International Journal of Machine Tools and Manufacture*, Vol. 36, No. 5, 1996, pp. 551-566.

[15] M. Guillot and A. El Ouafi, "On-Line Identification of Tool Breakage in Metal Cutting Processes by Use of Artificial Neural Networks," *Proceedings of ANNIE* 91, St Louis, 10 November 1991, p. 701.

[16] Y. I. Yao and X. Fang, "Assessment of Chip Forming Patterns with Tool Wear Progression in Machining via Neural Networks," *International Journal of Machine Tools and Manufacture*, Vol. 33, No. 1, 1993, pp. 89-102.

[17] S. Zurek, A. J. Moses, M. Packianather, P. Anderson and F. Anayi, "Prediction of Power Loss and Permeability with the Use of an Artificial Neural Network in Wound Toroidal Cores," *Journal of Magnetism and Magnetic Materials*, Vol. 320, No. 20, 2008, pp. 1001-1005.

[18] S. Y. Liang, *et al.*, "Machining Process Monitoring and Control: The State-of-the-Art," *Journal of Manufacturing Science and Engineering*, Vol. 126, No. 2, 2004, pp. 297-310.

[19] M. A. Davies, *et al.*, "On the Measurement of Temperature in Material Removal Processes," *Annals of CIRP*, Vol. 56, No. 2, 2007, pp. 581-604.

[20] X. Sun, R. Bateman, K. Cheng and S. C. Ghani, "Design and Analysis of an Internally-Cooled Smart Cutting Tool for Dry Cutting," *Journal of Engineering Manufacture*, 2011.

17

Location Results for Brace Spacing in Gap Tubular "K" Joints Using FEM

P. N. Jiki
Civil Engineering Department, University of Agriculture, Makurdi, Nigeria

ABSTRACT

The 8-node iso-parametric thin shell element was employed in the study of stress concentrations in the welded tubular "K" joint. Element equilibrium equations were derived using isoparametric formulation based on thin shell theory. After assembly, the resulting system equations were solved using existing fortran programs. Numerical experiments were conducted to isolate and locate ideal gap (positions) for the two braces of the "K" joint. The nominal stresses were calculated from which stress concentration factors were obtained. The resulting stress concentration factors were presented both as tables and as figures. A good agreement between our solutions and those for model joints in the literature is good and acceptable. It was found that the wider apart the brace spacing is, the weaker the strength of the joint. It was also found that the best location for the braces occurs when the stress level changes sign either from positive to negative or vice versa at a critical sampling point.

Keywords: Tubular Joint; Finite Element; Computer Simulation; Structural Joints; Effect of Bracing; Stress Concentration; Stress Distribution

1. Introduction

Lapped joints generally, and lapped tubular joints in particular are known to be stronger than gapped joints following a heuristic reasoning. However, for the purpose of bracing, it is not always feasible or even possible to brace a structural system with lapped joint throughout. Hence a strong case arises from using gapped tubular joints which are in many cases welded on site due to ease of construction and transportation. As usual, one of the problems with welded steel construction is the development of stress concentration or 'hot spot' stresses. The hot spots affect the strength of both lapped and gapped joints.

The problem of hot spot stresses or stress concentrations has been investigated by Kuang *et al.* [1] in their work on stress concentrations in tubular joints. Potuin *et al.* [2] also investigated stress concentrations in tubular joints. Gibstein [3] used a parametric study to investigate stress concentrations in T joints. Wordsworth and Smedly [4] investigated stress concentrations in offshore tubular joints without stiffners. However, joints of hollow square, rectangular and tubular sections are very often used in very important constructions such as large space roofs, offshore platforms [4] and trussed bridges [5,6] instead of their solid counterparts to reduce dead loads due to self weights. However, failure of these systems due to joint failures still pose engineering challenges. In order to understand why these joints still fail, So and So [7] carried out a study of "hot spot" stresses in a tubular "K" joint with combined in-plane and out-of-plane loadings. May be due to combined in-plane and out-of-plane loading, very high "hot spot" stresses were obtained from their model. Although they studied the effect of load variation, they did not study the effect of varying the gap between braces, on the strength of joint they studied. In order to understand the behaviour of these joints further, more studies were initiated by Paul *et al.* [8] to study the ultimate resistance of un-stiffened multi-planar tubular "T" and "K-K" joints. Parker [9] proposed the use of concrete filled hollow square section ("HSS") joints to improve the strength of such joints. The results of the finite element study on square hollow "K" joints by Jiki [10] seem to represent more realistically the behaviour of the joints but it was realised that the sharp corner edges of the section increased stress concentrations. This effect of geometry was also observed by Jiki [11] in the study of a rectangular hollow (RH) "K" joints. For the purpose of comparison using the finite element method to study the effect of section geometry on the distribution of stress concentrations in hollow joints, the same 8-node iso-parametric thin shell element is used to model these sections with different geometries. In all joints modelled by the finite element method, welded joint construction and the 8 node iso-parametric shell element were used through out to facilitate room for comparison.

It can be seen from the above review that the effect of stress concentrations on strength of tubular joints has been found to be a problem generally [5,6].

However, the particular problem of stress concentrations in lapped and gapped tubular joints with respect to effective location of the spacing of the braces has not hitherto been properly addressed.

The purpose of this study is to investigate the effect of effective location of braces (brace spacing) on the strength of the gapped tubular "K" joints. The effect of section geometry on the distribution of stress concentration factors in hollow section (HS) or tubular joints is also investigated. The finite element method of analysis was used in the investigation of the static stress concentrations.

2. Equilibrium Equations

The usual curved thin shell iso-parametric element was used for the present investigation (e.g. see Hinton and Owen [12]. The shape of the element is shown in **Figure 1**. Using potential energy formulation and the Rayleigh-Ritz process, the equilibrium equations were derived as follows:

The element total potential energy functional is given as :

$$\pi^e = U^e - W^e \tag{15}$$

where U^e is the strain energy of the element W^e is the external work done by the element during deformation.

In a compact formulation Equation (15) can be written as:

$$\pi^e = \iint_A [B]^T \{\delta\}^T [D][B][\delta]\mathrm{d}A - \int_b q\{\delta\}\,\mathrm{d}b \tag{16}$$

$$= \frac{1}{2}\{\delta\}^T \left[K^e\right]\{\delta\} - Q^e\{\delta\} \tag{17}$$

In which $\left[K^e\right]$ is the element stiffness matrix Q^e is the element boundary loads, such that:

$$\left[K^e\right] = \int_{-1}^{1}\int_{-1}^{1} [B]^T [D][B] \det\det[J]\,\mathrm{d}\xi\,\mathrm{d}\beta \tag{18}$$

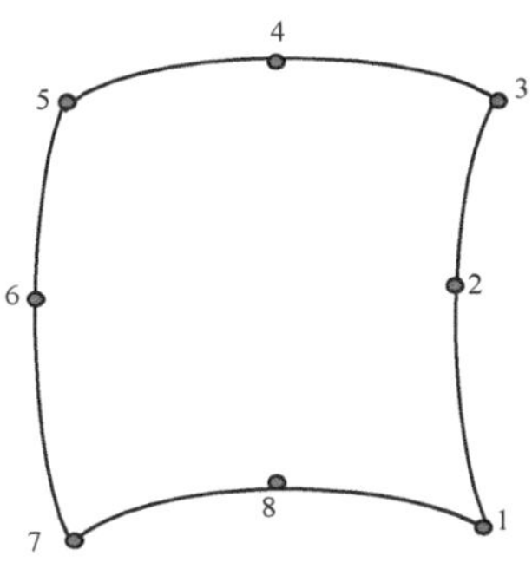

Figure 1. An 8-node isoparametric element.

$$Q^e = \sum_{l=1}^{n} q \tag{19}$$

After assembly we solve the system equations of Equation (20) for displacements.

$$\left[K^s\right]\{\Delta\} - \{Q\} = \{0\} \tag{20}$$

The finite element stresses are calculated from

$$\sigma_F = [D][B]\{\delta\} \tag{21}$$

in which

$$D = \frac{E}{\left(1-\upsilon^2\right)}\begin{bmatrix} 1 & \upsilon & 0 \\ \upsilon & 1 & 0 \\ 0 & 0 & \frac{1-\upsilon}{2} \end{bmatrix} \forall \text{ plane stress.} \tag{22}$$

3. Solution of System Equilibrium Equations

The system equilibrium equations for the present problem were solved after application of boundary conditions using the frontal solution algorithm and code developed by Hinton and Owen [12]. To check the validity of the program used we compare our results with those of So and So [7] for the case of in-plane loads as shown in **Table 1** herein. As we can see from that table our results or findings closely agree with theirs and is therefore acceptable. We have also investigated the effect of increasing the gap between braces on the stress concentration factors and our findings in this regard are shown in **Tables 2** and **3**. Having been satisfied with the performance of our program as a tool for the analysis, we have now studied the effect of variable brace spacing or gap on the strength of gap tubular "K" joints in details as presented in the tables and figures in this work.

4. Investigation of Stress Concentrations

The finite element discretization and loading of the joint for purpose of stress analysis is shown in **Figure 2**. Due to symmetry only half of the "K" joint used for the present study. After assembly we solve Equation (20) for nodal displacements and use Equation (21) to calculate stresses using Fortran code developed by Hinton and Owen [12]. After the finite element stresses were calculated, we also calculated nominal stresses using the usual expression for combined bending and axial effect as:

$$\sigma_N = \frac{P}{A} + \frac{My}{I} \tag{24}$$

and the nominal shear stresses are calculated from:

$$\tau_N = \frac{Q}{2\pi Rt} \tag{25}$$

Table 1. Comparison of S_{CFs} for "K" tubular joints subjected to inplane loads.

Load Type	S_{CF}					
	Obtained by Authors	Efthymiou and Durkin [7]	Obtained by So and So [7]	Authors	Efthymiou and Durkin [7]	Obtained by So and So [7]
In-plane bending	2.795	2.810	2.678	3.158	3.243	3.127

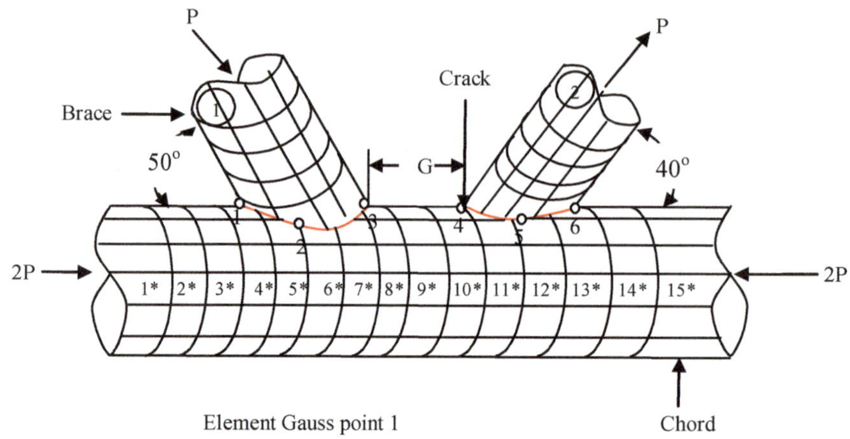

Figure 2. A finite element mesh for the K joint.

Table 2. Effect of increasing brace spacing (gap) on S_{CF} at corner of top chord (Brace angles = 50/40 deg).

Sampling point	50 mm gap	60 mm gap	70 mm gap
1	−0.99	−0.99	−0.83
2	−1.18	−1.18	−1.02
3	−0.70	−0.80	−1.92
4	−0.63	−0.86	−0.54
5	−1.22	1.73	+4.19
6	−0.69	−0.58	+2.59
7	−1.35	−0.96	−0.44
8	−0.88	−1.74	−0.99
9	−1.02	−0.83	−1.63
	Average = −0.962	Average = −0.960	Average = −0.066

Table 3. Effect of increasing brace spacing (gap) on S_{CF} at corner of top chord (Brace angles = 50/50 deg).

Sampling point	50 mm gap	60 mm gap	70 mm gap
1	1.19	1.17	−1.10
2	1.10	1.07	−1.15
3	0.72	0.73	−1.45
4	0.71	0.65	−0.64
5	1.26	1.10	+6.32
6	0.65	0.42	−1.15
7	1.39	1.44	−4.70
8	0.75	0.83	−1.16
9	0.98	1.12	−0.48
	Average = +0.972	Average = +0.950	Average = +0.612

in which Q is applied shear force, R is shell radius and t is shell thickness. The gap between the two braces in the first case was kept constant at 50 mm and stress concentration factor S_{CF} was obtained as

$$S_{CF} = \frac{\sigma_F}{\sigma_N} \forall \text{ flexure} \quad (26)$$

or

$$S_{CF} = \frac{\tau_F}{\tau_N} \forall \text{ shear} \quad (27)$$

where σ_F is finite element stress in the longitudinal axis of the chord τ_F is shear stress in the chord. σ_N is nominal stress given as $\sigma_N = \frac{P}{A} + \frac{My}{I}$.

5. Applications

Geometric parameters for the "K" joint.

Example 1.

For the purpose of comparison with our solution the dimensions of the "K" joint studied by So and So [7] are used here:

$$D = 232.85 \text{ mm}, \; d = 168.30 \text{ mm},$$

$$T = t = 12.7 \text{ mm}, \; \theta = 45^\circ.$$

We have also:

t/T = Brace of chord thickness ratio = 1.0,
D/T = Chord diameter to thickness ratio = 25.50,
d/D = Brace of chord diameter ratio = 0.52,
L/D = Chord length to diameter ratio = 6.18,
g/D = Brace separation to chord diameter ratio = 0.015

From g/D = ratio of 0.015, we calculate g to be $0.013 \times 323.85 = 4.86$ mm , say =5 mm.

And $L/D = 6.18$, we calculate L used by So and So [7]: $L = 6.18 \times 323.85 = 2001.4$ mm , say 2000 mm.

We next subject the joint with the above details to in-plane loading for the purpose of bending stress analysis. The result of analysis from our finite element code is compared with those of So and So [7] and Efthymiou and Durkin.

Our result for the chord is 0.53% lower than that of Efthymiou and Durkin and 4.37% higher than that of So and So [7], while for the chord; it is 2.37% lower that of Efthymiou and Durkin and 0.98% higher than that So and So [7]. In this case, we conclude that the comparison between our solution and that of the two published authors is overall less than 5% difference and is good and acceptable.

Example 2.

In this example we investigate the effect of damage and brace spacing/ separation on the strength of the "K" tubular joint. The geometric details of the joint is as follows:

Chord:

$D = 350$ mm, $t = T = 6.30$ mm, $L = 650$ mm; Load = 5T.

Brace:

$d = 250$ mm, $t = T = 6.30$ mm, $L = 650$ mm; Load = 2T.

The result of the peak (adverse) stress concentrations obtained mainly from top chord of the "K" joint are presented here in **Tables 2** and **3**.

6. Discussion of Results

Bracings in structural systems provide strength to the systems. However, there is no universal method of selecting locations for a selected bracing system. Usually the designer selects brace separations or gaps based on previous experience or heuristic reasoning. In this work we have used the finite element method as a tool to carry out some numerical experiments on locations of effective brace separations or spacings rather than use usual rule of thumb to select the desired spacings or gap between braces. By varying the gap between two braces and making computer runs produce a set of results. A close look at the results produced shows a trend in degradation of stresses as the gap is increased. This can give the analyst or the designer some idea of which brace separation to choose in the design.

In **Tables 2** and **3** we have presented the results of our study for three brace separations e.g. 50 mm, 60 mm and 70 mm respectively. We see in the tables that from 50 mm to 60 mm spacing/separation there is no alternating stress concentration factors S_{CFs} *i.e.* from positive to negative; that is, we have monotonic stress pattern and there is no stress reversal. However, as we increase the separation gap to 70 mm, we see the alternating pattern of stress concentration factors. This marks the limit of brace separation for the joint details above.

7. Conclusions

From our study of gapped tubular "K" joints we conclude as follows:

1) Brace separation distance for "K" tubular joint should be carefully selected so as to reduce over stress concentrations.

2) For best result, the joint should be lapped.

3) However, if the joint must be spaced, then based on our findings in **Tables 2** and **3** and **Figures 3** and **4**, 50 mm gap should be the limit of the spacing or gap. This is because after the 50 mm brace spacing or gap, adverse stress concentrations are observed when the spacing is increased beyond the 50 mm mark. In any case, the geometric parameters of the joint should be considered in order to avoid joint over stress.

4) We recommend or propose here that the finite element method be used as an ideal tool for the location of best brace spacing or gap (g).

5) Using our example joint studied here, we propose as a guide in the location result, that the ratio of the spacing

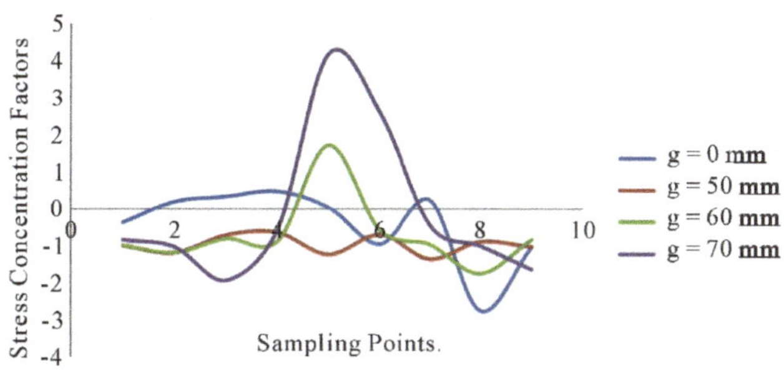

Figure 3. Effect of gap (g) on S_{CF} of top chord (centre).

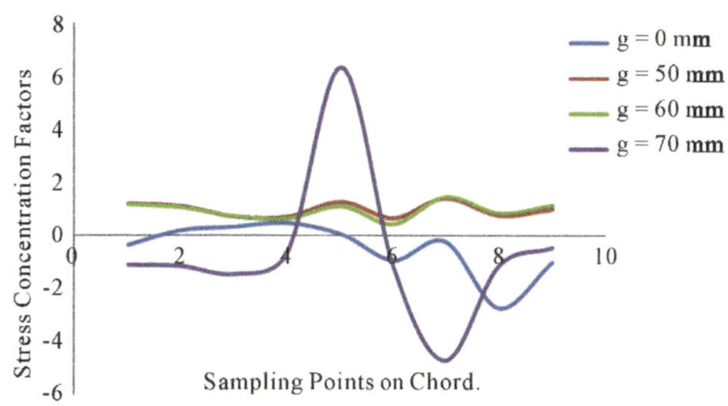

Figure 4. Effect of gap (g) on S_{CF} of top Chord (edge).

or gap g to the chord diameter D should not be greater than 1, that is, $g/D \leq 1$.

REFERENCES

[1] J. G. Kuang, A. B. Potuin and R. D. Leick, "Stress Concentration in Tubular Joints," *Offshore Technol Conference*, Houston, 4 May 1972, pp. 6-15.

[2] A. B. Potuin, J. G. Kuang, R. D. Leick and J. L. Kahlich, "Stress Concentration in Tubular Joints," *SPE Journal*, Vol. 17, No. 4, 1977, pp. 287-299.

[3] M. S. Gibstein, "Parametric Stress Analysis of T-Joints," Offshore Steel Seminar, Cambridge, 1978, pp. 3-26.

[4] A. C. Wordsworth and G. P. Smedley, "Stress Concentrations at Unstiffened Tubular Joints," Offshore Steel Seminar, Cambridge, 1978.

[5] S. Walbridge and A. Nussbaumer, "Probablistic Fatigue Analysis of Shop and Field Treated Tubular Truss Bridges," *Journal of Construction Steel Research*, Vol. 64, No. 2, 2008, pp. 156-166.

[6] A. Schumacher, and A. Nussbaumer, "Experimental Study on the Fatigue Behaviour of Welded Tubular K Joints for Bridges," *Journal of Engineering Structures*, Vol. 28, No. 5, 2006, pp. 745-755.

[7] A. K. Soh and C. K. Soh, "Hot-Spot Stresses of Tubular K Joints Subjected to Combined Loading," *Journal of Construction Steel Research*, Vol. 26, No. 2-3, 1993, pp. 125-140.

[8] J. C. Paul, Y. Makins and Y. Kurobane, "Ultimate Resistance of Unstiffened Multi Planar Tubular T T and K K Joints," *Journal of Structural Engineering, ASCE*, Vol. 120, No. 10, 1994, pp. 2853-2870.

[9] J. A. Parker, "Concrete Filled HSS Connections," *J. of Structural Engineering, ASCE*, Vol. 120, No. 3, 1995, pp. 458-467.

[10] P. N. Jiki, "Prediction of Stress Concentration Factors in Un-lapped Square Hollow 'K' Joints by the Finite Element Method," *Nigerian Journal of Technology*, Vol. 25, No. 1, 2006, pp. 5-14.

[11] P. N. Jiki, "Stress Analysis of Un-lapped Rectangular Hollow 'K' Joints by the Finite Element Method," *Botswana Journal of Technology*, Vol. 17, No. 2, 2009, pp. 46-53. (Botswana)

[12] E. Hinton and D. R. J. Owen, "Finite Element Programming," Academic Press, London, 1977, pp. 95-110.

18

The Influence Factors Study of Aerosol Particles

Tao Jiang[1], Wu Wei[2]
[1]Military Representative Office, Dalian, China
[2]China Ship Development and Design Center, Wuhan, China

ABSTRACT

The marine and coastal intake filtration device is mainly used to filter out the salt spray aerosol particles. So, the laboratory studies of the intake filtration device need to simulate the salt spray aerosols state of the ocean environment. But the distribution of ocean salt spray aerosol particle of diameter is very wide, so the simulation device of salt spray aerosol should produce a variety of salt spray aerosol particles. The paper is to mainly study the influencing factors of salt spray aerosol device to produce different particles diameter. And the investigation and study mainly include two aspects: numerical simulation and experimental study.

Keywords: The Generating Device of Salt Spray Aerosol; Nozzle; Atomization Mechanism; Numerical Simulation

1. Introduction

Since the marine atmosphere contains salt, which exists in the form of aerosol, the air intake of marine ship usually has the desalination requirements [1]. Currently, many large surface ships are equipped with gas turbine power, both at home and abroad. The desalination requirements of the air intake of gas turbine, which has a greater air intake, are more stringent. All of these require designers to do numerical simulation and experimental study to marine filtration device of desalination purposes in the appropriate ocean state. That can provide a basis for the design and manufacture of high-performance marine filter. The paper examines the factors of salt spray aerosol device of producing different diameter particle based on the actual development requirements of air intake filter and also conducts the mechanism study from numerical simulation and experimental study. The paper do the numerical simulation of the nozzle gas-liquid two-phrase flow field with a discrete phrase model (DPM) [2], taking the Reynolds average N-S equation as the control equation and the standard model as the turbulence model. In the boundary conditions, the inlet pressure of gas field, the initial diameter and initial velocity of droplets' inlet mass flow of a discrete phrase are given and also the outlet pressure of cylindrical space is given. Maintain the mass flow rate of the inlet droplets and study the influence law of salt spray aerosol particle diameter while changing the inlet pressure. Maintain the intake pressure and study the influence law of salt spray aerosol particle diameter while changing the mass flow rate of the inlet droplets.

The Mathematical Model of Nozzle Flow Field

First, to conduct the numerical simulation of the nozzle internal and external flow field, we establish a mathematical model of fluid flow in the nozzle. It includes turbulent flow model and laminar flow model [3].

Any liquidity issues meet the conservation of mass [4]. That's to say, the increase mass of fluid micro unit is equal to the inflow net mass of the micro unit in a period of time. Set the direction components of speed vector *u* are *u*, *v*, *w* in the direction of *x*, *y*, *z*. So, the mass conservation equation is

$$\frac{\partial p}{\partial t}+\frac{\partial(pu)}{\partial x}+\frac{\partial(pv)}{\partial y}+\frac{\partial(pw)}{\partial z}=0 \tag{1}$$

Supposing

$$\operatorname{div}(pu)=\frac{\partial(pu)}{\partial x}+\frac{\partial(pv)}{\partial y}+\frac{\partial(pw)}{\partial z} \tag{2}$$

The vector expression of Formula (1) is

$$\frac{\partial p}{\partial t}+\operatorname{div}(pu)=0 \tag{3}$$

This equation is the continuity equation of the flow.

Due to the water's little compressibility, it can be regarded as incompressible fluid. Provided the water's density ρ being constant, when the flow is in a steady state, ρ doesn't change with the time. So

$$\frac{\partial(pu)}{\partial x}+\frac{\partial(pv)}{\partial y}+\frac{\partial(pw)}{\partial z}=0$$

The momentum conservation equation in the direction of *x*, *y* and *z* are

$$\frac{\partial(pu)}{\partial t}+\operatorname{div}(puu)=-\frac{\partial p}{\partial x}+\frac{\partial\tau_{xx}}{\partial x}+\frac{\partial\tau_{yz}}{\partial y}+\frac{\partial\tau_{zx}}{\partial z}+F_x$$
$$\frac{\partial(pv)}{\partial t}+\operatorname{div}(pvu)=-\frac{\partial p}{\partial y}+\frac{\partial\tau_{xy}}{\partial x}+\frac{\partial\tau_{yy}}{\partial y}+\frac{\partial\tau_{zy}}{\partial z}+F_y \quad (4)$$
$$\frac{\partial(pw)}{\partial t}+\operatorname{div}(pwu)=-\frac{\partial p}{\partial z}+\frac{\partial\tau_{xz}}{\partial x}+\frac{\partial\tau_{yz}}{\partial y}+\frac{\partial\tau_{zz}}{\partial z}+F_z$$

where $\tau_{xx},\tau_{yz},\tau_{zx}$ are the component of viscous stress τ in the surface of the micro unit. *F* is the physical of the micro unit.

Equation (4) can also be written as follows:

$$\frac{\partial(pu)}{\partial t}+\frac{\partial(puu)}{\partial x}+\frac{\partial(pvu)}{\partial y}+\frac{\partial(puw)}{\partial z}$$
$$=\frac{\partial}{\partial x}\left(u\frac{\partial u}{\partial x}\right)+\frac{\partial}{\partial y}\left(u\frac{\partial u}{\partial y}\right)+\frac{\partial}{\partial z}\left(u\frac{\partial u}{\partial z}\right)-\frac{\partial p}{\partial x}+S_u$$
$$=\frac{\partial(pv)}{\partial t}+\frac{\partial(pvu)}{\partial x}+\frac{\partial(pvu)}{\partial y}+\frac{\partial(puw)}{\partial z}$$
$$=\frac{\partial}{\partial x}\left(u\frac{\partial w}{\partial x}\right)+\frac{\partial}{\partial y}\left(u\frac{\partial w}{\partial y}\right)+\frac{\partial}{\partial z}\left(u\frac{\partial w}{\partial z}\right)-\frac{\partial p}{\partial z}+S_w$$

The Formulas (3) and (4) are the momentum conservation equation. They can also be simplified as momentum equation, motion equation or Navier-Stokes equation.

Provided that the fluid's specific heat capacity is C_p, temperature is *T*, the heat transfer coefficient is *k*, the viscous dissipation is S_T, and its energy conservation equation can be expressed as

$$\frac{\partial(pT)}{\partial t}+\operatorname{div}(puT)=\operatorname{div}\left[\frac{ks}{C_p}\operatorname{gard}T\right]+S_T \quad (5)$$

Expanding to

$$\frac{\partial(pT)}{\partial t}+\frac{\partial(puT)}{\partial x}+\frac{\partial(pvT)}{\partial y}+\frac{\partial(pwT)}{\partial z}$$
$$=\frac{\partial}{\partial x}\left[\frac{k}{C_p}\frac{\partial T}{\partial x}\right]+S_T+\frac{\partial}{\partial y}\left[\frac{k}{C_p}\frac{\partial T}{\partial y}\right]+\frac{\partial}{\partial z}\left[\frac{k}{C_p}\frac{\partial T}{\partial x}\right]+S_T$$

From macroscopic aspect, laminar flow is regular viscous fluid motion. Laminar flow fluid is approximate the ideal fluid, so you can solve the equation directly from the N-S equation. Turbulence is the most ubiquitous fluid motion in nature and engineering. Its movement is very irregular, very unstable and the speed of its every point is immediately changeable [5]. Although the three-dimensional transient N-S equations apply equally, we still have to adopt the direct simulation method which has a high demand to computer if we solve the control equation of three-dimensional transient directly. However, this method doesn't used in the project currently. The method which is widely used in the engineering is the time averaging processing of the transient N-S equation. That is, take the turbulent motion as the superposition of two motions. One is the time average flow and the other is the instantaneous pulsating flow. Meanwhile [6], add the turbulent kinetic energy equation and turbulent dissipation rate equation which can reflect the turbulence characteristics. Here, *k* is to express turbulent kinetic energy and ε is to express turbulent dissipation rate. Taking the following equation:

$$k=\frac{\overline{u_i'v_j'}}{2}=\frac{1}{2}\left(\overline{u'^2}+\overline{v'^2}+\overline{w'^2}\right)$$
$$\varepsilon=\frac{u}{p}\overline{\left[\frac{\partial u_i'}{\partial x_k'}\right]\left[\frac{\partial u_i'}{\partial x_k'}\right]}$$

In the standard equation of $k-\varepsilon$, *k* and ε is the unknown quantity. The corresponding operation equation is:

$$\begin{aligned}&\frac{\partial(pk)}{\partial t}+\frac{\partial(pku_i)}{\partial x_i}\\&=\frac{\partial p}{\partial x_j}\left[\left(u+\frac{u_i}{\sigma_k}\right)\frac{\partial k}{\partial x_j}\right]+G_k+G_b+\rho\varepsilon-Y_M\\&+\frac{\partial(\rho\varepsilon)}{\partial t}+\frac{\partial(\rho\varepsilon u_i)}{\partial x_i}\\&=\frac{\partial p}{\partial x_j}\left[\left(u+\frac{u_i}{\sigma\varepsilon}\right)\frac{\partial\varepsilon}{\partial x_j}\right]+C_i\frac{\varepsilon}{k}\left(G_k+G_{3\varepsilon}G_b\right)\\&+G_{2\varepsilon}\rho\frac{\varepsilon^2}{k}+S_g\end{aligned} \quad (6)$$

In order to analyze the control equation and the solving of the various control equation with same procedures, the general forms of the basic control equation of fluid power are established [7].

$$\frac{\partial(\rho\phi)}{\partial t}+\operatorname{div}(\rho u\phi)=\operatorname{div}(T\operatorname{grad}\phi)+S \quad (7)$$

Expanding to

$$\frac{\partial(\rho\phi)}{\partial t}+\frac{\partial(\rho u\phi)}{\partial x}+\frac{\partial(\rho v\phi)}{\partial y}+\frac{\partial(\rho w\phi)}{\partial z}$$
$$=\frac{\partial}{\partial t}\left[\Gamma\frac{\partial\phi}{\partial x}\right]+\frac{\partial}{\partial y}\left[\Gamma\frac{\partial\phi}{\partial y}\right]+\frac{\partial}{\partial z}\left[\Gamma\frac{\partial\phi}{\partial z}\right]+S$$

2. Geometric Model and Mesh Division

Using Gambit to establish the simplified geometric model of the internal and external flow field of the nozzle (such as **Figures 1** and **2**).

The CFD method for solving control equation is actually to discrete the control equation in the space area and then obtain the discrete equations by solving them [8,9]. To discrete the control equation in the space domain, the grids must be used. So you have to grid the established the model, determine the size and type of the cell and also define the relevant boundary conditions before the solving of CFD [10] (such as **Figures 3** and **4**).

3. Numerical Simulation Results and Experiments

The paper mainly studies the influence law of the nozzle's atomized particle size when the pressure of the inlet air and the mass flow rate of the inlet droplets are changed. First, maintain the intake pressure and study the influence law of salt spray aerosol particle diameter while changing the mass flow rate of the inlet droplets. Second, maintain the mass flow rate of the inlet droplets and study the influence law of salt spray aerosol particle diameter while changing the inlet pressure (such as **Figures 5** and **6**).

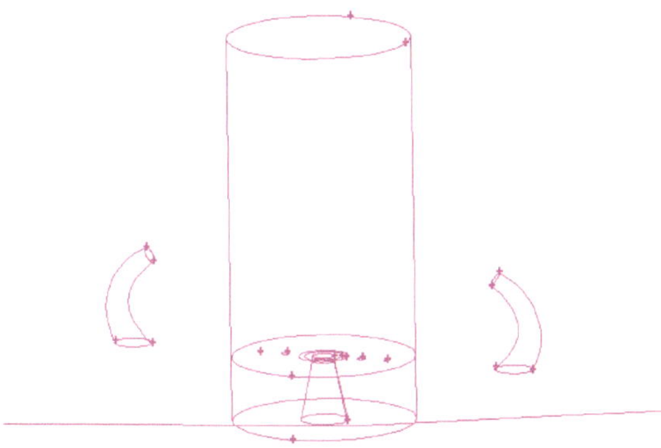

Figure 1. The partially enlarged view of internal and external flow field of the nozzle's simplified model.

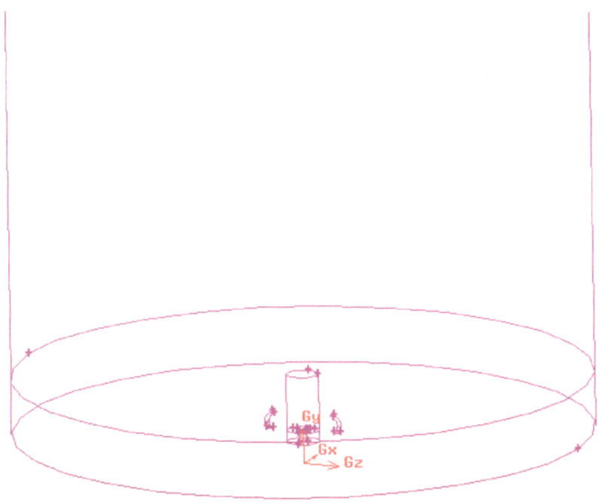

Figure 2. The simplified geometric model of the nozzle's internal and external flow field.

Figure 3. The contour model diagram.

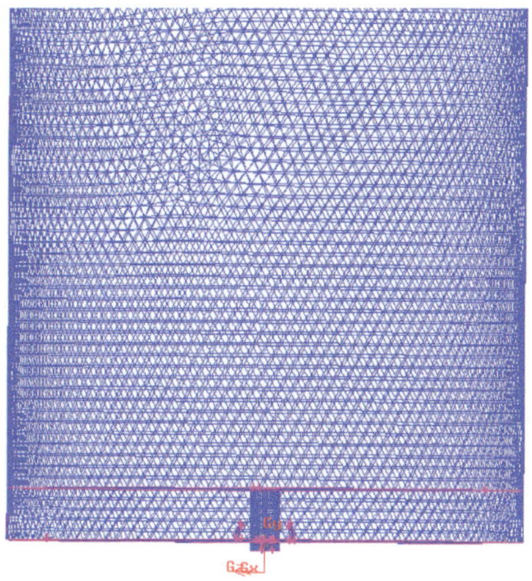

Figure 4. The grid map of the nozzle's internal and external flow field.

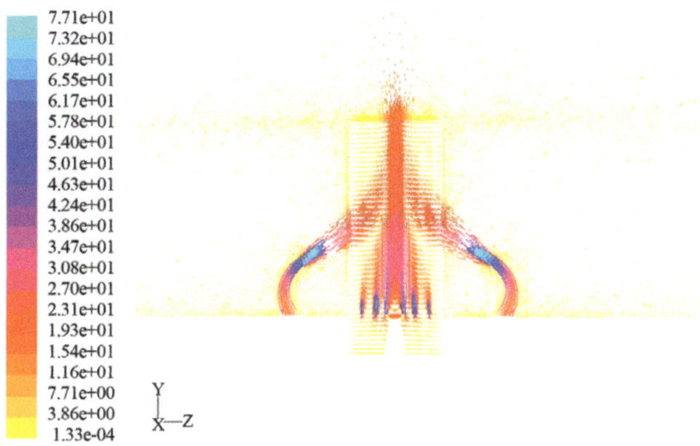

Figure 5. The spray velocity vector diagram.

Provided that the inlet air pressure is 105,325 Pa, the initial particle diameter is 0.03 mm, the initial particle velocity is 10 m/s, we can obtain the cross-sectional velocity vector diagram, the cross-sectional velocity contours diagram, the nozzle atomization particle size diagram when the mass flow rate of the inlet droplets are 0.6 g/s,

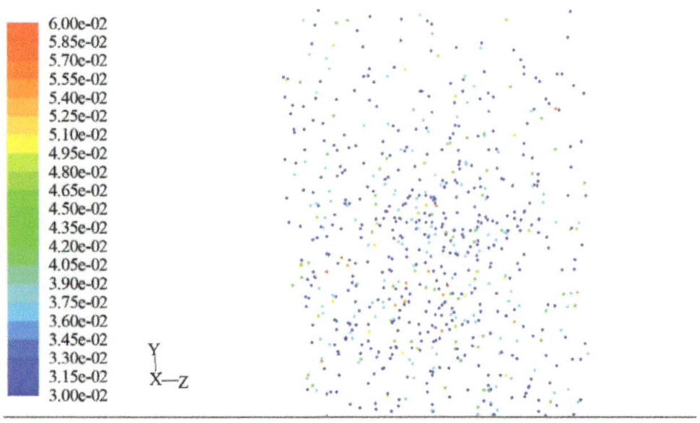

Figure 6. The spray particle size diagram.

0.8 g/s, 1 g/s, 1.5 g/s respectively. Finally, the influence law of atomization particle can be concluded about the mass flow rate of the inlet droplets.

As the **Figure 4** showed, the maximum speed of the spout gas can be 77.1 m/s when the mass flow of the inlet droplets is 0.6 g/s. The atomization particle diameter is among 0.03 mm and 0.06 mm as the **Figure 4** showed.

The maximum speed of the spout gas can be 77.1 m/s when the mass flow of the inlet droplets is 0.8 g/s and the atomization particle diameter is among 0.03 mm and 0.0723 mm. The maximum speed of the spout gas can be 77.1 m/s when the mass flow of the inlet droplets is 1 g/s and the atomization particle diameter is among 0.03 mm and 0.074 mm. When the mass flow of the inlet droplets is 1.5 g/s, the atomization particle diameter is among 0.03 mm and 0.0828 mm.

In summary, when the intake air pressure is constant, the spray particles diameter of the nozzle will increase with the increase of the mass flow of the inlet droplets.

As the **Figure 7** showed, the maximum speed of the spout gas can be 66.8 m/s when the intake air pressure is 104,325 Pa and the atomization particle diameter is among 0.03 mm and 0.102 mm, as the **Figure 8** showed.

The maximum speed of the spout gas can be 77.1 m/s when the intake air pressure is 105,325 Pa and the atomization particle diameter is among 0.03 mm and 0.0723 mm. The maximum speed of the spout gas can be 108 m/s when the intake air pressure is 109,325 Pa and the atomization particle diameter is among 0.03 mm and 0.0624 mm.

In summary, when the mass flow of the inlet droplets is constant, the outgas velocity increases, the atomization effects become well and the particle diameter decrease with the increase of the intake air pressure.

The environment parameters of the experiment are as follows: the atmospheric pressure is 99.50 kPa and the environment temperature is 30.6 degrees Celsius.

The spray particle diameter and particle velocity distribution diagram are showed as **Figure 9** when the mass

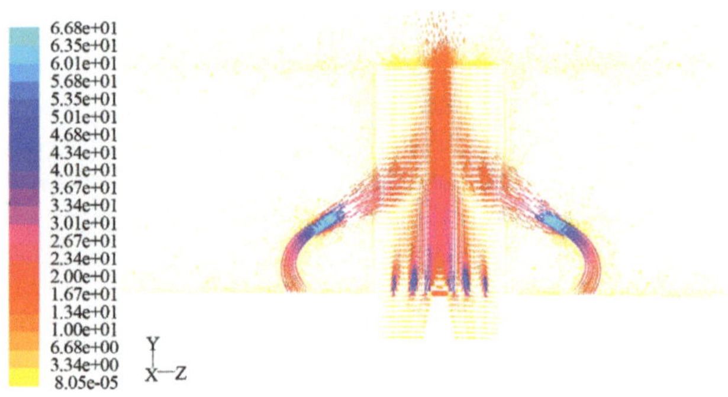

Figure 7. The velocity vector diagram.

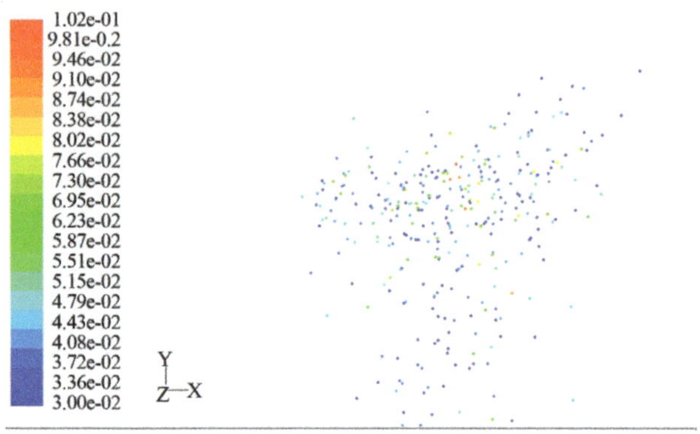

Figure 8. The particle atomization diameter diagram.

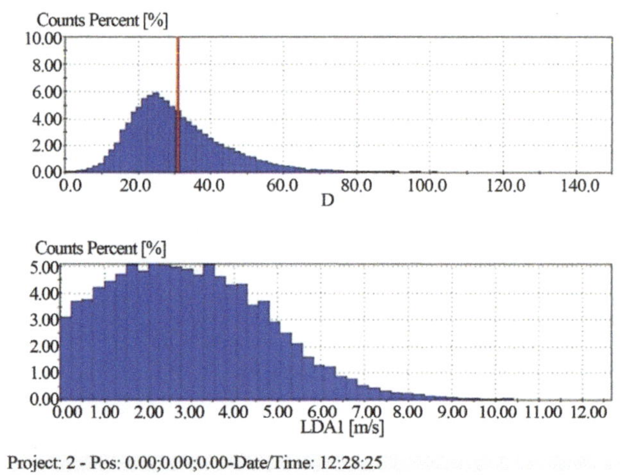

Figure 9. The spray particle diameter and particle velocity distribution diagram.

flow of inlet droplets is 0.013 kg and the intake air pressure is 111.23 kPa.

The spray particle diameter and particle velocity distribution diagram are showed as **Figure 10** when the mass flow of inlet droplets is 0.033 kg and the intake air pressure is 111.30 kPa.

The spray particle diameter and particle velocity distribution diagram are showed as **Figure 11** when the mass flow of inlet droplets is 0.041 kg and the intake air pressure is 111.69 kPa.

The spray particle diameter and particle velocity distribution diagram are showed as **Figure 12** when the mass flow of inlet droplets is 0.060 kg and the intake air pressure is 332.18 kPa.

The spray particle diameter and particle velocity distribution diagram are showed as **Figure 13** when the mass flow of inlet droplets is 0.67 kg and the intake air pressure is 462.81 kPa.

The data also can be expressed as the **Table 1** showed.

From the **Table 1**, the following conclusion can be drawn. That's, from the first three data sets, the average particle diameter increases with the increase of the mass flow of the inlet droplets when the intake air pressure is constant. And from the latter three data sets, the average particle diameter decrease with the increase of the intake air pressure when the mass flow of the inlet droplets is constant.

4. Conclusions

The paper mainly studies the factors of the salt spray

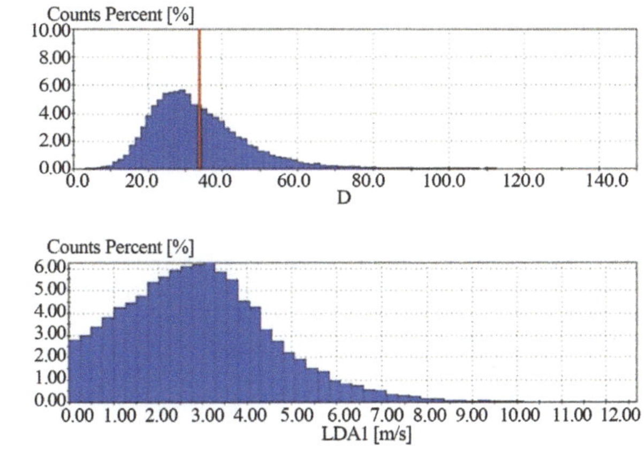

Project: 4 - Pos: 0.00;0.00;0.00-Date/Time: 12:31:10

Figure 10. The spray particle diameter and particle velocity distribution diagram.

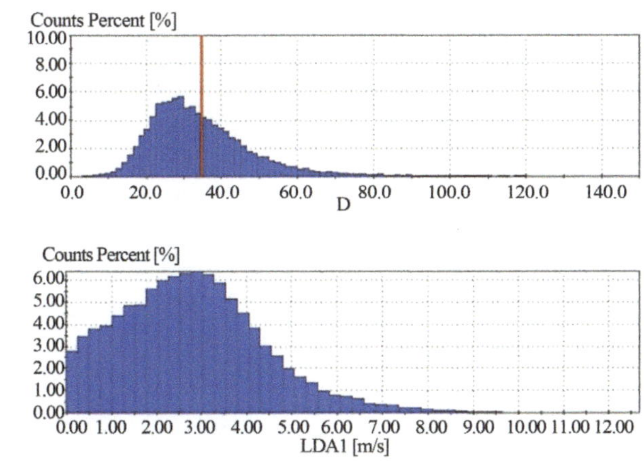

Project: 5 - Pos: 0.00;0.00;0.00-Date/Time: 12:31:45

Figure 11. The spray particle diameter and particle velocity distribution diagram.

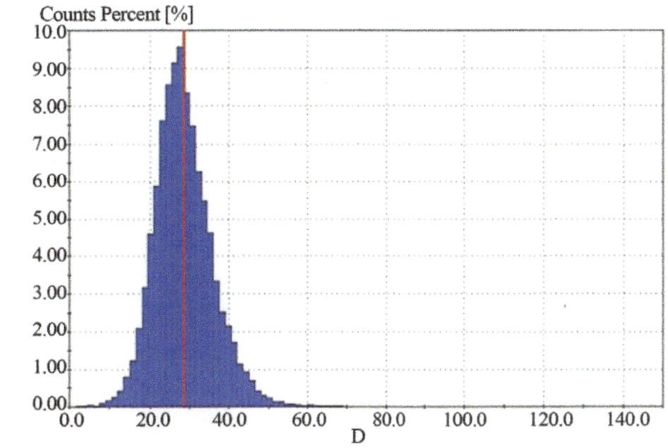

Project: 6 - Pos: 0.00;0.00;0.00-Date/Time: 12:40:26

Figure 12. The spray particle diameter and particle velocity distribution diagram.

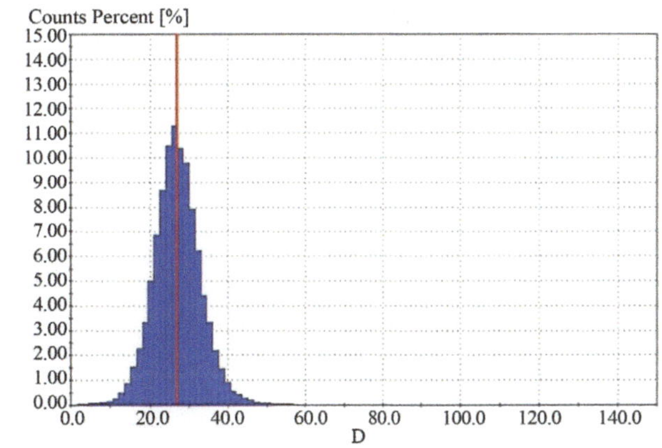

Project: 8 - Pos: 0.00;0.00;0.00-Date/Time: 12:42:18

Figure 13. The spray particle diameter and particle velocity distribution diagram.

Table 1. The distribution table of spray particle diameter.

The flow rate of the inlet droplets (kg)	The intake air pressure (kPa)	The average particle size (μm)
0.013	111.23	44.6
0.033	112.3	50.7
0.041	110.69	54.5
0.06	332.18	32.6
0.064	404.36	30.5
0.067	462.81	29.9

aerosol device to produce different diameter particles. It mainly includes two aspects (the mass flow of the inlet droplets and the intake air pressure). On the basis of the analysis of the nozzle atomization mechanism and the learning of the existing nozzle, the paper conducts the numerical simulation to a given nozzle through the software. By the simulation of the nozzle's internal and external flow field, the flow and air flow characteristics of the nozzle's internal and external flow field are obtained. By the analysis of the cross-sectional velocity vector diagram, the cross-sectional velocity contours diagram, the nozzle atomization particle size diagram, the influence law of salt spray aerosol device to produce different diameter particles can be concluded about the mass flow rate of the inlet droplets and the intake air pressure. Finally, the influence laws of the two factors are verified by the combination of the experiment summary. That's,

1) The nozzle's spray particles diameter increases with the increase of the mass flow of the inlet droplets when the intake air pressure is constant;

2) when the mass flow of the inlet droplets is constant, the outgas velocity increases, the atomization effects become better and the particle diameter decreese with the increase of the intake air pressure.

REFERENCES

[1] P. Aeroso, "Measurement, Priciples, Technologies, and Applications," Chemical Industry Press, Beijing, 2007.

[2] Z. Z. Han, "FLUENT Fluid Simulation Examples and Application," Beijing Institute of Technology Press, Beijing, 2004.

[3] F. J. Wang, "Computational Fluid Dynamics Analysis," Tsinghua Press, Beijing, 2004.

[4] L. J. Wang, "Study of High Pressure Water Mist Nozzle and Fire Extinguishing Performance," Huazhong University of Science and Technology, Wuhan, 2005.

[5] D. F. Che, "A Number of Flow and Its Application," Xi'an Jiao Tong University Press, Xi'an, 2007.

[6] J. Cai, "Air filter ABC," China Building Industry Press, Beijing, 2002.

[7] D. X. Cheng, "Mechanical Design Manual," Chemical Industry Press, Beijing, 1969.

[8] Q. L. Hui, "Wind Tunnel Testing," National Defence Industry Press, Beijing, 2000.

[9] L. Y. Hou and X. C. Hou, "Nozzle Technology Handbook," Sinopec Press, Beijing, 2002.

[10] C. Wu, X. Y. Zhou and G. W. Wu, "Eddy Pressure Nozzle and Its Application in Spray Granulation in Fluidized Bed," *Chemical Equipment Technology*, Vol. 24, No. 2, 2003, pp. 4-7.

19

Measurement of Effective Reading Distance of UHF RFID Passive Tags

Yuan-Ping Luh, Yin-Chang Liu
Department of Mechanical and Electrical Engineering, National Taipei University of Technology, Chinese Taipei

ABSTRACT

Currently, for manufacturing UHF RFID passive tags, the online inspection mechanism is very simple. This mechanism tries to read tag IDs in the near field, and hence validates tag's usability. For UHF RFID applications, tag usability is a very rough indicator and cannot characterize the performance of a tag accurately. In practice, effective reading distance is the key performance index of a tag. This study proposes chip's turn-on power approach to characterize the effective reading distance of UHF RFID passive tags. The experimental results presented in this paper demonstrated the feasibility of this approach. Moreover, in comparison with the large-scale setup, this mini-scale setup produces a smaller error in the estimation of the effective reading distance. Using a mini anechoic chamber, the mini-scale setup can be adopted in practice for online tag performance inspection to grade tag's compliance with effective reading distance.

Keywords: RFID; Turn-On Power; Passive Tag; Performance

1. Introduction

Most tag machinery manufacturers, such as Muehlbauer and Bielomatik [1], provide fundamental online mechanism that tries to read tag IDs in the near field, and hence validates tag's usability. For UHF RFID applications, tag usability is a very rough indicator and cannot characterize the performance of a tag accurately. In practice, effective reading distance is the key performance index of a tag because most online proofed tags are inadequate for implementation due to the insufficient reading distance.

For tag manufactures, to characterize the effective reading distance of an UHF RFID passive tag, the simplest approach is to gradually position the sampled RFID tag away from interrogator's antenna in a large space until 50% or less of signal is received. However, such approach implies low productivity because it is a time consuming process. Moreover, in an opening area, environmental electromagnetic interference may significantly influence the measurement accuracy on effective reading distance [2-4].

To eliminate the environmental electromagnetic interference, the present state-of-the-art approach is to measure the effective reading distance inside a large anechoic chamber [5-7]. However, the use of large anechoic chamber implies high capital investments. In addition, such approach is still a time consuming process.

Since current approaches to characterize the effective reading distance of UHF RFID passive tags are costly and time consuming, an economic approach to accelerate the characterization process is needed. To propose such an innovative approach, this study reviews the traditional approach and the fundamental background of electromagnetic wave propagation in advance. Afterward, the theoretical equations for this innovative approach are derived. Finally, an experimental study is conducted to demonstrate the feasibility of this innovative approach.

2. Theoretical Background of Current Approach

The template is used to format your paper and style the text. All margins, column widths, line spaces, and text fonts are prescribed; please do not alter them. You may note peculiarities. For example, the head margin in this template measures proportionately more than is customary. This measurement and others are deliberate, using specifications that anticipate your paper as one part of the entire journals, and not as an independent document. Please do not revise any of the current designations. To characterize the effective reading distance of an UHF RFID passive tag, the current approach is shown in **Fig-**

ure 1. In this approach, the power of interrogator output (P_T) is the same as the setup during operations, and the initial distance between RFID tag and interrogator's antenna is 0.9 meter. After setup, the interrogator transmits electromagnetic (radio frequency) signal to energize the tag every 500 msec for 30 times in a reading cycle, and therefore receives the ID number returned by the tag. In a reading cycle, if more than 50% of returned signal is received by the interrogator's antenna, the distance between RFID tag and interrogator's antenna will be increased by 0.1 meter for next cycle, until 50% or less of signal is received. The final distance between RFID tag and interrogator's antenna is the effective reading distance of a tag. Note that the approach in **Figure 1** is a time consuming process.

In an opening area, the multi-path propagation effects may be caused by electromagnetic interference reflection. Therefore, the effective reading distance may not be corrected measured [8]. To eliminate the environmental electromagnetic interference and measure the effective reading distance correctly, the RFID tag and interrogator's antenna are better installed inside a large anechoic chamber. The environment inside an anechoic chamber is so-called an ideal free space. Note that the use of large anechoic chamber implies high capital investments. The strength of electromagnetic (RF) signal will be attenuated with the distance through wave propagation [9]. After long distance propagation, the power of electromagnetic (RF) signal may be too weak to energize RFID tags. This causes the failure of reading tag's ID numbers. Theoretically, the amount of power attenuation can be calculated. In an ideal free space, the relation between power attenuation (in dB, P_{Loss}) of electromagnetic (RF) signal with propagation distance (in meter, *d*) is as follows:

$$P_{Loss} = 20 \cdot \log \frac{4\pi \cdot d}{\lambda} \qquad (1)$$

where λ = the wavelength of electromagnetic (RF) wave (in meter) [10].

The transmission path of electromagnetic (RF) signal from an interrogator to a tag is shown in **Figure 2**. The power of electromagnetic (RF) signal received by the tag (P_{Tag}) can be estimated from the power of interrogator output (P_T) as follows:

$$P_{Tag} = P_T - P_{CL} + G_A - P_{Loss} \qquad (2)$$

where P_{CL} = the cable loss

G_A = the interrogator's antenna gain

P_{Loss} = the path loss as calculated by Equation (1).

Additionally, the electromagnetic (RF) signal strength returned from the tag to the interrogator's antenna, as characterized by the Received Signal Strength Indicator (RSSI), can be estimated theoretically as follows:

$$RSSI = P_T - P_{CL} + G_A - P_{Loss} - P_{Loss} = P_{Tag} - P_{Loss} \qquad (3)$$

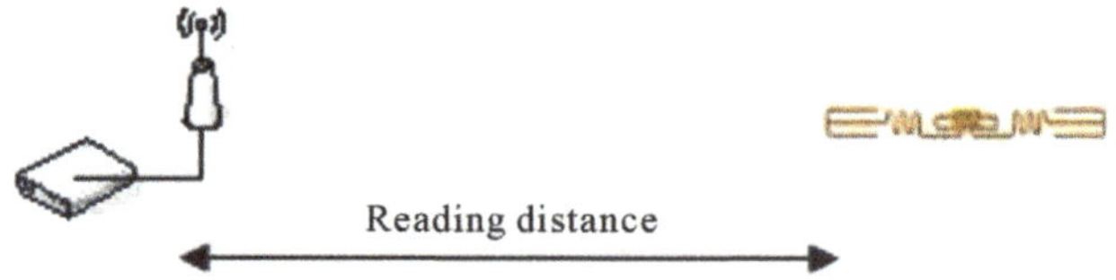

Figure 1. The setup of the traditional inspection.

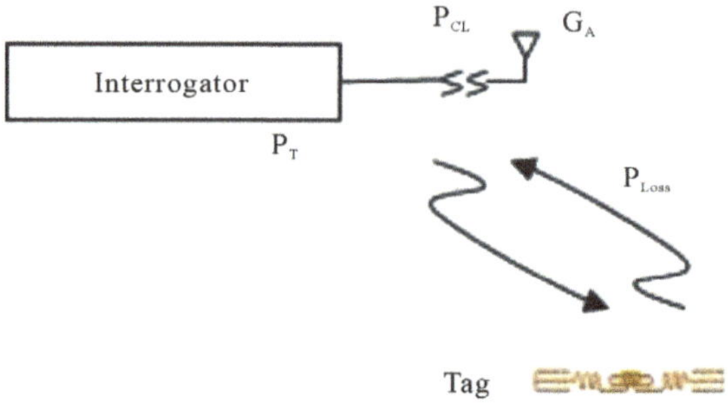

Figure 2. RF propagation path of the RFID UHF Tag. Note. That under the circumstance of reading rate nearly 50% and more, the power of electromagnetic (RF) signal received by a tag is defined as the minimum turn-on power of a tag ($P_{Turn\text{-}on}$).

Equation (3) shows the value of $RSSI_{Tag}$ is reduced through wave transmission. Therefore, the RSSI is often used to estimate the distance between a tag and a interrogator's antenna practically [11,12]. As illustrated in this section, the traditional approach to characterize the effective reading distance of tags is a time consuming process. In addition, the use of large anechoic chamber for characterization implies high capital investments. These two shortcomings result in the traditional approach not suitable for routine inspection. For developing online inspection equipment, space reduction and process simplification are the key issues.

3. Theoretical Background of This Approach

To propose this approach to characterize effective reading distance of a tag, the required space for measurement should be minimized, and the characterization process should be simplified. For space reduction, the static test method prepared by EPCglobal [13] gives enlightenment. Theoretical study in previous section shows that when the distance between RFID tag and interrogator's antenna equals "effective reading distance", the power of electromagnetic (RF) signal received by a tag is approximately $P_{Turn\text{-}On}$. To determine the value of $P_{Turn\text{-}On}$, EPCglobal publishes the static test method as the industrial standards. In accordance with the EPCglobal static test standard, a tag is placed on the platform and facing to the interrogator antenna, as shown in **Figure 3**. Specially, rather than gradually positioning the RFID tag away from interrogator's antenna, a power attenuator is used to simulate the power loss caused by transmission distance.

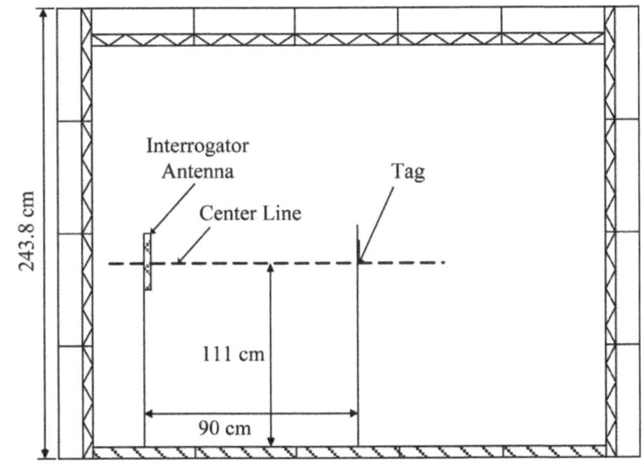

Figure 3. The setup of EPCglobal static test method.

Hence, the distance between tag and interrogator's antenna is fixed to 0.9 meter during experiment. Even though 0.9 meter is still not small enough for building on-line inspection equipment; however, the use of power attenuator seems to be a feasible way for space reduction.

Physically, a tag can be placed on any location as close to interrogator's antenna as possible. However, if the distance between tag and interrogator's antenna is too short, the phenomenon of near field electromagnetic interference will be occurred. On the basis of distance to radiation source, the electromagnetic field is classified into three regions: reactive near field, radiating near field and far-field [14]. Theoretically, the radiation wavelength (λ) and the antenna width (D) determine the range of near field. For the distance to radiation source less than $\lambda/2\pi$ or $2D^2/\lambda$, the near field reactive or radiating effects will occur. Practically, to avoid the near field interference, the distance to radiation source longer than one wavelength is suggested. The relation between wavelength (λ) and frequency (f) is as follows:

$$\lambda = \frac{c}{f} \tag{4}$$

where c = the speed of light (3×10^8 m/sec).

In most cases, the frequency of 915 MHz is adopted by the UHF RFID passive applications. The wavelength for 915 MHz RF signal is 0.32 meter. Therefore, to avoid the near field interference, the distance between UHF RFID tag and interrogator's antenna should be more than 0.32 meter. In this study, 0.35 meter is selected for further study. Such size is smaller enough for building on-line inspection equipment. To characterize the effective reading distance of tags in such mini space, a power attenuator is used to simulate the power loss caused by long transmission distance. The power gap (P_{gap}) induced by transmission distance and power attenuator can be expressed as follows:

$$P_{gap} = P_{Loss0.35} + P_{att} = 22.6 + P_{att} \tag{5}$$

where $P_{Loss0.35}$ = the path loss for d = 0.35 meter = 22.6 dB by Equation (1).

P_{att} = the magnitude of power attenuation.

With power attenuator, by modifying Equations (2) and (3), the P_{Tag} and RSSI can be estimated theoretically as follows:

$$P_{Tag} = P_T - P_{CL} + G_A - 22.6 - P_{att} \tag{6}$$

$$RSSI_{0.35\,m} = P_{Tag} - 22.6 - P_{att} \tag{7}$$

Note. That under the circumstance of reading rate nearly 50% and more, the power of electromagnetic (RF) signal received by a tag is defined as the minimum turn-on power of a tag ($P_{Turn\text{-}On}$). Based on Equation (6), the circumstance of $P_{Turn\text{-}On}$ is achieved by adjusting P_{att}, and the corresponding P_{gap} can be calculated from Equation (5). By substituting P_{gap} to P_{Loss} in Equation (1), the equivalent distance (d) for P_{gap} is obtained. Such equivalent distance is the tag's effective reading distance as characterized in 0.35 meter ($RD_{0.35\,m}$).

Even though the size for characterization is reduced with the assistance of using attenuator, it is still a time consuming process to adjust power attenuation, P_{att}, to achieve the circumstance of reading rate nearly 50% and more. Therefore, for on-line inspection, the value of P_{att} has to be unchanged. Such setup is feasible for online characterization because customers only care about the effective reading distance guaranteed by manufacturers instead of the maximum reading distance exactly. For the declared effective reading distance by a manufacturer, the value of P_{att} is assigned in prior to tag production. Theoretically, an interrogator can send electromagnetic (RF) signal 1000 times per second, and can energize a tag within 0.1 second. Therefore, the inspection time for a tag can be reduced to less than 1 second practically.

In this section, the minimum required space for measuring effective reading distance is determined, and the simplified characterization process is also specified. Such approach is proposed for developing on-line inspection equipment. However, the correctness of using power attenuator in such limited space to measure effective reading distance is not confirmed yet. Hence, further experimental study is designed for verification.

4. Design of Experimental Study

To verify the feasibility of proposed innovative approach, an experimental study has been designed. In the experimental study, the effective reading distance of a tag will be characterized by two setups. One uses the large-scale setup, as shown in **Figure 1**. The other uses the mini-scale setup with a power attenuator, as shown in **Figure 3**. All experimental works are performed in an anechoic

Table 1. Traditional test in large-scale.

Tag #	DogBone		ShortDipole	
	RD_{Max} (m)	RSSI (dB)	RD_{Max} (m)	RSSI (dB)
1	7.25	−62.9	6.65	−62.4
2	7.15	−63.3	6.55	−61.7
3	7.25	−63.5	6.65	−62.3
4	7.20	−62.8	6.60	−61.6
5	7.20	−63.7	6.55	−61.7
6	7.10	−62.4	6.50	−62.1
7	7.25	−63.6	6.40	−61.7
8	7.15	−63.3	6.55	−61.7
9	7.10	−62.8	6.60	−61.9
10	7.15	−63.2	6.65	−62.3
Avg	7.18	−63.15	6.57	−61.94
Dev	0.06	0.41	0.08	0.31

chamber to eliminate the environmental electromagnetic interference. If the effective reading distances of a tag as determined by different setups are close enough, the feasibility of proposed innovative approach can be approved.

In the experimental study, two different types of UHF RFID passive tags are chosen for testing. One is the UPM DogBone and the other is the UPM ShortDipole. Both types use the Impinj Monza 3 chip. The minimum sensitivity of the Impinj Monza 3 chip is −15 dB [15]. The maximum reading distance declared by UPM is 9.2 meters for DogBone, and is 8.8 meters for ShortDipole [16,17]. The quantity of each tag type for experiment is 10. The adopted RFID UHF interrogator for experiment is Alien9780 [18]. The power of interrogator output (P_T) is set to 30 dB throughout the experiment. The interrogator antenna used in experiment is Alien ALR9610-BC [19]. The antenna gain (G_A) of Alien ALR9610-BC is 5.73 dB. The cable length to connect interrogator and antenna for experiment is 3 meter, and the power loss induced by cable transmission (P_{CL}) is 2.55 dB approximately.

By the large-scale setup, the effective reading distance of a tag and the associated RSSI are measured directly. The initial distance between RFID tag and interrogators antenna is 2 meters. After setup, the interrogator transmits electromagnetic (radio frequency) signal to energize the tag every 500 msec for 30 times in a reading cycle, and therefore receives the ID number returned by the tag. In a reading cycle, if more than 50% of returned signal is received by the interrogator's antenna, the distance will be increased by 0.5 meter for next cycle, until 50% or less of signal is received. After attaining such critical condition, the effective reading distance can be further determined by reducing the distance by 0.05 meter per cycle, until almost exact 50% of signal is received. By the mini-scale setup, the distance between RFID tag and interrogator's antenna is fixed to 0.35 meter. In addition, the power attenuation of attenuator (P_{att}) is set to 25 dB initially. The adopted attenuator for experiment is KAY 1/839 [20]. After initial setup, the interrogator transmits electromagnetic (radio frequency) signal to energize the tag every 500 msec for 30 times in a reading cycle, and therefore receives the ID number returned by the tag. In a reading cycle, if more than 50% of returned signal is received by the interrogator's antenna, the power attenuation of attenuator (P_{att}) will be increased by 1 dB for next cycle, until 50% or less of signal is received. After attaining such critical condition, the final value of P_{att} can be further determined by reducing the power attenuation by 0.1 dB per cycle, until almost exact 50% of signal is received. With P_{att}, the P_{gap} and $P_{Turn\text{-}On}$ are determined by Equations (5) and (6). Moreover, the effective reading distance of a tag and the associated RSSI are derived by Equation (1) with P_{gap} and Equation (7), respectively.

5. Experimental Results and Analysis

The traditional test results are shown in **Table 1**. The experimental results for two different types of UHF RFID passive tags are shown in **Tables 2** and **3**. As shown in **Table 1**, the effective reading distance of a tag (RD_{Max}) and the associated RSSI are measured directly by the large-scale setup. Differently, by the mini-scale setup, the effective reading distance of a tag ($RD_{0.35\ m}$) and the associated RSSI ($RSSI_{0.35\ m}$) are derived from the measured power attenuation of attenuator (P_{att}) by Equations (1), (5) and (7).

In **Tables 2** and **3**, the mean value of $RD_{0.35\ m}$ is 7.13 meters for the tags of UPM DogBone type, and is 6.50 meters for the tags of UPM ShortDipole. Furthermore, the mean value of $RSSI_{0.35\ m}$ is −64.36 dB for the tags of UPM DogBone type, and is −62.74 dB for the tags of UPM ShortDipole. In comparison with the large-scale setup, the results by the mini-scale setup only have less than 0.7% error on RDMax and less than 1.9% error on RSSI for DogBone type. And in comparison with the large-scale setup, the results by the mini-scale setup only have less than 1.1% error on RDMax and less than 1.3% error on RSSI for ShortDipole type. The experimental result shows the correctness of proposed innovative mini-scale setup can be approved. It is noteworthy that the $RD_{0.35\ m}$ and $RSSI_{0.35\ m}$ has less deviation than RD_{Max} and RSSI, as shown in **Tables 2** and **3**. Such consequence implies that the measurement consistency is better by the mini-scale setup than by the large-scale setup.

Table 2. Experimental results for DogBone tags.

Tag #	Mini-Scale (0.35 m)				
	P_{att} dB	P_{gap} dB	$P_{Turn\text{-}On}$ dB	RD m	RSSI dB
1	26.3	48.9	−15.38	7.09	−64.28
2	26.3	48.9	−15.38	7.09	−64.28
3	26.4	49.0	−15.48	7.18	−64.48
4	26.4	49.0	−15.48	7.18	−64.48
5	26.3	48.9	−15.38	7.09	−64.28
6	26.4	49.0	−15.48	7.18	−64.48
7	26.3	48.9	−15.38	7.09	−64.28
8	26.3	48.9	−15.38	7.09	−64.28
9	26.4	49.0	−15.48	7.18	−64.48
10	26.3	48.9	−15.38	7.09	−64.28
Avg	26.3	-	-	7.13	−64.36
Dev	0.05	-	-	0.05	0.10

Table 3. Experimental results for ShortDipole tags.

Tag #	Mini-Scale (0.35 m)				
	P_{att} dB	P_{gap} dB	$P_{Turn\text{-}On}$ dB	RD m	RSSI dB
1	25.6	48.2	−14.68	6.56	−62.88
2	25.5	48.1	−14.58	6.47	−62.68
3	25.6	48.2	−14.68	6.56	−62.88
4	25.5	48.1	−14.58	6.47	−62.68
5	25.5	48.1	−14.58	6.47	−62.68
6	25.5	48.1	−14.58	6.47	−62.68
7	25.4	48.0	−14.48	6.40	−62.48
8	25.5	48.1	−14.38	6.47	−62.68
9	25.6	48.2	−14.68	6.56	−62.88
10	25.6	48.2	−14.68	6.56	−62.88
Avg	25.5	-	-	6.50	−62.74
Dev	0.07	-	-	0.06	0.13

By the mini-scale setup, the minimum turn-on power of a tag ($P_{Turn\text{-}On}$) is derived from the measured power attenuation of attenuator (P_{att}) by Equation (6). Theoretically, the power magnitude to energize a tag is limited by the sensitivity of the chip used inside a tag. Due to some tag package limitations, such as material impedance and characteristic differently, the power magnitude to energize a tag will be higher than the chip's sensitivity in general. Hence, for UPM ShortDipole cases in **Table 3**, it seems reasonable that the calculated $P_{Turn\text{-}On}$ are slightly higher than the chip's sensitivity (−15 dB). In **Table 2**, the calculated $P_{Turn\text{-}On}$ are slightly lower than the chip's sensitivity (−15 dB). Nevertheless, such experimental result is still reasonable since tag's antenna gain effect and bonding loss are not considered for calculating the $P_{Turn\text{-}On}$ in this study. Tag's antenna gain can enlarge the transmitted power form interrogator to energize a tag. In general, the designed antenna gain of tag is from 0.8 dB to 1.2 dB.

6. Conclusions

To measure the effective reading distance of an UHF RFID passive tag, the present state-of-the-art approach is to gradually position the RFID tag away from interrogator's antenna in a large anechoic chamber until 50% or less of signal is received. However, such approach is costly and time consuming. This study proposes a new approach to accelerate the characterization process. In this approach, rather than gradually positioning the RFID tag away from interrogator's antenna, a power attenuator is used to simulate the power loss caused by transmission distance. With power attenuator, this study claims that the distance between tag and interrogator's antenna can be reduced to 0.35 meter. Also, by experimental study, the equilibrium between mini-scale setup and large-scale setup is verified. Therefore, the feasibility of proposed new approach is confirmed. Using a mini anechoic chamber associated with the simplified characterization process, this mini-scale setup can be adopted in practice for online tag performance inspection to grade tag's compliance with effective reading distance. The mini-scale layout setup is shown in **Figure 4**.

The IOT network in the future, there will be many passive sensors need to build. These sensors in the manufacturing and assembly process will face the read performance QC issues. Great potential in this study can be

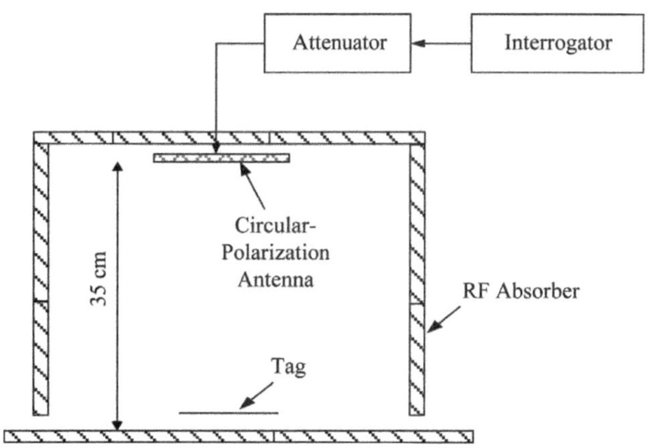

Figure 4. The setup of the inspection.

used to assist the industry to solve such problems.

REFERENCES

[1] ABI research, "The RFID Passive Label Market," 2007, pp. 1-8.

[2] X. M. Qing, C. K. Goh and Z. N. Chen, "Impedance Characterization of RFID Tag Antennas and Application in Tag Co-Design," *IEEE Transactions on Microwave Theory and Techniques*, Vol. 57, No. 5, 2009, pp. 1268-1274.

[3] F. Fuschini, C. Piersanti, L. Sydanheheimo, L. Ukkonen and G. Falciasecca, "Electromagnetic Analyses of Near Field UHF RFID Systems," *IEEE Transactions on Antenna and Propagation*, Vol. 58, No. 5, 2010, pp. 1759-1770.

[4] K. Weigelt, M. Hambsch, G. Karacs, T. Zillger and A. C. Hubler, "Labeling the World: Tagging Mass Products with Printing Processes," *IEEE Pervasive Computing*, Vol. 9, No. 2, 2010, pp. 59-63.

[5] L. Cabria, J. A. Garcia, E. Malaver and A. Tazon, "A PHEMT Frequency Doubling Active Antenna With BPSK Modulation Capability," *IEEE Antennas and Wireless Propagation Letters*, Vol. 3, 2004, pp. 310-313.

[6] A. Cataldo, G. Monti, E. D. Benedetto, G. Cannazza, L. Tarricone and L. Catarinucci, "Assessment of a TD-Based Method for Characterization of Antennas," *IEEE Transactions on Instrumentation and Measurement*, Vol. 58, No. 5, 2009, pp. 1412-1419.

[7] G. Cerri, V. M. Primiani, C. Monteverde and P. Russo, "A Theoretical Feasibility Study of a Source Stirring Reverberation Chamber," *IEEE Transaction on Electromagnetic Compatibility*, Vol. 51, No. 1, 2009, pp. 3-11.

[8] X. H. Cao and H. B. Xiao, "Propagation Prediction Model and Performance Analysis of RFID System under Metallic Container Production Circumstance," *Microelectronics Journal*, Vol. 42, No. 2, 2011, pp. 247-252.

[9] C. Floerkemeier and S. Sarma, "RFIDSim—A Physical and Logical Layer Simulation Engine for Passive RFID," *IEEE Transactions on Automation Science and Engineering*, Vol. 6, No. 1, 2009, pp. 33-43.

[10] T. Rappaport, "Wireless Communications," Prentice Hall, Upper Saddle River, 1996.

[11] C. C. Chen, D. C. Mao and C. Lee, "ART: A Novel Localization Method for Managing Uncertainty Region in Wireless Fading-Signal Sensor Networks," *Ubiquitous Computing and Communication Journal*, Vol. 2, No. 3, 2007, pp. 30-41.

[12] H. S. Ahn and W. Yu, "Environmental-Adaptive RSSI-Based Indoor Localization," *IEEE Transactions on Automation Science and Engineering*, Vol. 6, No. 4, 2009, pp. 626-633.

[13] EPCglobal Inc., "Static Test Method for Applied Tag Performance Testing Rev 1.9.4," 2008.

[14] H. Fan, "Using Radiating near Field Region to Sample Radiation of Microstrip Traces for Far Field Prediction by Genetic Algorithms," *IEEE Microwave and Wireless Components Letters*, Vol. 19, No. 5, 2009, pp. 272-274.

[15] Impinj Inc., "UHF Gen 2 RFID Tag Chip Monza™ 3 Product Brief," 2008.

[16] UPM Raflatac Inc., "DogBone EPC Class 1 Gen 2 Data Sheet: 3001571," 2010.

[17] UPM Raflatac Inc., "ShortDipole EPC Class 1 Gen 2 Data Sheet: 3001488," 2010.

[18] Alien Technology Corp., "ALR-9780 Hardware Setup Guide, Rev. D," 2005.

[19] Alien Technology Corp., "ALR-9800 Hardware Setup Guide, Rev. B," pp.3, 2006.

[20] https://www.valuetronics.com/Used_Kay_1_839.aspx

The Development of Decision-Making Maps for the High-Speed Machining (HSM) of Rhyodacite (Carijó Basalt) in Milling-Based Processing Applications

Hélio Dorneles Etchepare, Wilson Kindlein Júnior
PPG3M, Federal University of Rio Grande do Sul, Porto Alegre, Brazil

ABSTRACT

The demand for innovative products that overcome competitive pricing and conquer market shares can be met by interdisciplinary approaches that bridge product design, engineering and technology. Ornamental and covering stones stand out among the materials in need of innovation and are commonly used in architecture and construction as coverings in urban elements, funerary art, and art and decor. In spite of the many applications of ornamental stones, products with low added value are typically observed. Although Brazil displays significant geological potential for ornamental stones, its actual market participation is as a supplier of un- or semi-processed stones. Even with the market's visible expansion and Brazil's increased representation in the market, the technological advances used in the final processing of stones are restricted to improve the tools used for cutting, especially the durability of diamond-coated tools, and for polishing, which, in the short-term, meets the demands for marketing, producing and distributing semi-finished products. Accordingly, a little exploited aspect of ornamental stones appears, especially for "carijó basalt" (rhyodacite), that is, the inclusion of new processing and value-adding technologies or the inclusion of non-conventional processes in the sector. The goals of this work are to study and develop HSM (high speed machining) CNC (computer numerically controlled) milling processes that are applicable to rhyodacite, specifically to determine the milling parameters that give the best results for machining time with respect to tool wear and material abrasion and to replicate the results obtained in the samples. The results show that the raw material used in the present study, because it is natural, presents significant variations in composition and hardness, which prevent specific milling parameters from being determined. However, using a post-processor specifically developed for this study, it was possible to draft a decision-making map that aids in the execution of this process. Additionally, equipment failure occurred during every attempt and with every adjustment of the milling process. This indicates that the application of multipoint tools to rhyodacite milling is difficult in this industry. Thus, the use of single-point tools should be the natural path to follow for a potential practical industrial application in Brazil.

Keywords: Technology; Rocks; Processing; Multidisciplinarity

1. Introduction

According to Brazil's Ministry for Mines and Energy (MME) [1], in a report from 2001, the country's covering rock sector can be divided into 18 productive groups composed of entrepreneurial activities in 7 states and 80 municipalities. The main product in the state of Rio Grande do Sul is basalt. The report states that Brazil has 11.500 companies in this sector, providing 120.000 direct jobs and possessing facilities capable of processing from 40 to 50 m^3/year, which represents a production of 6.0 million tons/year. Basalt stands out, with approximately 7% of this total. According to Bizzi (2003), the commercial transactions in the sector, including machinery, equipment and raw materials, corresponded to approximately US$ 2.5 billion in 2003 [2].

Carijó basalt has been used as a covering material in construction and is extracted mainly from a region in the northeast of the Rio Grande do Sul state, known as the "Basalt Region," due to the significant number of extraction and processing companies present there. This "Basalt Region" encompasses, according to the Basalt Magazine (Syndicate of the Quarry Exploitation Industry of Nova Prata and Region, 2000) [3], the municipalities of André da Rocha, Guabijú, Nova Araçá, Nova Bassano, Nova Prata, Paraí, Protásio Alves, São Jorge, Veranópolis, Vila Flores and Vista Alegre do Prata (**Figure 1**).

According to data made available by the Chamber of Industry and Commerce (Câmara de Indústria e Comércio—CIC) on the Basalt Region, the gross revenue generated from extracting and processing this material is 501.3 million Brazilian Reais, of which approximately 52.7 million Brazilian Reais are paid as taxes and 37 million Brazilian Reais are paid as salaries. Additionally, the an export volume is approximately 48 million Brazilian Reais, all of which make this industry one of the most important industries in the region due to its significant participation in the state- and country-level economies [4].

There are 300 mining areas in the region, according to data published in the journal Socio-Economical Hierarchy of the Basalt Region. In the municipality of Nova Prata alone, also known as the National Basalt Capital, there are 104 registered mining and 89 basalt processing companies, which represent 10% of the municipality's economy and provide approximately 3000 jobs, both direct and indirect. **Table 1** shows the main products of basalt processing in Nova Prata, as well as their volumes and corresponding fractions [4].

Given the latent potential of using computer numerically controlled techniques in rock processing applications, in the present article, we seek to systematically study the viability of using HSM milling in this context and to determine the best machining parameters for this purpose.

2. HSM Milling

According to Dieter (1981) [5], in general, two distinct manufacturing classes are utilized in the solid state, including deformation processes, in which the desired shape is produced through plastic deformation of the materials while preserving their initial volume, and machining processes, in which part of the material is removed through specific processes to produce the target shape. Machining processes are used to produce parts with high dimensional tolerance, good surface finishes and relatively complex shapes. Currently, significant technological developments have been achieved in machining processes, particularly with the addition of computer numerically controlled (CNC) and high speed machining (HSM) techniques, which guarantee precision and repeatability of movement. One of the machining areas that has benefited most significantly from these developments is, according to Faller *et al.* (2006), the field of milling processes [6].

In HSM milling, the machine (**Figure 2**) interprets

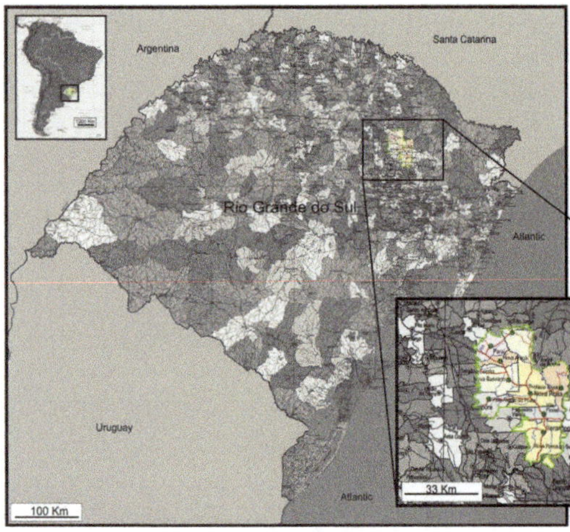

Figure 1. The Basalt Region in Rio Grande do Sul.

Table 1. The products obtained from basalt.

Main products	*Volume produced* (m^3*/year*)	(%)
Paving stone	30,000	50
Slabs	6000	10
Irregular pieces of stone paving	12,000	20
Foundation stones	6000	10
Other	6000	10
Total	60,000	100

Adapted in 08/01/2011 from the Chamber of Industry and Commerce, Basalt Region, 2010.

Figure 2. The CNC Digimill 3D milling centre. Adapted in 11/28/2012 from http://www.tecnodrill.com/prod_03.htm.

digital instructions and activates a set of servos and stepper motors that control each axis of the machine (X, Y and Z in the case of 3-axis milling machines), moving the table or the tool in an automatic and interpolated manner (movement can be made simultaneously along all axes). Some or all of the numerical data are converted during the process, e.g., distances, angles, speeds, etc., which allows almost any shape to be conceived/manufactured in real applications. The arrival of this technology was only possible after the necessary tools, automation and informatics were invented.

Any machining process depends critically on the material from which the tool is made and the interaction of the machining forces with the workpiece, where the mechanical properties of the tool, especially its hardness, should be superior to those of the raw material (the workpiece to be machined).

The development of machining tools has generally been accompanied by technological developments, both with respect to the specific characteristics of the raw materials and the specialisation of manufacturing processes. Machining tools generate surfaces in two ways: by using a profile tool, the shape of which resembles the shape of the surface to be produced, or by advancing the tool along the part's length. Profile tools are relatively more expensive but are important in line production applications. Machining parameters, such as the cutting depth, the X-, Y- and Z-speeds, and the spindle[1] rotation speed, all of which vary according to the material and tool, as noted by Freitas (2006) [7], should be considered to prevent excessive tool and/or equipment wear. Most cutting tools are made of cast steel or hard metal, according to Dieter (1981), but there are also tools made of fused alloys and others possessing diamond coatings, with the latter being more suited in the processing of ornamental stones.

Figueira (2006) [8] points out that diamond-coated tools are the most commonly employed tools in the blasting, cutting and polishing processes for ornamental rocks, ceramics and non-ferrous materials. According to Stemmer (2008) [9], the extraordinary hardness of diamond, which exceeds that of any other material, can vary between 5000 and 7000 on the Knoop scale, or 56 to 102 GPa (giga-Pascal), depending on the crystal's orientation. The high thermal conductivity of diamond can hinder the performance of polycrystalline tools because of the premature tearing of the abrasive grains due to the absorption by the metallic matrix of the heat generated from the friction between the workpiece and the tool. The potential usefulness of numerical control in stone processing has prompted the discussion of applying HSM milling to cutting applications for this type of material to verify the best conditions for its application.

[1]device that provides the tool with rotation.

3. Materials and Methods

According to the Brazilian Geological Service, CPRM [10], the stones used in this study, extracted from mining areas in the Basalt Region, are part of the Geological Province of Paraná flood basalts, the Caxias Facies, which formed in the Cretaceous period between 65 and 135 million years ago. This facies consists mainly of massive basalt flood of intermediate to acidic composition, mafic and with fine granular texture and lamellar structures, which explains why extraction is performed in the horizontal orientation.

According to Motoki (2004) [11], the plateau that is stratigraphically called the Serra Geral mountain range, spanning across the states of Santa Catarina and Rio Grande do Sul, is composed mainly of basalt lava with high iron and low silica contents. However, at the top of these mountains, the basaltic lava is covered by rhyodacite, a volcanic rock with low iron and high silica contents, which is often mistaken for and commercialised as basalt [12].

Three plates of this material were acquired to better assess its characteristics and to obtain an overall better understanding of its nature: one measuring 95 × 60 × 2 cm^3, labelled "control," and two smaller plates, labelled "random," one measuring 74 × 11 × 2 cm^3 and the other measuring 45 × 15 × 2 cm^3. The plates were obtained from different quarries in the Basalt Region. The control plate, which was visually darker than the other plates, was mapped and subdivided into 425 specimens measuring 3 × 3 × 2 cm^3. The random plates, both exhibiting a red tone, were subdivided into 93 samples of 3 × 3 × 2 cm3 without control of specimen position. The test bodies were washed to remove surface dust and were then dehumidified for 24 hours (**Figure 3**).

A set of 22 samples was selected to characterise the rock with respect to hardness using Knoop's test and chemical composition using X-ray diffraction analysis. Of these samples, 11 were taken from the control plate and 11 from the random plates, specifically, samples B2, B9, B17, B24, I5, I13, I21, P2, P9, P17 and P24 from the control plate and samples 1, 9, 17, 25, 33, 49, 57, 65, 73, 81 and 89 from the random plates. For the X-ray fluorescence assay, a subset of 11 samples was taken from the aforementioned set, namely, samples I7, B2, B17, I5, I13, I21, P9, P17, and P24 (control) and 1 and 9 (random).

The results from the hardness tests, X-ray diffraction and fluorescence analyses and petrography indicated that it was not necessary to classify the samples prior to the machining tests given their homogeneity. For the rhyodacite tests, the machining parameters used for steel were taken as the initial reference values because of the significant hardness and well-known parameters of this material. However, because diamond-coated (polycrystal-

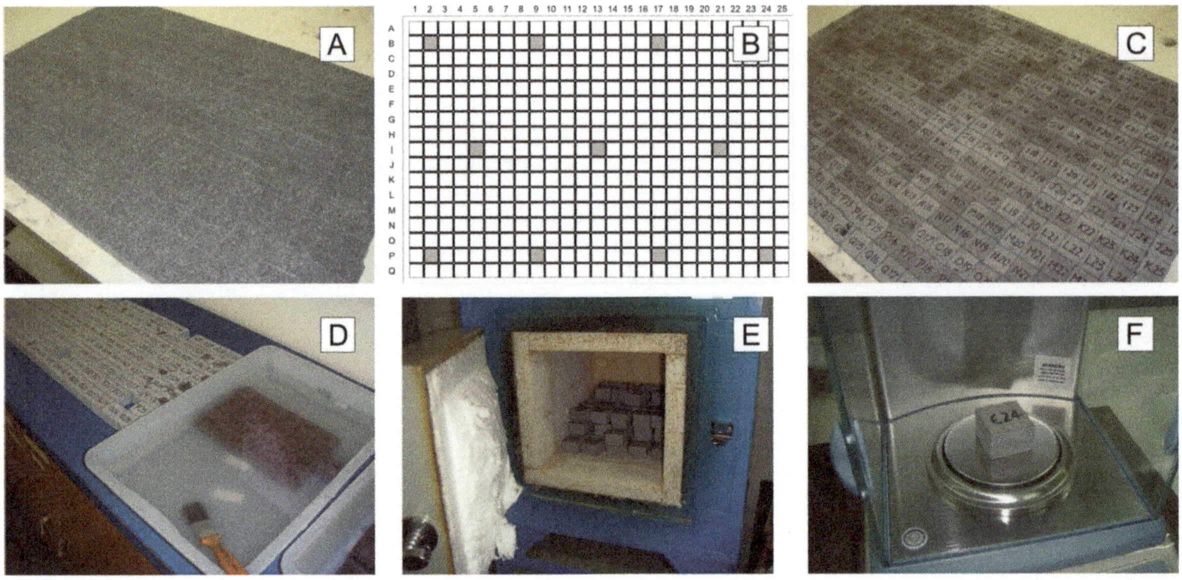

Figure 3. The control plate (A) was mapped (B) and subdivided into 425 specimens of 3 × 3 × 2 cm^3 (C). The samples indicated on the map (B) were analysed for hardness using Knoop's test, X-ray diffraction and fluorescence and petrography. All samples were washed (D) and dehumidified by evaporation for 24 h (E), and their masses were assessed using a digital analytical scale (F).

line with nickel matrix) tools were intended for use, the spindle rotation was altered to 23,000 rpm.

Additionally, to minimise possible variations, 100 tools of the KG Sorensen® brand model 3018 batches 2266 and 5359 were purchased. The tools were classified according to their diameter, measured with a Mitutoyo® digital calliper accurate to the 0.001 mm. The diameters of the tools varied between 2.766 and 3.005 mm, with an average diameter of 2.911 mm and a standard deviation of 0.04 mm. By statistically considering 95% of these measurements to be accurate, tools with diameters smaller than 2.81 mm were discarded. The mass of each tool was measured using a precision scale and was found to vary between 0.3193 g and 0.3372 g, with an average weight of 0.3219 g. Measurements of the mass and length of each tool were used to determine their wear relative to the established machining parameters.

4. Rhyodacite Milling Tests

To initiate testing, all rhyodacite samples were calibrated according to their planeness, and a fixing support, levelled with the help of a dial gauge, was adapted to the milling centre to allow the samples to be quickly exchanged. Additionally, two ISO 15488-A collet chucks were purchased; these chucks were capable of holding tools with shanks between 1 and 2 mm in diameter. The goal of such procedures was to optimise the setup times. All proposed patterns were milled according to the parameters given in **Figure 4**.

All tests were performed using water-soluble mist coolant (*Unix Solúvel* 100 for light machining) diluted at the ratio of 1/19, as recommended by the manufacturer. For all patterns except number 4, the tests were performed with one tool each, and no loss of mass was detected; for pattern 4, however, it was necessary to use 4 tools in separate tests, for which the speed was gradually reduced to 6.38 mm/sec, 25% below the standard speed used in the steel tests and other rhyodacite tests. None of these procedures was successfully concluded. Because of the lack of success for pattern 4, due to the excessive wear of the lower end of the tools being used, which was caused by the significant number of plunges into the sample, an alternative strategy was tested, namely, *raster* machining, originally from the X-axis to the Y-axis. This change in the machining direction reduced the number of vertical, *i.e.*, Z-axis, displacements of the tool from 486 to 126, thus significantly decreasing the impact on the sample. Maintaining the same parameters as that used for steel, the machining time based on this new strategy, as estimated by the ArtCAM® software, would be 57 min and 53 sec, which corresponds to a reduction of 9 min and 58 sec in the total machining time.

New assays were performed, at first excluding the displacement variables in X and Y, to determine the speed along the Z-axis, which wears the tool the least. It was possible to conclude through these tests that the Z-axis speed should be decreased and possibly that the speeds along the X- and Y-axes should be increased to optimise the processing time and minimise tool wear. This conclusion led to a new proposal for speed control, namely, independent control along the X-, Y- and Z-axes, which was not possible in the ArtCAM® and EdgeCAM® CAM

Description	Plaster Manufacturing	"Carijó Basalt" Manufacturing
Pattern 1 – flat and smooth geometry - strategy - raster - final depth - 3 mm - horizontal step - 0.2 mm - vertical step - 0.5 mm - feed speed - 12 mm/seg. - plundge speed - 4 mm/seg. - rotation speed - 23000 RPM - machining time - 49 min 37 seg. - material removed - 2700 mm³	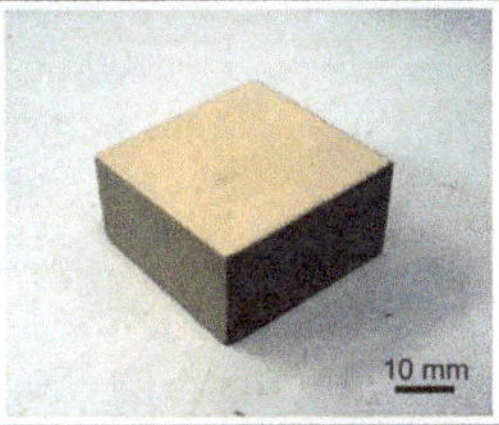	
Pattern 2 – tilted profile - strategy - raster - final depth - 3 mm - horizontal step - 0.2 mm - vertical step - 0.5 mm - feed speed - 8.5 mm/seg. - plundge speed - 4 mm/seg. - rotation speed - 23000 RPM - machining time - 55 min 17 seg. - material removed - 1350 mm³		
Pattern 3 - concave profile - strategy - raster - final depth - 3 mm - horizontal step - 0.2 mm - vertical step - 0.5 mm - feed speed - 8.5 mm/seg. - plundge speed- 4 mm/seg. - rotation speed - 23000 RPM - machining time - 55 min 08 seg. - material removed - 885,67 mm³		
Pattern 4 - square profile with well defined vertical walls and corners - strategy - raster - final depth - 3 mm - horizontal step - 0.2 mm - vertical step - 0.5 mm - feed speed - 6.4 mm/seg. - plundge speed - 1.5 mm/seg. - rotation speed - 23000 RPM - machining time - 1 H 16 min 51 seg. - material removed - 1575 mm³		
Pattern 5 - wave profile - strategy - raster - final depth - 3 mm - horizontal step - 0.2 mm - vertical step - 0.5 mm - feed speed - 8.5 mm/seg. - plundge speed - 4 mm/seg. - rotation speed - 23000 RPM - machining time - 59 min 58 seg. - material removed - 879.68 mm³		
Pattern 6 - free profile representing variations of complex geometry - strategy - raster - final depth - 3 mm - horizontal step - 0.2 mm - vertical step - 0.5 mm - feed speed - 8.5 mm/seg. - plundge speed - 4 mm/seg. - rotation speed - 23000 RPM - machining time - 1 H 38 min 34 seg. - material removed - 1444.96 mm³		

Figure 4. A finishing comparison between geometric patterns based on the dissertation of Gustavo Freitas (2006) [7] machined in plaster and "carijó basalt."

software packages being used.

Because of the difficulties encountered in finding a software package that permits the independent control of each speed, it was necessary to develop a post-processor to allow control over the basic speed in the experiment. To do this, the post-processor analyses the CNC program generated by the CAM software, in this case a file with. tap extension, and individually sets specific values for the X- (or Y-, depending on the strategy adopted) and Z-axis speeds, thus allowing precise control and machining time optimisation. Aside from independent X-, Y- and Z-axis speed controls, the post-processor allows the user to change the feed speeds without having to redo the programs in the CAM software based on the analysis of the trigonometric coordinates in X (or Y) and Z in the program. It also allows free control of these speeds at the user's discretion, as well as of the negative Z-axis displacement, by considering the angle of the surface to be machined, which can be constrained to an operator defined range.

A total of 65 tests were performed using the different machining patterns introduced in **Figure 4**. The parameters used to execute these machining processes for pattern 1 (smooth and flat geometry), 2 (tilted profile), 3 (concave profile), 5 (wave profile) and 6 (free profile) achieved efficiencies between 85% (pattern 1) and 66% (pattern 5). The machining process of pattern 4 had a success rate below 10%, making it the hardest pattern to execute.

All tests considered satisfactory had their repeatability verified in a new specimen by using a new tool and the same criteria and standards. Even with rigorous control over the advancing speed in all three movement axes of the tool, when using the post-processor, it was not possible to determine the set of optimal parameters that would allow the procedure to be repeated without constant intervention in the various established parameters.

Among the many variables inherent in any machining process, such as variations in the equipment and tools, greater control exists for those variables that are related to the tools, which are easier to measure and classify according to a pattern. However, the raw material used in this study, rhyodacite, which is a natural material, presents variations that directly interfere with the tool's wear pattern, preventing uniformity and the determination of a specific set of parameters that can be used in general HSM milling processes with this type of equipment, cooling system and tool.

However, because success was obtained under certain conditions, we can trace a decision-making map to assist with this type of job. The map considers, as its initial reference, the pre-established surface patterns used in this study and the actions taken to perform the tests until a positive result was achieved.

The development of the decision-making map is divided into two steps: in the first step (decision making map 1), only the procedures that are possible without using a post-processor are considered; in the second step (decision-making map 2), parameter control is performed in a way that can only be accomplished when using the post-processor. All patterns from **Figure 4** were machined without using the post-processor. Its use, however, optimises and makes the procedures easier, faster and less likely to damage the tools.

Initially, in decision-making map 1, the fundamental parameters are established based on the parameters that yielded successful rhyodacite milling events. For patterns 1, 2, 3, 5 and 6, the feed and plunge speeds (FS and PS, respectively), as well as the horizontal and vertical steps (HS and VS, respectively) are equivalent because the forces to which these tools are subjected are similar. For pattern 4, however, the feed and plunge speeds were reduced to minimise these forces. A significant fraction of the surface textures used in commercial products can be fitted, as previously discussed, to one of the six patterns or a combination thereof (**Figure 5(A)**).

Decision-making map 1 starts with the milling of a rhyodacite surface using pre-established parameters. After the initial procedure is completed, the success of the procedure is ascertained. If it has not been successful, the suggestion is to use a different cutting tool and restart the process with the same parameters (**Figure 5(B)**). Again, the success of the process is ascertained. If it was once more not successful, the tool's wear is determined, which may occur in one of two moments, either during the process, *i.e.*, as the tool is actually machining, or during entry, *i.e.*, as the tool plunges into the surface at the start of the process. The operator is ultimately responsible for observing under which conditions the wear occurred (**Figure 5(C)**).

In the case of wear during the process, it is suggested that the advance speed be reduced by 10% (or 50% for pattern 4). In both cases, this speed reduction is relative to the initial value of the parameter. Afterwards, it is suggested that the tool be exchanged and that the procedure be restarted. If the procedure was not concluded and no visible wear has occurred, then the tool is likely damaged (broken or buckled), and a new speed reduction of 10% (or 50% for pattern 4) should be applied, along with a 0.1 mm reduction in the vertical step to minimise axial stresses. The process should then be started with a new tool (**Figure 5(D)**).

Should the tool continue to wear or break despite all of these recommended actions, the surface to be machined should be zoned, *i.e.*, divided into areas that vary according to the dimensions of the tested samples. For each sector, a new tool must be used, and the process must be restarted. At the end of the procedure in the first area, the success is ascertained (**Figure 5(E)**).

If success is not attained, the tool must be checked for damage at its tip or buckling. This wear can occur in two situations, either when the tool enters the rhyodacite at the start of the machining process or during the machining process when the tool moves along the X- or Y- and Z-axes. Here again, the operator's attention is necessary to determine in which situation the damage has occurred.

If damage occurs during the machining process, it is suggested that the feed speed be reduced by 10% (or 50% for pattern 4), which minimises the axial stresses and favours processing. Conversely, if wear is observed during entry, the plunge speed must be reduced by 10%, which minimises premature wearing of the tools. In either case, the speed reduction is relative to the initial parameter. After these changes have been implemented, the tool must be exchanged, and the procedure is restarted. If the procedure is not concluded and no visible wear has

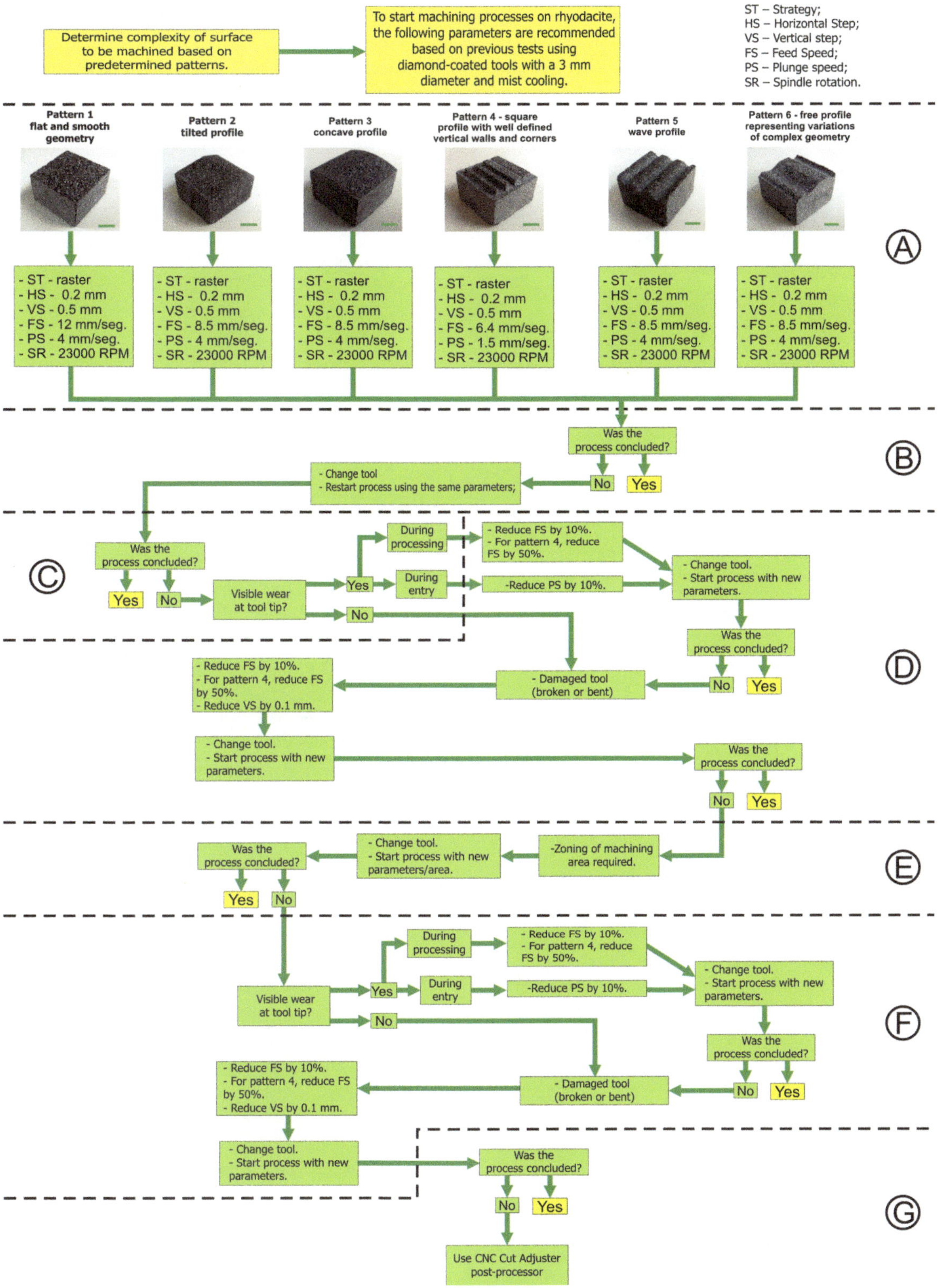

Figure 5. Decision-making map, without the use of the CNC Cut Adjuster software for adjusting the machining parameters for rhyodacite.

occurred, the tool is damaged (bent or broken), and a new reduction in feed speed of 10% (or 50% for pattern 4) with a 0.1 mm reduction in the vertical step, relative to the previously used parameters, must be applied to minimise the axial stresses during machining. The process must be restarted at the area being worked with a new tool (**Figure 5(F)**). The success of the procedure is once more ascertained. If even with all of these interventions there is no conclusion and no improvement in the machining conditions and/or reduction with respect to tool damage, then it is not be possible to finalise the procedure (**Figure 5(G)**).

In decision-making map 2 (**Figure 6**), the process occurs after all initial parameters have been reset. This

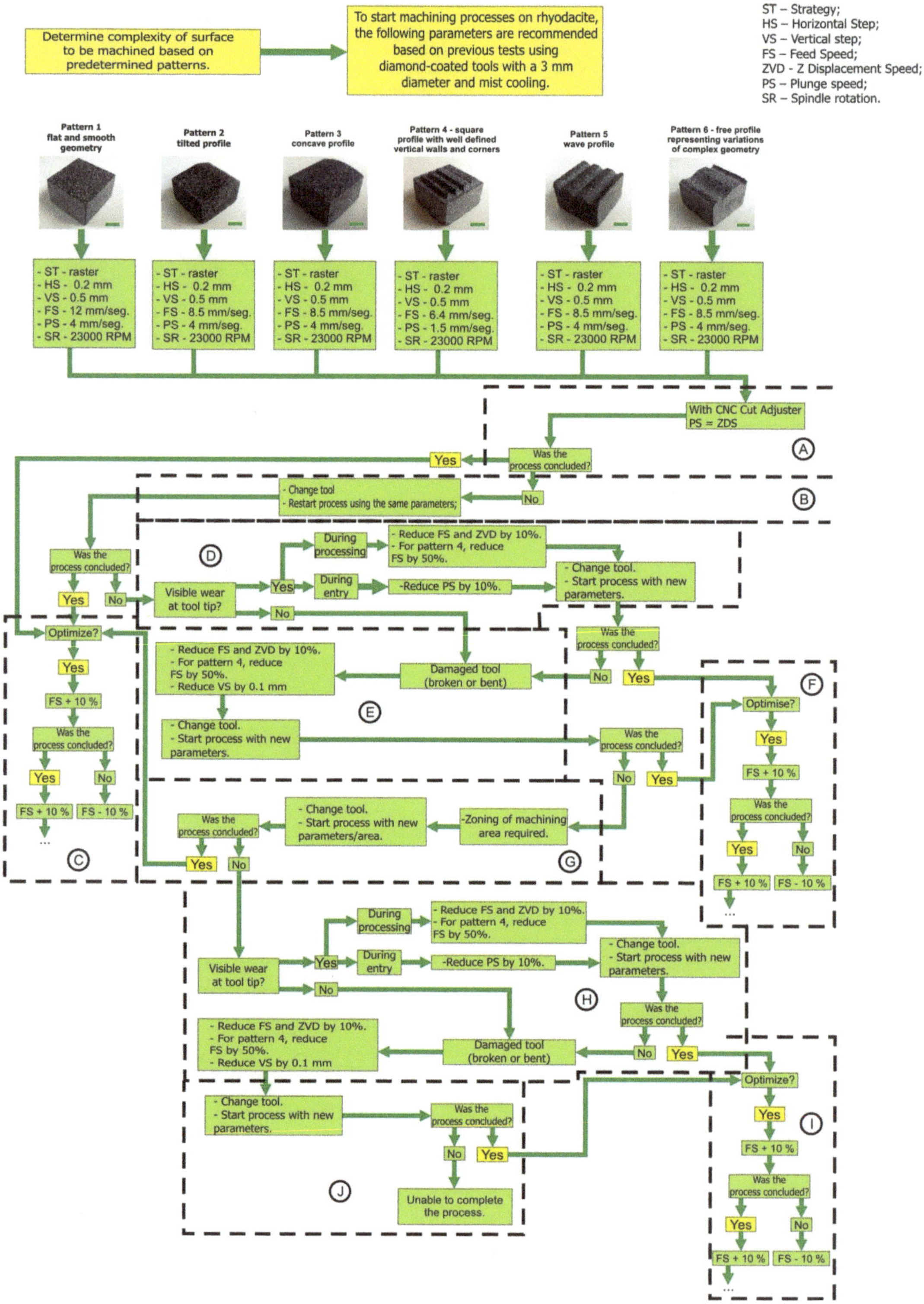

Figure 6. Decision-making map, taking into account the use of the CNC Cut Adjuster software to adjust the machining parameters for rhyodacite.

procedure is suggested so that the starting point is well known. Using the CNC Cut Adjuster post-processor developed in this work, it is suggested that the vertical displacement speed in Z (ZS) be made equal to the plunge speed (PS) defined in the CAM software. This way, all descending displacements in Z are reduced in speed, which favours tool preservation (**Figure 6(A)**). Changing the advance, Z displacement and plunge speeds is not possible without this software. After the process is finished, its success is ascertained. In the case of failure, it is suggested that the tool be changed and that the procedure be restarted using the same parameters (**Figure 6(B)**). The success of the procedure is once more ascertained. In case of success, it may be possible to optimise the procedure by increasing the feed speed by 10% and repeating the procedure. If success is obtained again, the feed speed may once more be increased by 10%; in the case of failure, the previous increase is retracted (**Figure 6(C)**). If the process failed to achieve completion, the tool's wear must be determined, which may occur in one of two moments, as shown in decision-making map 1, either during the process, *i.e.*, as the tool is moved by the milling machine, or during entry, *i.e.*, as the tool plunges into the surface at the start of the process. The operator must observe under which conditions the wear has occurred (**Figure 6(D)**).

For wear during the process, the suggestion is to reduce the feed speed and Z-axis displacement speed by 10% (or 50% for pattern 4); however, for wear during entry, the suggestion is to reduce the plunge speed by 10%. In both cases, the speed reduction is relative to the initial parameter. Next, it is suggested that the tool be exchanged, and the procedure is restarted. If the procedure fails to conclude and no visible wear occurs, the tool is likely damaged (buckling or breakage), which requires a new reduction of 10% (or 50% for pattern 4) for the feed and Z-axis displacement speeds with a 0.1 mm reduction of the vertical step to minimise the axial stresses, after which the process should be restarted with a new tool (**Figure 6(E)**). Because it is easier to control the machining parameters, it is possible, when the process concludes successfully, to optimise the procedure by gradually improving the X-speed of the tool in 10% increments. As the feed speed is increased, the chance of damage is also increased. If the tool becomes damaged, then it is necessary to restart the procedure with the most recent feed speed for which success was obtained (**Figure 6(F)**). After implementing all of these suggestions, if tool wear or damage continues to occur, it becomes necessary to zone, *i.e.*, to divide, the surface to be machined according to the dimensions of the samples tested. For each sector, a new tool must be used, and the process is restarted. At the end of the procedure in the first area, the success of the procedure is ascertained (**Figure 6(G)**).

In the case of failure, whether the tool incurred any damage, such as wear at the tip or buckling, must be determined. This wear, as previously mentioned, can occur in one of two situations, either during the tool's entry into the rhyodacite, *i.e.*, at the beginning of the machining process, or during the machining process as the tool moves along the X- or Y- and Z-axes. The determination of the situation in which damage occurred requires accurate observation by the operator. Should the observed wear occur during the machining process, it is suggested that both the feed and Z-axis displacement speeds be reduced by 10% (or 50% for pattern 4). This minimises the axial stresses and favours processing. However, if the wear is observed during entry, it is necessary to reduce the plunge speed by 10%, thus minimising premature wear of the tool. In either case, the reduction in speed is relative to the initial parameter. After these changes, it is necessary that the tool be changed, and the procedure is restarted. If the procedure was not completed and no visible wear has occurred, the tool is damaged (bent or broken), and a new reduction of 10% (or 50% for pattern 4) in the feed and Z-axis displacement speeds must be made to minimise the axial forces during machining. It is necessary to restart the process in the zone being worked with a new tool (**Figure 6(H)**). Once again, the success of the process is ascertained. If the machining process concludes, then it is possible to optimise the procedure as previously described (**Figure 6(I)**). With all of these interventions, if there is no conclusion and no improvement in the machining conditions and/or reduction in tool damage, then it is not be possible to finalise the procedure (**Figure 6(J)**).

5. Conclusions

In this article, we present a characterisation study of the material behaviour of rhyodacite ("carijó" basalt), a stone material, including its processing by high speed machining (HSM). Our approach enhanced our understanding of the necessary processing conditions for rhyodacite and achieved a deeper understanding of this material's characteristics. The technological contributions of this study can be found in the systematisation of the HSM milling procedure and the development of decision-making maps to increase the possibilities of successfully processing this stone material.

The original contributions of this study are related to the use of low-cost diamond-coated polycrystalline tools for rhyodacite processing while correlating the rock properties with tool wear, with the goal of determining the applicable parameters.

The results of this study have shown that the low-cost diamond-coated polycrystalline tools with nickel matrix, obtained from the market, were not able to withstand the forces to which they were subjected. This made it necessary, here in Brazil, to elaborate a specific post-processor to obtain more control over the parameters provided by CAM software packages. However, it was extremely difficult to maintain the machining parameters, and countless failure problems were encountered with the multi-point tool.

This study suggests that adopting the use of HSM tools for the industrial-scale processing of rhyodacite in Brazil would require technological developments towards low-cost diamond-coated tools, the cost of which should be compatible with the commercial value of this rock; the recommendation is to start with monocrystal diamond tools or single-point tools fabricated with high hardness ceramics, such as tungsten carbide, cubic boron nitride and PCD.

REFERENCES

[1] Brazil—Ministry for Mines and Energy, "Development of Studies for Drafting the Duodecenial Plan (2010-2030) of Geology, Mining and Mineral Transformation," 2009.

[2] A. Motoki, "Petrographic Description of Igneous Rocks," State University of Rio de Janeiro, Rio de Janeiro, 2004

[3] L. A. Bizzi, C. Schobbenhaus, R. M. Vidotti and J. H. Gonçalves, "Tectonic Geology and Mineral Resources of Brazil," CPRM—Brazilian Geological Service, 2003.

[4] Basalt Magazine, "Nova Prata: Syndicate of the Quarry Exploitation Industry of Nova Prata and Region," 2000.

[5] CIC, "Socio-Economical Hierarchy of the Basalt Region," Chamber of Industry and Commerce—CIC, Nova Prata, 2010.

[6] G. E. Dieter, "Mechanical Metallurgy," 2nd Edition, Guanabara Dois, Rio de Janeiro, 1981.

[7] R. Faller, F. Silva and W. Kindlein, "The Use of Very High Speed Machining with Four Axes in the Product Design," VII P&D Design, Curitiba, 2006.

[8] G. Freitas, "Methodology and Applicability of LASER 3D Digitalisation of Molds for Shoes and Components," UFRGS, Porto Alegre, 2006.

[9] M. Filgueira and D. G. Pinatti, "Development of a New Processing Route for Diamond-Coated Tools: Metallurgy of Powder and Rotary Forging," *Revista Matéria*, Vol. 11, No. 3, 2006, pp. 181-188.

[10] C. E. Stemmer, "Cutting Tools II: Drills, Reamers, Taps, Milling Cutters, Broaches, Grinding and Abrasive Wheels," UFSC, Florianópolis, 2008.

[11] CPRM, Company of Research on Mineral Resources/ Brazilian Geological Service, 2010. http://www.cprm.gov.br

[12] A. Motoki, L. Zucco, T. Vargas, J. L. Neves, "Basalt of the Serra Gaúcha Mountain Range and its Relation to Regional Development," 2nd Brazilian Conference of University Extension, Belo Horizonte, 2004.

Energy Efficiency Evaluation in Pumping System

Sharad P. Jagtap[1], Anand N. Pawar[2]
[1]Mechanical Engineering Department, Government Polytechnic, Murtizapur, India
[2]Mechanical Engineering Department, Government Polytechnic, Amravati, India

ABSTRACT

Like 3M's (Man, Machine and Material), energy is one of the major inputs for economic development of the country. In case of developing countries like India, the energy sector assumes critical importance in the view of the ever increasing energy needs, requiring huge investments for new power plant erection. About 85% of total primary energy need comes from fossil fuels. Particularly, oil contributes to 31% primary energy sources and the existence of it from domestic source in India may last for only about 20 years only at the current reserve to production (R/P) ratio. Similarly, coal contributes to 54% primary energy sources and may last for about 114 years at the current reserve to production (R/P) ratio. Water Pumping System is one of the major utilities in SME's as well as power & process industry, and consumes about 10% - 15% electrical power in Industrial and Agricultural sector. As per TERI Energy data directory yearbook, Industrial sector and Agricultural sector contribute to 45% and 7% of the total energy needed respectively in India. A measurable amount of Industrial production is delivered from small and medium industries operated with inefficient equipment, where it has been difficult to implement efficiency improvements. The objective of Water Pumping System in most cases is either to transfer liquid from one reservoir to another or to circulate liquid across the heat exchanger around a system. Since worldwide, centrifugal pumps account for the majority of electricity used by pumps, this paper focuses on energy saving in centrifugal pumping system with a case study work carried out in an industry.

Keywords: Energy Audit; Energy Saving; Pumping System; Performance Evaluation of Pumping System

1. Introduction

The pumps come in a various sizes for wide applications. It can be classified according to their operating principles as dynamic or displacement type. Dynamic pumps can be sub-classified as centrifugal and special effect pumps and Displacement pumps can be classified by rotary and reciprocating pumps. Out of all types of pumps, reciprocating pump is more energy efficient but increases maintenance cost, that's why worldwide, centrifugal pumps accounted for major electricity using equipment in palmping system [1].

Due to ever increasing in economic growth for developing countries, product variation and its production rate in process industry are continuously being changed; hence pump is dynamic device which is not delivered to the fixed volume of liquid as its pressure head changes. Therefore, to attain the best performance from pumping system, it is to be analyzed critically. Overall energy efficiency of plant would be achieved when energy intensity in a specific product, process or area of production or consumption is reduced without affecting output or comfort levels. Nowadays, such plant energy efficiency evaluation is carried out in "Energy Audit" projects.

2. Energy Conservation

Under the provisions of the Energy Conservation Act, 2001 The, Government of India setup Bureau of Energy Efficiency (BEE), a statutory body under the ministry of power on 1st March 2002, with the primary objective of reducing energy intensity of the Indian economy. This will be achieved with active participation of all stakeholders, resulting in accelerated and sustained adoption of energy efficiency in all sectors [2]. Pumping system is used in wide application in industry as well as agriculture sector. By the survey of few Energy Efficiency Projects in India, there are many ways for enhancing energy efficiency in pumping system which is directly depends on its operation philosophy and capacity uses.

Energy efficiency study (Energy Audit) means the verification, monitoring and analysis of use of energy including submission of technical report containing recommendations for improving energy efficiency with cost benefit analysis and action plan to reduce energy consumption. Energy audit helps to understand more about the ways energy is used in the industry, and helps to identify the area where energy waste & scope for improvement can occur. It is carried out by following way.

Preliminary Audit/Walk through audit: By using existing data (energy bills and invoices) and general observations through walk through, establishes energy consumption pattern and its baseline or reference point as well as identify areas for more detail study/measurements.

Targeted Energy Audit: It is often results from Preliminary EA and provides detail analysis of targeted project like pumping system, compressed air system, steam and boiler system, lighting system, etc.

Detail Energy Audit: It is comprehensive energy study of the plant constitute energy balance, detail measurement and calculations for performance evaluation of kind of systems in the plant. Identification of Energy Conservation Opportunities (ENCON), Cost benefit analysis and preparation of action plan for implementation of ENCON measures is carried out. After implementation of ENCON measures monitoring and periodic review also carried out [3].

Detail energy efficiency study of a chemical industry was carried out in India. In this paper energy efficiency analysis of chilled water pumping system for Vapor Absorption Machine (VAM) has analyzed. The total connected load of the entire plant was around 8700 kW. Studied plant was a chemical industry therefore most of the processes energy load were critical (affects production) for power failure from electricity board hence power supply for such process were catered by two gas engines. **Figure 1** shows the average electrical energy demand of entire plant. Energy generation per month for installed gas engines ware studied from history recorded, the average energy generation per month was 2.41 lakhs units (kWh) and simultaneously energy bills from electricity board also analyzed which was shown average energy consumed per month was 0.9 lakhs units (kWh).

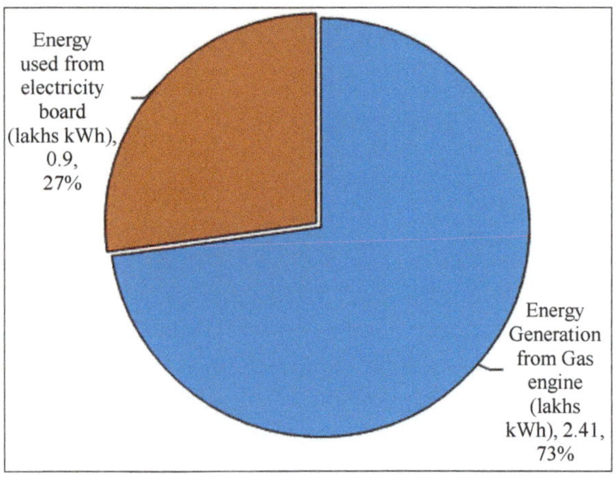

Figure 1. Total Energy Consumption of the Plant.

3. Energy Conservation in Chilled Water Pumping System of VAMs

Performance analysis and finding of suitable measures for pumping system of VAM (Vapor Absorption Machine) is carried out as an actual case study work. Two nos. of hot water driven Vapor Absorption Machines were installed. Each VAM is dedicated with individual Gas Engine jacket heat. Both the VAM operations are as per plant demand. Radiator fans for Gas Engines were available to reject the heat from hot water when the chiller is not in operation. The installed lithium bromide-Water pair VAM as shown in **Figure 2** consists of five basic components as per following.

1) *Generator*: The purpose of the generator is to deliver the refrigerant vapor to the rest of the system. It accomplishes this by separating the water (refrigerant) from the lithium bromide-and-water solution. In the generator, a high-temperature energy source, hot water from diesel Engine jacket cooling flows through tubes that are immersed in a dilute solution of refrigerant and absorbent. The solution absorbs heat from the warmer water, causing the refrigerant to boil (vaporize) and separate from the absorbent solution (Li-Br). As the refrigerant is boiled away, the absorbent solution becomes more concentrated. The concentrated absorbent solution returns to the absorber and the refrigerant vapor migrates to the condenser.

2) *Condenser*: The purpose of condenser is to condense the refrigerant vapors. Inside the condenser, cooling water flows through tubes and the hot refrigerant vapor fills the surrounding space. As heat transfers from

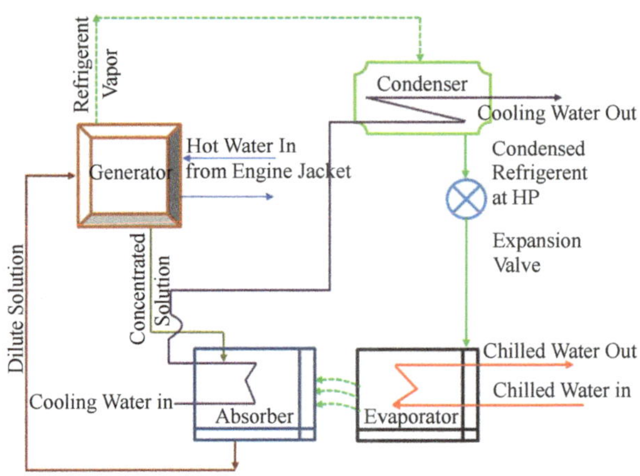

Figure 2. Components of VAM.

the refrigerant vapor to the water, refrigerant condenses on the tube surfaces. The condensed liquid refrigerant collects in the bottom of the condenser before traveling to the expansion device. The cooling water system is typically connected to a cooling tower. Generally, the generator and condenser are contained inside of the same shell.

3) *Expansion Valve*: From the condenser, the liquid refrigerant flows through an expansion device into the evaporator. The expansion device is used to maintain the pressure difference between the high-pressure (condenser) and low-pressure (evaporator) sides of the refrigeration system by creating a liquid seal that separates the high-pressure and low pressure sides of the cycle. As the high-pressure liquid refrigerant flows through the expansion device, it causes a pressure drop that reduces the refrigerant pressure to that of the evaporator. This pressure reduction causes a small portion of the liquid refrigerant to boil off, cooling the remaining refrigerant to the desired evaporator temperature. The cooled mixture of liquid and vapor refrigerant then flows into the evaporator.

4) *Evaporator*: The purpose of evaporator is to cool the circulating water (Chilled water). The evaporator contains a bundle of tubes that carry the system water to be cooled/chilled. High pressure liquid condensate (refrigerant) is throttled down to the evaporator pressure (typically around 6.5 mm Hg absolute). At this low pressure, the refrigerant absorbs heat from the circulating water and evaporates. The refrigerant vapors thus formed tend to increase the pressure in the vessel. This will in turn increase the boiling temperature and the desired cooling effect will not be obtained. So, it is necessary to remove the refrigerant vapors from the vessel into the lower pressure absorber. Physically, the evaporator and absorber are contained inside the same shell, allowing refrigerant vapors generated in the evaporator to migrate continuously to the absorber.

5) *Absorber*: Inside the absorber, the refrigerant vapor is absorbed by the lithium bromide solution. As the refrigerant vapor is absorbed, it condenses from a vapor to a liquid, releasing the heat it acquired in the evaporator. The absorption process creates a lower pressure within the absorber. This lower pressure, along with the absorbent's affinity for water, induces a continuous flow of refrigerant vapor from the evaporator. In addition, the absorption process condenses the refrigerant vapors and releases the heat removed from the evaporator by the refrigerant. The heat released from the condensation of refrigerant vapors and their absorption in the solution is removed to the cooling water that is circulated through the absorber tube bundle. As the concentrated solution absorbs more and more refrigerant; its absorption ability decreases. The weak absorbent solution is then pumped to the generator where heat is used to drive off the refrigerant. The hot refrigerant vapors created in the generator migrate to the condenser. The cooling tower water circulating through the condenser turns the refrigerant vapors to a liquid state and picks up the heat of condensation, which it rejects to the cooling tower. The liquid refrigerant returns to the evaporator and completes the cycle.

Figure 3 shows the chilled water pumping arrangement of VAMs. Primary chilled water pumps were transferring water from primary chilled water storage to evaporator of VAMs and get more chilled and stored it in secondary chilled water storage, from where secondary chilled water pumps transfer it to plant departments.

At most of the plants facility, it was observed that without understanding the correct pressure drop in pumping system causes oversized selection of pumps, resulting into running of pumps far away from duty point (best efficiency point). Thus due to dynamic flow-head-efficiency characteristics of centrifugal pumps, leads to substantial drop in operating efficiency. Besides, worn-out parts and increased clearances over the period of time also cause de rating in operating efficiency. As a thumb rule, operating cost of a centrifugal pump is approximately 15 times higher than the initial cost. Due to these reasons, study of pumping system generally offers maximum energy saving potential with attractive payback.

Theoretically capacity requirement for present pumping system is analyzed *i.e.* actual required volume of pumping water (m^3/hr) with respective to required task (*i.e.* energy load of heat exchanger, evaporator) and pressure head (m) through its elevation from pump reference. This theoretical study helped to compare the actual system parameters of pumping system. Pump always de-

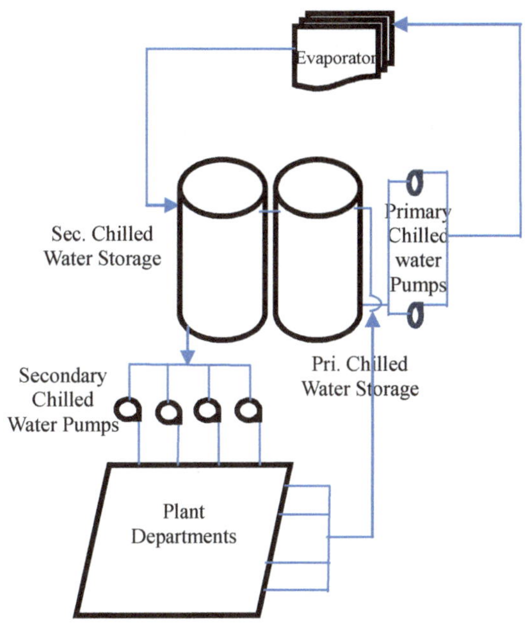

Figure 3. Chilled water pumping arrangement.

signs at certain operating point (flow rate in m^3/hr and pressure head in meter) at which if it is operated, gives maximum energy efficiency. Theoretical required pressure head study based on

- Desired volume (m^3/Hr) of liquid
- It static head (Delivery elevation from reference of Pump)
- Dynamic head (Constitute Major and Minor pressure Losses)

Major dynamic pressure head losses are the friction loss in length of pipe line and pressure loss across the equipment (heat exchanger, surface condensers, evaporators, etc.) and minor dynamic pressure losses are caused by fittings, bends, valves, etc.

Objectives of the work: The major objective of the study is summarized hereunder.

- Performance evaluation of process cooling water pumping system
- Identification of Energy Conservation Opportunities (ENCON)

Operating Philosophy of Primary chilled water pumping system:

When both the VAMs are in operation, a higher head primary chilled water pump (*MEGA-G*-50-250) was running while smaller head primary chilled pump (*MEGA-G*-50-160) is running when only one VAM is under operation. Separate secondary chilled pumps are installed for individual section where chilled water is demanded.

3.1. Installation Details

Specifications of hot water driven (from gas engine jacket cooling) of both VAM in the plant were as per **Table 1**.

As per chilled water arrangement shown in **Figure 3**, there were two separate pumping water network (primary and secondary). The installation details are shown in **Table 2**.

For performance analysis of chilled water & hot water pumps and capacity measurement of VAMs, the simultaneous measurements of water flow, pressure , power of respective pump and temperature at various points of VAMs is carried out by using ultra-sonic flow meter, digital pressure gauge, three phase load analyzer and digital temperature meter respectively.

3.2. Performance Analysis and Recommendations for VAMs

Performance analysis of VAM-I: Basic analogy for heat exchangers analysis is used by referring [4] and the actual measurements and performance evaluation for different parameters for VAM-I was carried out is shown in **Table 3**.

Performance analysis of VAM-II: Similarly actual measurements and performance evaluation for different

Table 1. Design details for VAM's.

Make	Thermax
Model	B511S
Capacity (TR)	95
COP	0.66
Refrigerant	Water
Nos.	2
Design chilled water flow (m^3/hr)	57.46
Design chilled water in/out temperature (°C)	12/7
Design Chilled water pressure drop (Kg/cm^2)	1
Design cooling water flow (m^3/hr)	168
Design cooling water in/out temperature (°C)	32/36.2
Design Cooling water pressure drop (Kg/cm^2)	1
Design Hot Water flow (m^3/hr)	44
Design hot water in/out temperature (°C)	88/78

Table 2. Design details for Chilled water pumps.

S.N.	Particular	Make	Type/Model	Flow (m^3/hr)	Head (m)	Motor (KW)	Speed (rpm)
			Primary Chilled water pumps: 02 nos.				
1	Primary Pumps	KSB	MEGA-G-50-250	60	60	22	2945
		KSB	MEGA-G-50-160	60	32	7.5	2880
			Secondary Chilled water pumps: 04 nos.				
1	Secondary Pump-A	Beacon	50/160	30	40	7.5	2900
2	Secondary Pump-B	Beacon	50/160	50	35	7.5	2865
3	Secondary Pump-C	Beacon	50/160	30	40	7.5	2900
4	Secondary Pump-D	Kirloskar	DB 50/20	60	43	15	2900

parameters for VAM-II was carried out is shown in **Table 4**.

Recommendation for both the VAMs: As per **Tables 3** and **4**, both the VAM's heat input were less compared to design level as shown in **Figure 4**. Similarly it was also observed that refrigeration effect of both the VAM's was less to their design. Hence the Co-efficient of Performance (COP) of both VAMs-I and VAM-II is reduced to expected design value of 0.66 as shown in **Figure 5**. Therefore it was recommended to clean the hot water and cooling water heat exchangers of both the VAMs to improve the heat input rate (H1) and refrigeration Capacity (H2), and hence consequently the COP of both VAMs would be improved.

Table 3. Performance analysis of VAM-I.

Chilled water flow (m^3/Hr.)	43.7	Difference	
Chilled water temp. diff. (°C)	12.2	7.4	4.8
Chilled water pressure drop (kg/cm^2)	2.8	1.6	1.2
Cooling water flow (m^3/Hr.)		135	
Cooling water temp. diff. (°C)	29.2	32.9	3.7
Cooling water pressure drop (kg/cm^2)	1.4	0.5	0.9
Hot water flow (m^3/Hr.)		50	
Hot water temp. diff. (°C)	78.5	71.7	6.8
Hot water pressure drop (kg/cm^2)	2.2	1.6	0.6
Actual Capacity (TR)		**69.37**	
Heat Input: H1 (Kcal/Hr.)		**340,000**	
Heat absorbed through chilled water : H2 (Kcal/Hr.)		209,760	
COP		**0.62**	

Table 4. Performance analysis of VAM-II.

Chilled water flow (m^3/Hr.)	42	Difference	
Chilled water temp. diff. (°C)	14.7	9.5	5.2
Chilled water pressure drop (kg/cm^2)	2.5	1.3	1.2
Cooling water flow (m^3/Hr.)		125	
Cooling water temp. diff. (°C)	28.2	31	3.7
Cooling water pressure drop (kg/cm^2)	1.0	0.4	0.6
Hot water flow (m^3/Hr.)		54	
Hot water temp. diff. (°C)	73.6	67	6.6
Hot water pressure drop (kg/cm^2)	2.4	1.0	1.4
Actual Capacity (TR)		**72.22**	
Heat Input : H1 (Kcal/Hr.)		**356,400**	
Heat absorbed through chilled water: H2 (Kcal/Hr.)		218,400	
COP		**0.61**	

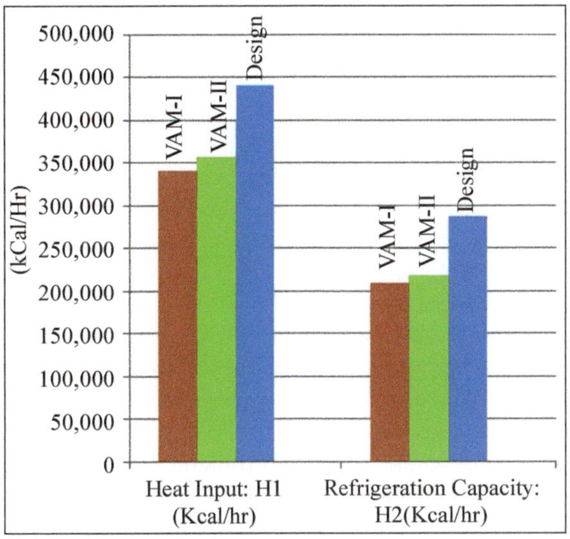

Figure 4. Heat rate analysis for both VAMs.

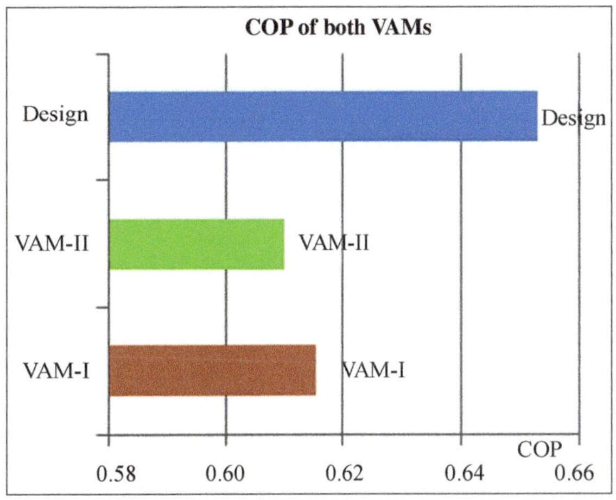

Figure 5. Comparison of actual COP with design COP.

3.3. Performance Analysis and Recommendations for Chilled Water Pumps

3.3.1. Performance Analysis and Recommendations for Primary Chilled Water Pumps

With the help of ultrasonic flow meter and digital pressure gauge and electrical power analyzer chilled water flow rate (m^3/Hr.), pressure (m) and electrical power (kW) respectively were measured. The measured values are plotted in following **Table 5**.

Two pumps of same flow rate (60 m^3/Hr) but different design pressure head (60 m and 32 m) were installed. The higher pressure head pump was used when both the VAMs were in operation for catering chilled water circulation in evaporators of both VAMs. The necessary pressure head, flow rate and their electrical power consumption were measured with the help of respective measuring instrument and summarized in **Table 5**.

During site survey and taking measurements, it was observed that, when low design head Primary Chilled Water pump (MEGA-G-50-160) in operation, the primary pump discharge pressure at 43 m^3/Hr flows is 33 m and the pressure at the inlet of VAM m/c is 28 m, which indicated 6 m of pressure drop in supply line. Similar was the case observed with high design head capacity pump operation (MEGA-G-50-250 as per operating philosophy of VAM), When both the chillers are in operation, the primary pump discharge pressure at capacity of 56 m^3/hr indicated 10 m of pressure drop in supply line. This was also counterchecked with existing 100 mm diameter pipe and proposed retrofitted 150 mm diameter pipe with the help of Hazen-Williams Formula [5] as shown in **Table 6**.

Hd = [6.1961 × 103 × (Q × δ/60)1.852]/[100 × C1.852 × **Dh** 4.87]

Where,

Hd Pressure head drop (m)

C Design coefficient determined for the type of pipe or tube—the higher the factor, the smoother the pipe or tube

Q Flow rate (m^3/Hr)

δ Density of liquid (for Water—1000 kg/m^3)

Dh Inside hydraulic diameter (mm)

From the analysis of Pressure head loss for 100 mm and 150 mm diameter pipeline of Primary chilled water pumping system as shown in **Figure 6**, It was recommended laying of 150 mm diameter pipeline (instead of 100 mm installed) with insulation. Also it was recommended to replace of both primary pumps with following rating.

Desired pressure head losses in pipeline after retrofitting of correct sized pipeline laying:

Desired total head: (Static Head) + (Chilled water pressure drop in evaporator) + (Discharge line friction) + (#Minor Losses)

(# Minor pressure head losses were very less compares to other components in present VAM system, hence neglected.)

Proposed pump specification and power Consumption for MEGA-G-50-160 model

Static head	: 5 m
Chilled water pressure drop in evaporator for small pump operation	: 12 m
Discharge line friction for small pump	: 1 m
Total Pressure head for Small pump	: 18 m
Required Capacity as per VAM	: 45 m^3/Hr
Efficiency (from market survey)	: 72%
Assume Electrical Motor Eff.	: 85%
Power consumption for above pump	: 3.61 kW
Proposed pump specification and power Consumption for MEGA-G-50-250 model	
Total Pressure head for Big pump	: 36 m
Required Capacity as per VAM	: 90 m^3/Hr
Efficiency (from market survey)	: 78%
(Refer Annexure-II)	
Power consumption for above pump	: 13.32 kW
Existing power consumption	: (20.9 + 8.12)/2*
	: 14.51 kW
*For calculation of saving potential, single VAM and two VAM's operation at a time is considered 50% - 50% as per operation history.	
Proposed power consumption	: (3.61 + 13.32)/2
	: 8.46 kW

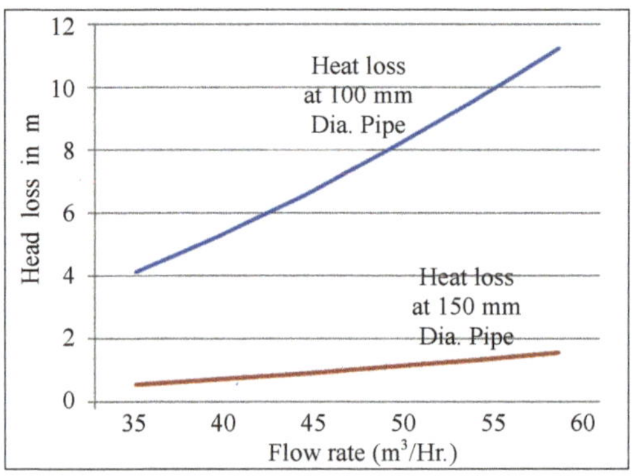

Figure 6. Analysis of Head Loss in pipeline.

Table 5. Actual Measurements and performance evaluation for Primary Chilled Water Pumps.

S.N.	Pump no.	Capacity (m^3/hr)	Total Head (m)	Power (kW)	Efficiency (%)
1	MEGA-G-50-250	56.2	66	20.9	56.9*
2	MEGA-G-50-160	43.7	33	8.12	56.9

*Efficiency = Total Head (m) × Flow rate (m^3/hr)/{367 × Eff.Motor × Input Electrical Power (kW)}.

Table 6. Pressure Head loss evaluation for different diameter pipeline.

Particular	Low design head pump	Higher design head pump
Length of pipe line (m)	140	140
Flow in pipe line (m^3/hr)	43	56
Pipe line diameter (mm)	150	150
Friction factor (.C) (Refer Annexure-I)	100	100
Press. Head Loss when 100 mm pipe diameter (m) line (mm)	6.058	9.880
Press. Head Loss when 150 mm pipe diameter (m) line (mm)	0.84	1.37

3.3.2. Performance Analysis and Recommendations for Secondary Chilled Water Pumps

With the help of measuring instruments, actual parameters of secondary chilled water pumps are measured and analysis for its performance was carried out as shown in **Table 7**.

From above analysis, it is clear that the efficiencies of Secondary Pump-A & Secondary Pump -B are very poor and recommended to replace with energy efficient new pump. Hence it was recommended to replace the existing pump with following proposed rating.

Secondary pump-A:

Capacity (m^3/hr) 15
Pressure (m) 45
Efficiency (%) 55
Proposed power consumption (kW): 3.93

Secondary pump-B:

Capacity (m^3/hr) 65
Pressure (m) 22
Efficiency (%) 77
Proposed power consumption (kWh): 5.95

Proposed power consumption in Secondary chilled water pumps = 3.93 + 5.95 + 6.2 + 10.4
= 26.48 kW

3.3.3. Cost-Benefit Analysis

Overall analysis of power consumption reduction was carried out by comparing existing and proposed measures for both primary and secondary chilled water system which is summarized in **Table 8**. Overall about 21% powers could be reduced by proposed measures as shown in **Table 8**.

Operational hours per day: 16
Operating days per annum: 350
Total operating hours per annum: 5600
Energy cost (Rs./kWh): 6.25
Energy Saving potential (Rs./year): 321,650
Tentative Investment (Rs.): 450,000
Payback (months): 17

Table 7. Actual Measurements and performance evaluation for Secondary Chilled Water Pumps.

S.N.	Particular	Pump Identity	Capacity (m^3/Hr)	Total Head (m)	Power (kW)	Efficiency (%)
1	Secondary Pump-A	PA-639B	13.4	41	5	35.2
2	Secondary Pump-B	PA-638A	60.3	19	8.02	45.8
3	Secondary Pump-C	PA-637B	37	33	6.2	63.1
4	Secondary Pump-D	PA-634A	38	52	10.4	60.9
		Existing total power consumption			29.62	

Table 8. Power consumption reduction in Chilled water system.

S.N.	Pump	Existing Power (kW)	Proposed Power (kW)	Power Saving (kWh)	Power Saving (%)
1	Primary Chilled water	14.51	8.46	6.05	20.82
2	Secondary Chilled water	29.62	26.48	3.14	
	Total	44.13	34.94	9.19	

4. Result and Discussion

Through detail study, performance analysis of chilled water pumping system for VAMs of the industry was indicating remarkable saving opportunities in installed pumps. Installed primary and secondary chilled water pumps were not selected at proper duty point and also pipe diameter design in pipeline network was undersized. Overall chilled water pumping system consumes 44.13 kWh electrical units while that of proposed pumps would consume only 34.94 electrical units. Almost 21% electrical power could be saved. Also in VAMs observation, the pressure drop across the evaporator is 1.2 kg/cm^2 compares to manufacturers design it should be 1.0 kg/cm^2. This indicates evaporator tubes to maintain cleaned periodically.

5. Conclusion

The present study is honestly tried elaborate effort to examine the performance of a pumping system of VAM of a chemical industry. Energy audit as per Energy Conservation Act 2002 is an effective tool in identifying and perusing a comprehensive energy management program. A systematic approach to comprehensive understanding of any type of industry with their different processes can give the industry a plan with which it can effectively manage the industrial energy with its minimum cost. This approach could be useful for an industry in combating essential energy cost and also tapping many other benefits like improved production cost, higher profit and most important satisfaction for contributing in world energy saving.

REFERENCES

[1] Bureau of Energy Efficiency, "General Aspects of Energy Management and Energy Audit," Guide Books-01.

[2] Bureau of Energy Efficiency, "Energy Efficiency in Electrical Utilities" Guide Books-03, Chapter-01,02, Chapter-06,04.

[3] N. R. Kumbhar and R. R. Joshi, "An Industrial Energy Auditing: Basic Approach," *International Journal of Modern Engineering Research* (*IJMER*), Vol. 2, No.1, pp. 313-315.

[4] Çengel Y.A. and Boles M.A., "The Second Law of Thermodynamics," In: *Thermodynamics an Engineering Approach*, 7th Edition, Chapter No. 06, McGraw Hill Companies Publication Book, New York.

[5] S. W. Bedford, "Fluid Mechanics," 9th Edition, Tata McGraw Hill Edu. Pvt. Ltd., New Delhi.

Annexure-I

Table. Hazen-Williams's coefficients.

Pipe Material	Recommended C Value	
	New Pipes@	Design Purpose
Unlined Metallic Pipes		
Cast Iron, Ductile Iron	130	100
Mild Steel	140	100
Galvanized Iron above 50 mm dia.#	120	100
Galvanized Iron 50 mm dia and below used for house service connections.#	120	55
Centrifugally Lined Metallic		
Cast Iron, Ductile Iron and Mild Steel Pipes lined with cement mortar or Epoxy Up to 12,000 mm dia	140	140
Above 1200	145	145
Projection Methose Cement Mortar Lined Metallic Pipes		
Car Cast Iron, Ductile Iron and Mild Steel Pipes	130*	110**
Non Metallic Pipes		
RCC Spun Concrete		
Pre stressed Concrete		
Up to 1200 mm dia	140	140
Above 1200 mm dia	145	145
Asbestos Cement	150	140
PVC, GRP and other Plastic Pipes.	150	145

Note: @ The C values for new pipes included in the Table 6.1 are for determining the acceptability of surface finish of new pipelines. The user agency may specify that flow test may be conducted for determining the C value of laid pipelines. #the quality of galvanizing should be in accordance with the relevant standards to ensure resistance to corrosion throughout its design life. *For pipes of diameter 500 mm and above; the range of C values may be from 90 to 125 for pipes less than 500 mm. **In the absence of specific data, this value is recommended. However, in case authentic field data is available, higher values up to 130 may be adopted.

Annexure-II

Pump performance Data sheet			
Customer	:	Quote Number	:
Customer Reference	:	Size	: CE50/16
Item No	: 02	Stage	: 01
Uses Territory	:	Based on Curve No	: HD18811CAD Rev 1
Quantity	: 01	Date Last Saved	: 10 May 2011, 7:07 AM
Operation Conditions		Liquid	
Flow Rated	:75.00 m^3/Hr	Liquid Type	:
Differential Head/Pressure Rated (required)	: 35 m	Additional liquid description	: Water
Differential Head/Pressure Rated (actual)	: 35.1 m	Solids diameter Max.	: 0.00 mm
Suction pressure rated/Max	: 0.00/0.00 bar g	Temperature Max.	: 20.00 deg C
NPSH available, rated	: Ample	Fluid density rated/Max	: 1.000/1.000 SG
Frequency	: 50 Hz	Viscosity rated	: 1.00 cST

Continued

		Vapor pressure rated	: 0.00 bar a
Performance		Material	
Speed rated	: 2900 rpm	Material requested : 01	
Impeller diameter Rated	: 164 mm	MOC C1260(012)/C1260(012)/CS40C8(053)	
Impeller diameter Maximum	: 174 mm	Material Selected : 01	
Impeller diameter minimum	: 135 mm	MOC C1260(012)/C1260(012)/CS40C8(053)	
Efficiency	: 79%	Pressure Data	
NPSH required/ margin required	: 4.32/0.50 m	Maximum working pressure	: 3.73 bar g
nq (imp. Eye flow)/ S (imp. Eye flow)	: 28/141 metric units	Maximum allowable working pressure	: 5.59 bar g
Minimum continuous safe flow (MCSF)	: 8.80 m^3/Hr	Maximum allowable suction pressure	: 1.96 bar g
Head Maximum, rated diameter	: 38.01 m	Maximum test pressure	: 5.59 bar g
Flow Best Efficiency Point 9BEP)	: 81.01 m^3/Hr	Driver and power data	
Flow ratio (rated/BEP)	: 92.54%	Driver sizing specification	: Rated Power
Diameter ratio (rated/max)	: 94.25%	Margin over specification	: 10%
Head ratio (rated dia./max dia.)	: 85.71%	Service factor	: 1.00 (used)
Cq/Ch/Ce [ANSI/HI9.6.7-2004]	: 1.00/1.00/1.00	Power hydraulic	: 7.14 kW
Selection status	: Acceptable	Power rated	: 9.09 kW
		Power Maximum, rated diameter	: 12.68 kW
		Minimum recommended Motor rating	: 11.00 kW/14.75 hp

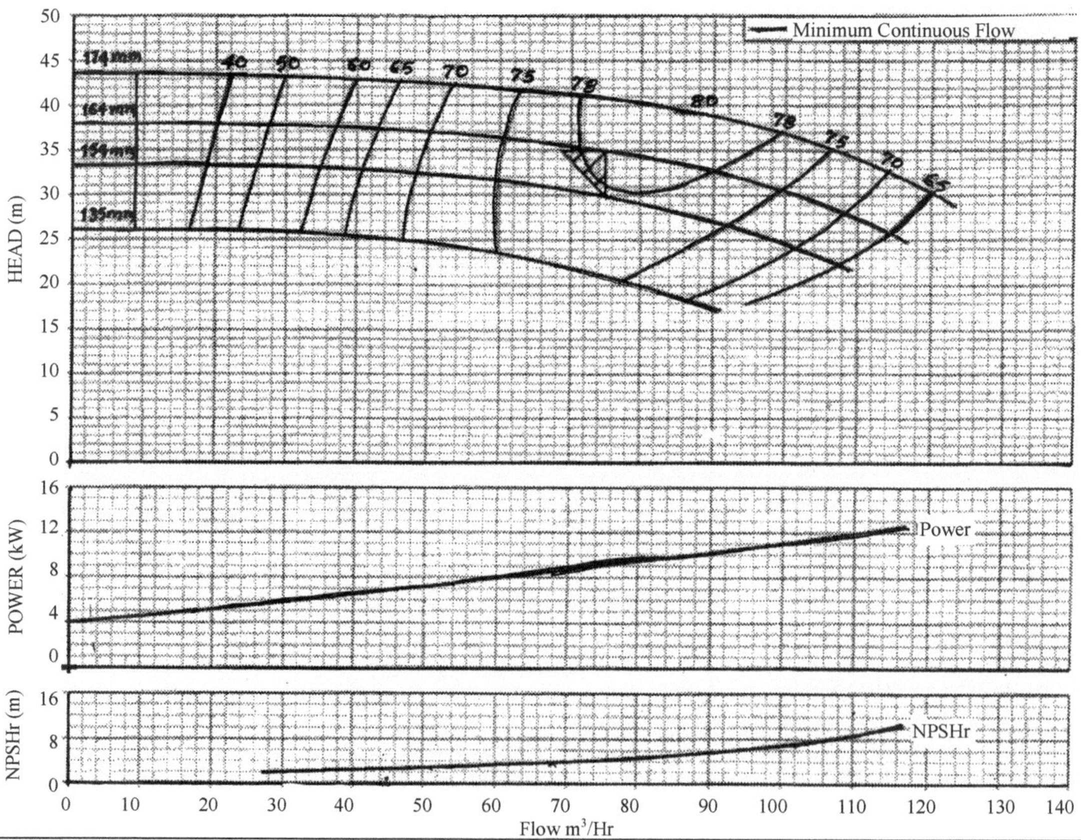

22

Thermal Activation of Asymetrical Composites for Vibration Control

Axel Imbert[1,2], Gildas L'Hostis[1*], David Rigel[2], Fabrice Laurent[1], Bernard Durand[1]
[1]Laboratoire de Physiques et Mécanique, Ecole Nationale d'Ingénieur de Sud Alsace, Mulhouse, France
[2]Centre Technique des Industrie Mécanique, Centre Recherche Matériaux Textiles, Mulhouse, France

ABSTRACT

The CBCM (Controlled Behaviour Composite Material) is a thermal active composite, which has been developed for morphing applications. The thermal activation is made by a source of heating generated within the composite structure. The coupling between the induced thermal field and the thermomechanical properties of the various components of the composite structure leads to the change of the structure shape. The heat source is generated by Joule effect, Carbon yarns inserted in the composite, are connected to a power supply. The application field of CBCM technology is the domain of shape modification and active assembly. The objective of this work is to illustrate the capabilities of CBCM in the domain of vibration control. We will study several reference plates with different constitution. The influences of these different constitutions, of the CBCM effect and the loss of stiffness for the matrix will be highlighted, for two boundary conditions, free/free and embedded/embedded.

Keywords: Composite Material; Vibration Control; Smart Materials; Active Structure

1. Introduction

Smart materials can be defined as materials which are able to change their form, their mechanical or any physical properties to adapt to their background. They are more and more used for technical applications and particularly in the field of controlling structural deformations. In this case, main actuators use piezoelectric and shape memory alloys materials. To make an active composite, these actuators may be used at the scale of the fiber or the matrix [1,2], at the scale of the interface between the successive layers in a laminate [3,4] or at the scale of the linkage between the actuator and a passive composite structure [5,6]. However, the interface strength between the actuator and the composite plays a crucial role. It is a limit of such a technology for the time life of the structure, especially when the rigidity of the composite structure is high, because to deform the structure the actuator has to fight against this rigidity. The use of a thermal actuation, based on the bimetallic strip effect can be a solution to overcome the problems of bonding between the actuator and the structure. Indeed, the whole structure can be considered as an actuator and the problem of interface cohesion is not concentrated at the interface actuator/structure but distributed all along the interfaces of the laminated composite. For composite structures, the bimetallic strip effect can be obtained by coupling an internal heat source with an unsymmetrical laminate.

The concept of a thermal activation based on a bimetallic effect, CBCM effect [7,8] has been studied for morphing structure, the objective of this work is to show the ability of this technology in the field of vibration control of composite structures. For the control of vibration the main actuator remains the piezoelectric, for example the Macro Fiber Composite [9,10]. However in the literature, some application using SMA actuators can be found. The particular behavior of the SMA actuators inserted in the composite or fixed on the composite surface, leads to two technics of vibration control which lead to a frequency shift for different eigenvalue. The ASET technic (Active Strain Energy Tuning) [11], use the shape memory property of the SMA. A non-activated shape and an activated shape are programmed for the SMA actuator by a thermo mechanical cycle. SMA actuator is inserted in the composite have to be maintained during the manufacturing of the composite. During actuation the shape recovery of the SMA induces a stress field and modifies

*Corresponding author.

the dynamic behavior of the composite. To modify the dynamic behavior, the APT technic [12,13] uses the change of the SMA behavior during the phase transformation (variation of the Young's modulus between austenitic and martensitic phase.

A CBCM combines these two actions, stress field increase and change for the behavior of the matrix, especially a loss of stiffness during heating.

After a presentation of the CBCM principle, the constitution of the different tested plates, the experimental procedure, boundary conditions and experimental apparatus will be totally described. In a first par for two kinds of boundary condition, free/free and embedded/embedded and for two CBCM, the influence of the plate rigidity on their dynamical response will be studied: for non-activated and activated plates. The shift of frequency for the different eigenvalue will be highlighted. In the second part of this work, the influence of the loss of stiffness for the matrix will be studied for a CBCM plate and a symmetrical plate. An embedded/embedded boundary condition will be use and from the experimental date, the shape of the two first modes will be reconstructed.

2. Material and Methods

2.1. How CBCM Works

The bending or the warping is a well-known drawback of a dissymmetrical composite when submitted to temperature variations.

Two different manners to have a bending effect by heating the CBCM have been identified: one called "single effect" or "temperature effect", another called "double effect" or "gradient effect". The characterization between "single" or "double" is a reference to the possibility for CBCM plates to bend in one or two directions. The single effect is obtained thanks to an asymmetric laminate made of layers with different coefficients of thermal elongation in a given direction. The different coefficients can be obtained either by different materials or by different orientations of the reinforcement in the layers.

The double effect can be obtained with any composite structure containing an insulating layer, for instance a sandwich structure. If only one side of the composite is heated, the other side being insulated, a gradient of temperature appears within the structure. The result is, as previously shown, the bending of the plate **Figure 1**.

The thermal response time of the CBCM and consequently the response time of the structure is the same order as the SMA one [14]. This response time is directly linked to the intensity of the current in the active layer.

2.2. Description of the CBCM Plates

The composite plates (395 × 125 × 2 mm^3), prepared by compression moulding, are made of seven layers: two layers ($2D^G$) of plain weave Glass fabric (196 g/m^2), two layers (90^G) of Glass unidirectional (588 g/m^2), an active layer (A_L) and two layers ($2D^A$) of plain weave Aramid fabric (173 g/m^2). Three constitutions of plates are studied: two CBCM plates SL = $\{(2D^G)_2/(0^G)_2/A_L/(2D^A)_2\}$ and ST = $\{(2D^G)_2/(90^G)_2/A_L/(2D^A)_2\}$where 0° is along the longitudinal direction of the plate. The difference between the two CBCM plates becomes to the orientation of the main reinforcement, the glass unidirectional. The latest constitution is symmetrical SYM = $\{(2D^G)_1/(90^G)_1/(2D^A)_1/AL/(2D^A)_1/(90^G)_1/(2D^G)_1\}$. The active layer is made of four parallel carbon yarns (6000 filament of 7 μm diameter) along the longitudinal direction of the plate. The yarns are connected to a DC generator and provide the internal heat source of the structure. The resin used for the matrix is an epoxy resin (Epolam 20 - 25 from Axson Corporation). The curing cycle followed is recommended by the manufacturer to obtain a glass transition temperature Tg of 130°C.

2.3. Experimental Procedure

In this work two experimental procedures are used. The first procedure consists to characterize the influence of

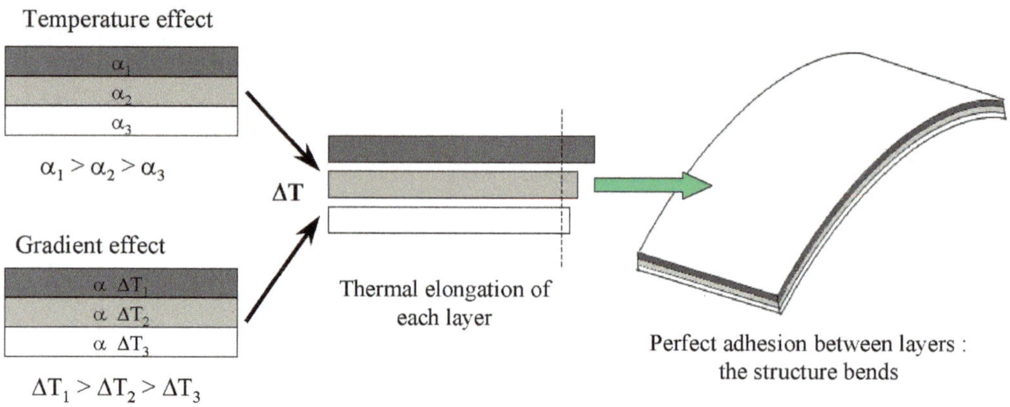

Figure 1. CBCM effects.

the internal stress field induced by CBCM effect on the whole dynamic response of plates with different rigidity (SL and ST plates). Two kinds of boundary conditions are considered: free/free and embedded/embedded. To perform the dynamic analysis, the measuring systems is composed by a Brüel & Kjær acquisition system, an impact hammer (stress sensor PCB 086C05 # 13381), a piezoelectric accelerometer (Brüel & Kjaer 4393# 1803841, sensitivity 0319 pC/ms^{-2}), an amplifier (Endevco 133 # BD69, gain 10 mV/ms^{-2}) and a Pulse acquisition module with I/O 4/2 way 3109. For the value of a modal frequency, this set gives accuracy equal to ±2 Hz.

The second experimental procedure consists to highlight the influence of change of rigidity of the plates due to the mechanical softening of the matrix function due to the temperature increase. The plate is embedded on line A and B **Figure 2**. The measurement area (240 × 125 mm^2) is discretized in 20 measurement nodes. The dynamic analysis is performed using a system acquisition, Brüel & Kjaer, an impact hammer (with force sensor PCB 086C05 # 13381), a front acquisition Pulse (Controller Module 7539 A). The speed of the plate is measured using a laser vibroscope (Ometron DV-500-D) in the node 1 as the reference node. During the test, the 20 nodes are successively affected by the hammer. The modal frequency accuracy is equal to 0.2 Hz. After treatment, in order to construct the shape of the different modes, experimental data are transferred to ME'scope VES software.

The temperature of the plate is measured by thermocouple on the upper side of the plate and a DAQ NX acquisition system of National Instrument.

3. Results and Discussion

3.1. Influence of the Plate Rigidity

The difference of rigidities between SL and ST plates is due to the orientation of the two glass unidirectional layers. The longitudinal rigidity is greater for the SL plate than the ST one; which is the opposite for the transversal rigidity.

For the free/free boundary conditions, **Figure 3** shows the responses of the two plates at ambient temperature equal to 20°C (without the activation of CBCM effect). The smaller longitudinal rigidity of the ST plate leads to a decreasing of the modal frequency. This result is in accordance with those obtained during the static characterization of plates for a free point bending test. For the SL and ST plates, the flexural modulus is 20 GPA and 16 GPA respectively. It can be noted that the morphing properties free displacement and blocking force given by the unconstrained and constrained test respectively are higher for the ST plate [14].

For a temperature of activation T_A = 100°C, **Table 1** summarizes the results obtained for the first seven modes

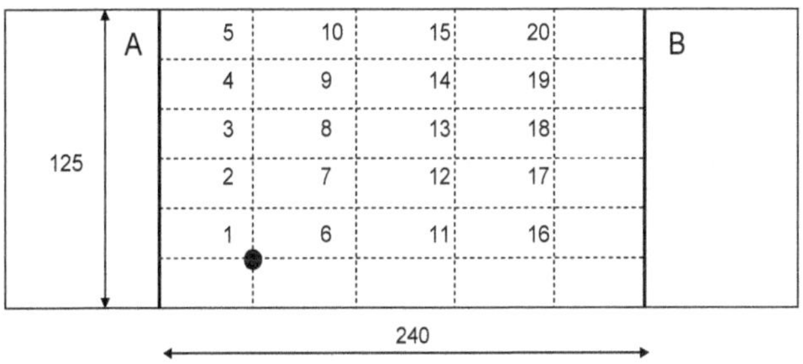

Figure 2. Discretized plate.

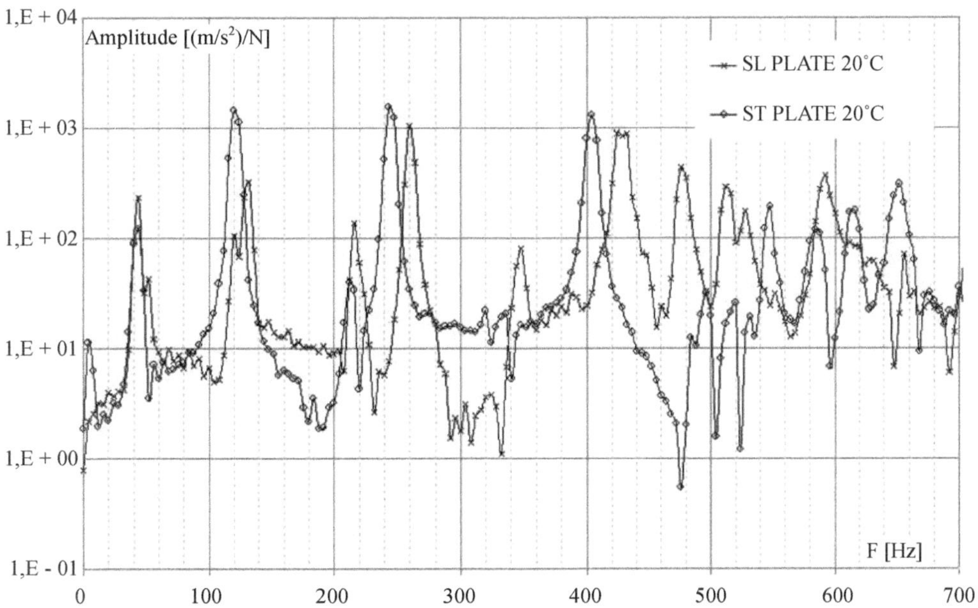

Figure 3. SL and ST plates dynamical response at 20°C (free/free boundary conditions).

of the plates. δ is the difference (in %) between the frequency modal values of the non-activated and activated plates. the non-indicated values correspond to a lower value than the accuracy of the measurement system (4 Hz). For the two plates, the CBCM effect affects the same values of modal frequency. The maximal values of δ are observed for the ST plate. For mode 1 and 3 no change of frequency may be observed.

For the free/free boundary conditions and due to the morphing property of the plates, the direction of curvature changes between the non-activated and activated plate **Figure 4**.

For the embedded/embedded test the change of curvature is cancelled by the boundary condition. At ambient temperature, the values of modal frequency **Table 2** show the difference of dynamical response of the two plates. Contrary to the precedent results, the ST plate is more rigid than the SL one. This change of rigidity becomes to the boundary conditions, because due to the clamping an internal stress field appears in the plates, higher in the ST plates due to the higher value of the initial curvature.

For T_A = 100°C the CBCM effect can be observed for the SL plate **Figure 5** and the ST plate **Figure 6**. The same kinds of previously results are obtained: no changes for certain frequency modal value and for the modes a lower value for the activated plates. After activation, the rigidity of the ST plate remains higher than the SL plate (**Table 2**), however the maximum values of δ are obtained for the ST plate. Depending of the glass unidirectional orientation, the CBCM effect not affects the same

Table 1. Frequency modal value for free-free boundary conditions (Hz).

	SL			ST		
Mode	20°C	100°C	δ (%)	20°C	100°C	δ (%)
1	44	44	-	4	6	-
2	120	104	13	44	36	18
3	132	128	-	120	120	-
4	216	192	11	212	180	15
5	260	232	11	244	236	3
6	348	312	10	332	296	11
7	428	416	3	404	388	4

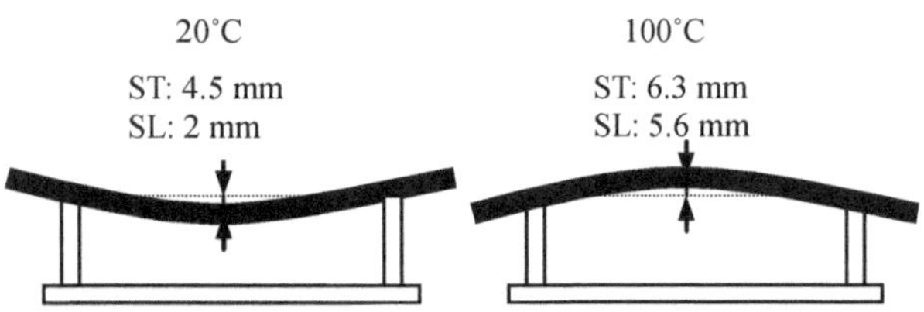

Figure 4. Curvature change for SL and ST plates.

Table 2. Frequency modal value for embedded/embedded boundary conditions (Hz).

	SL			ST		
Mode	20°C	100°C	δ (%)	20°C	100°C	δ (%)
1	32	32	-	32	32	-
2	60	60	-	92	92	-
3	108	104	-	128	112	13
4	148	136	8	252	248	-
5	292	284	3	304	276	9
6	496	472	5	492	480	2
7	648	604	7	660	604	8

nodal frequency values, and the influence of the CBCM effect is different depending on the frequency. The first three modes are unchanged for the SL plate; the third mode of the ST plate has the maximum value of δ. An opposite result is obtained for the fourth mode of the plate SL. This difference of dynamical behavior between SL and ST plates may be induced by the initial stress field due to the embedded boundary conditions. But the thermal activation has an influence on the mechanical property of the matrix and induces a loss of stiffness.

Thereby, when the plate is heating three phenomena are in competition. The initial stress field and the stress field induce by the CBCM effect. Linking to the asymmetry of the composite, these fields are higher for the ST plate and lead to an increase in stiffness along the longitudinal direction of the plate. The third phenomenon is the loss of stiffness and for the ST plate, due to the orientation of the glass unidirectional; this phenomenon's effect is higher for the longitudinal direction.

3.2. Influence of the Matrix Loss Stiffness

For the embedded/embedded boundary conditions, the dynamic behavior of the thermal active plates is influenced by three coupled phenomena. The initial and thermal stress fields due to the curvature of the plate and the CBCM effect respectively, lead to an increase of the stiffness. On the opposite, the latest phenomenon induces a loss of rigidity linking to the matrix loss stiffness. To isolate this phenomenon, two plates are studied the ST plate and a symmetrical plate SYM. The dynamic analysis is made of two steps. The first is the determination of the modal frequency but only for the two first modes. The second step consists to reconstruct the associated modal shape of the modes.

For node 1, the reference point **Figure 2**, a change of modal frequency is observed for the SYM plate **Figure 7**. The value of δ between the plate at T_{amb} and the activated plate at T_A = 100°C is δ = 12% for the first mode (10.3

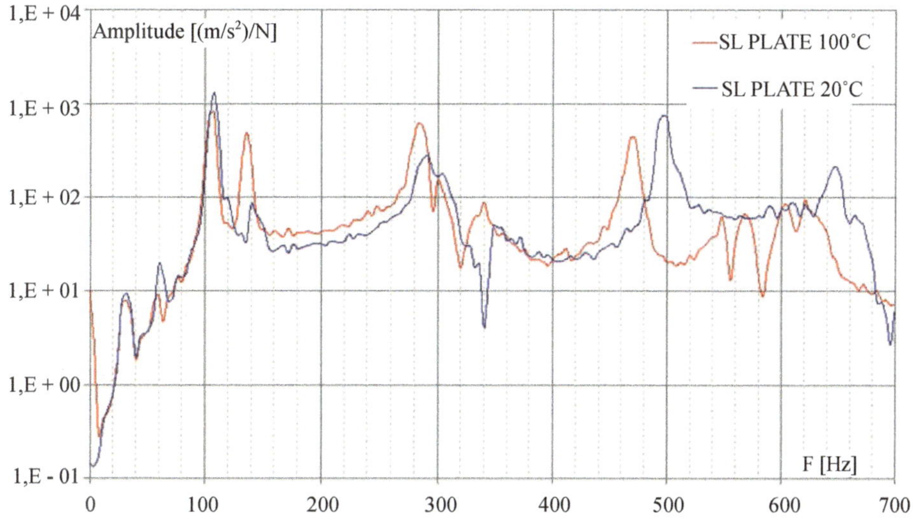

Figure 5. SL plate dynamical response (embedded/embedded boundary conditions).

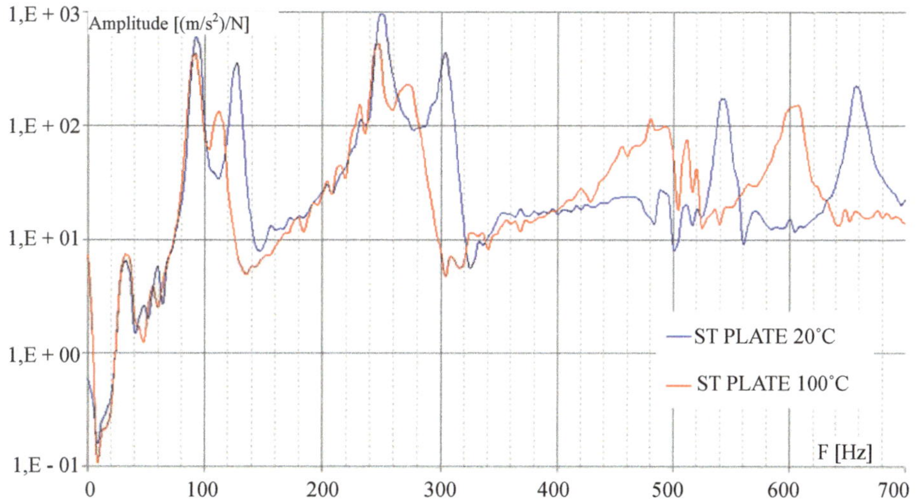

Figure 6. ST plate dynamical response (embedded/embedded boundary conditions).

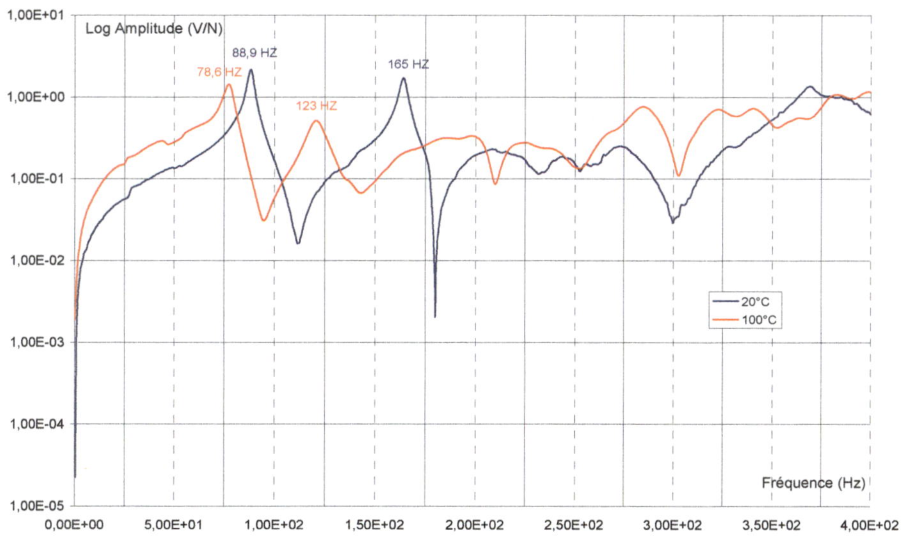

Figure 7. Dynamic response, SYM plate, embedded/embedded, at 20°C and 100°C.

Hz), and δ = 25% for the second mode (42 Hz). The corresponding damping coefficients are respectively, 1.72% and 1.15% for the non-activated plate and 3.22% and 3.35% for the activated plate.

The previous results obtained for the ST plate are confirmed **Figure 8**. The value of δ between the plate at T_{amb} and at T_A = 100°C is δ = 3% for the first mode (4 Hz) and δ =12% for the second mode (20 Hz). The corresponding damping coefficients are respectively, 1.35% and 1.14% for the non-activated plate and 2.48% and 2.39% for the activated plate.

For the two plates, the thermal activation leads to a decrease in the value of the frequency, and an increase of the damping coefficients. This result is consistent with the variation of the behavior of the epoxy depending to the temperature increase: loss of rigidity and increase of viscous properties.

However, for the asymmetrical plate ST, the changes of modal frequency and damping factor depending to the temperature are smaller than for the symmetrical plate. For the symmetrical plate, the changes are only due to the loss of stiffness of the epoxy resin depending to the temperature. For the ST plate, during the thermal activation, the competition between CBCM effect, and the loss of epoxy stiffness remains. For the two modes of the ST plate, the coupling between these opposite phenomena acts differently. For the first mode, the loss of stiffness is partly compensated by the CBCM effect.

This result may explain the dynamical behavior observed in Section 3.1 (**Tables 1** and **2**). For certain mode the loss of stiffness is totally compensated by the CBCM effect, and the value of the modal frequency remains the same. For the other modes the loss of stiffness is not compensated and a change of frequency is obtained. So depending on the shape of the mode, the coupling acts differently.

From experimental data, the shape of the two modes is reconstructed. For the SYM plate **Figure 9**, at ambient temperature the two modes are for the first a (2,1) mode and a (1,2) mode for the second. After the thermal activation, the same modal shapes are observed. The mode most affected by the loss stiffness (δ = 25%) is the second mode (1,2). This result is consistent with the orientation of the main reinforcement, the Glass unidirectional, along the transversal direction of the plate.

For ST plate **Figure 10**, the shape of the modes is conserved between the non-activated and activated structure. The modes are: (1,1) mode for the first and (1,2) for the second. As previously for the SYM plate, the second mode, the same that the SYM plate, is more affected by the loss stiffness (δ = 12%). However the first mode remains quasi constant (δ = 3%). The bending effect along the longitudinal direction of the plate due to CBCM effect affect mainly this first mode composed by a flexural mode along the longitudinal direction. Along this direction, the CBCM effect balances the loss of stiffness.

4. Conclusions

For the all modes, the softening of the matrix depending of the temperature increase induces an important change for their modal frequency. This result has been also observed for inserted SMA, for their temperature of activation. The loss of rigidity of the matrix is accompanied by an increase in its viscous properties. For a SYM plate, a thermal active layer may be used to change its dynamic behavior, but all modes will be modified by this thermal

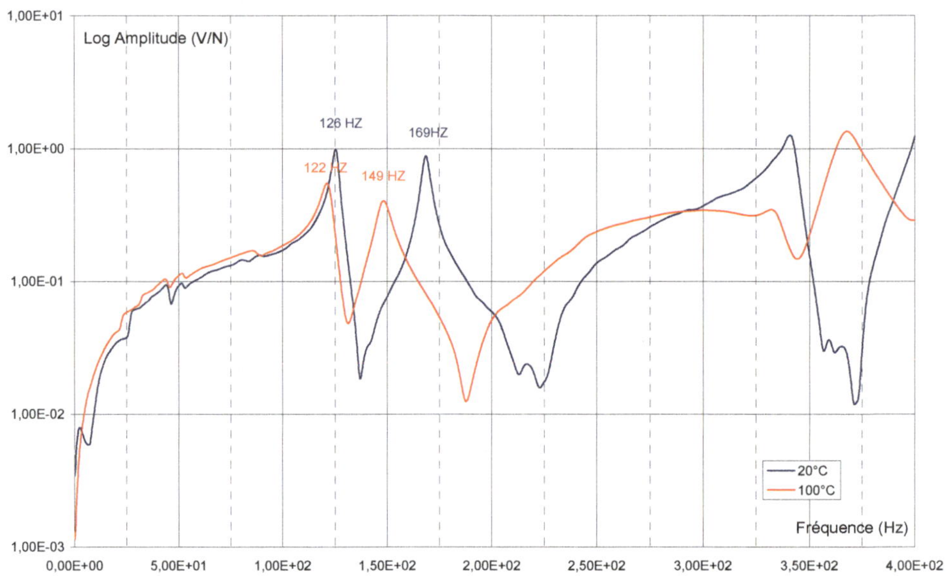

Figure 8. Dynamic response, ST plate, embedded/embedded/ at 20°C and 100°C.

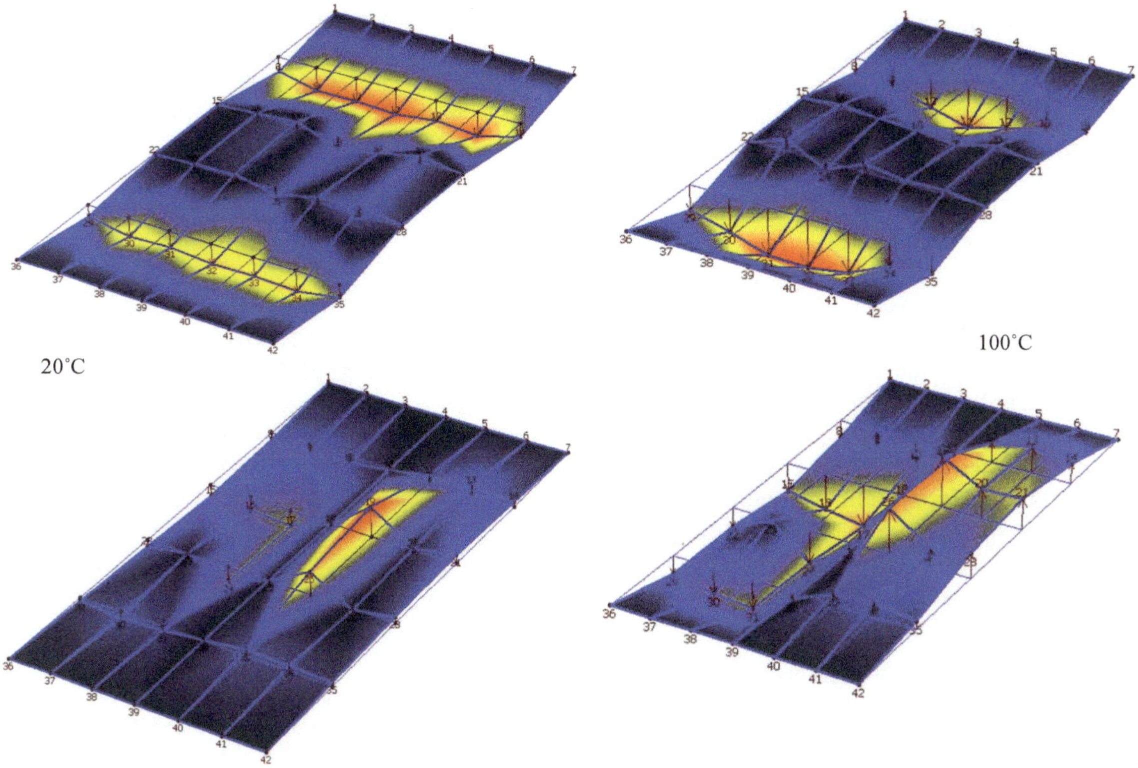

Figure 9. Modal shape, SYM plate.

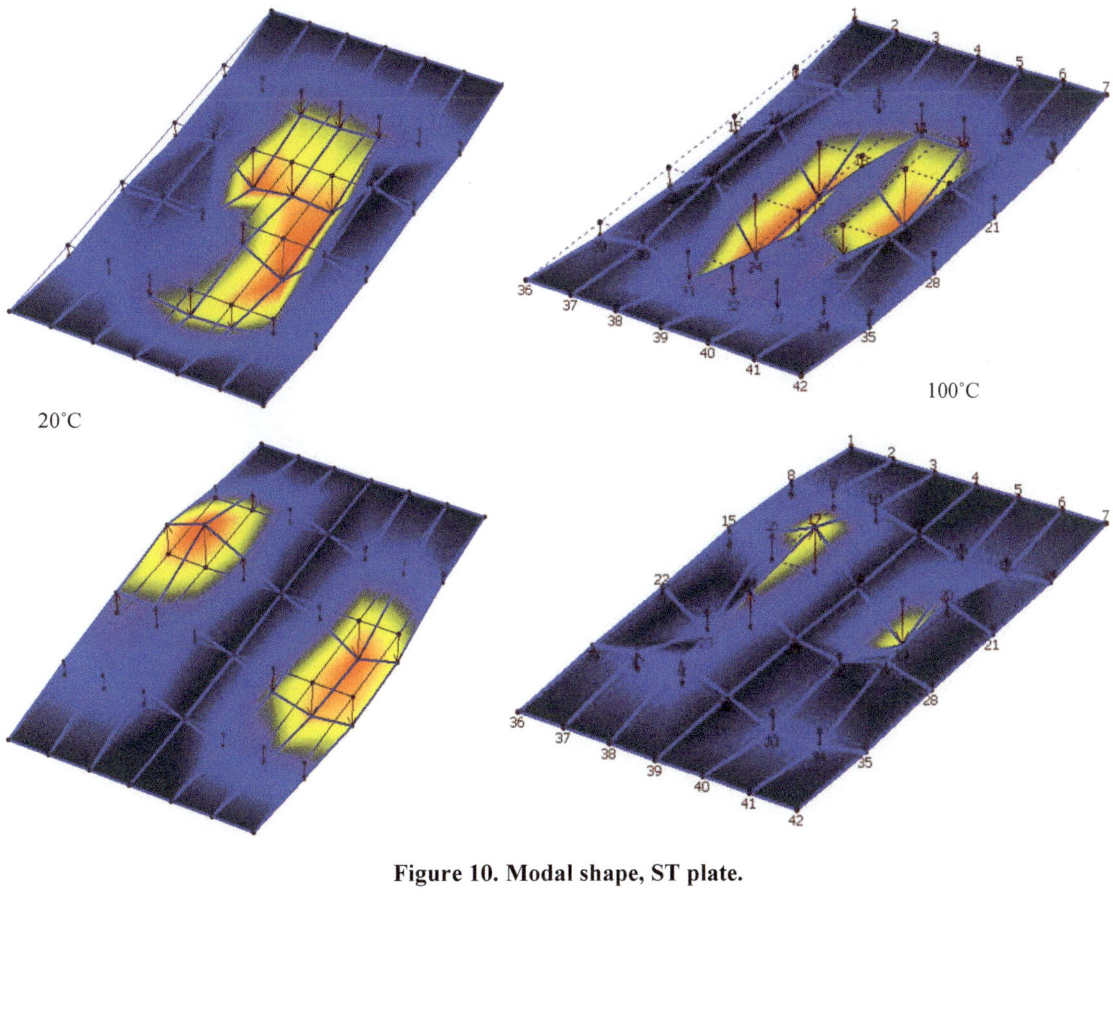

Figure 10. Modal shape, ST plate.

activation. For the asymmetrical plate, the CBCM effect is added and for certain modes can be compensated the softening effect.

The softening is induced by the behavior of the matrix and is linked to the heat field in the composite plate. Depending to this thermal field, the softening can be local or global at the entire plate. The CBCM effect is also linked to the heat field; it can be local or global. However the CBCM effect depends also to the organization of the structure nature and orientation of the reinforcement. Therefore the coupling depending to the thermal field, between the loss of rigidity and the CBCM effect gives a wide field of possibility for the adaptation of the composite structures dynamic behavior.

Finally, one of the main advantages of the thermal activation compared to SMA or piezoelectric technologies is its simplicity of implementation in the process of composite structures manufacturing and this for any kinds of structure geometry.

REFERENCES

[1] G. C. Psarra, J. Parthenios and C. Galiotis, "Adaptative Composites Incororating Shape Memory Alloy Wires," *Journal of Material Science*, Vol. 36, No. 3, 2001, pp. 535-546.

[2] M. V. Donadon, S. F. M. Almeida and A. R. de Faria, "Stiffening Effects on the Natural Frequencies of Laminated Plates with Piezoelectric Actuators," *Composite Part B*, Vol. 33, No. 5, 2002, pp. 335-342.

[3] C. S. Zhang and Q. Q. Ni, "Bending Behaviour of Shape Memory Polymer Based Laminates," *Composites Structures*, Vol. 78, No. 2, 2005, pp. 153-161.

[4] D. M. Elzey, A. Y. N. Sofla and H. N. G. Wadley, "A Shape Memory-Based Multifunctional Structural Actuator Panel," *International Journal of solids and Structures*, Vol. 42, No. 7, 2005, pp. 1943-1955.

[5] K. J. Yoon, K. H. Park, H. C. Park and D. Perreux, "Thermal Deformation Analysis of Curved Actuator LIPCA with a Piezoelectric Ceramic Layer and Fiber Composite Layers," *Composite Science and Technology*, Vol. 63, No. 3-4, 2003, pp. 501-506.

[6] M. R. Schultz, M. W. Hyer, R. B. Williams, W. K. Wilkie and D. J. Inmann, "Snap-Through of Unsymmetric Laminates Using Piezocomposite Actuators," *Composites Science and Technology*, Vol. 66, No. 14, 2006, pp. 2442-2448.

[7] H. Asanuma, O. Haga, J. I. Ohira, G. Hakoda and K. Kimura, "Proposal of an Active Composite with Embedded Sensor," *Science and Technology of Advanced Material*, Vol. 3, No. 2, 2002, pp. 209-216.

[8] H. Drobez, G. L'Hostis, K. Gautier, F. Laurent and B. Durand, "A New Active Composite," *Journal of Smart Material and Structures*, Vol. 18, No. 2, 2009, Article ID: 025020.

[9] A. Kovalovs E. Barkanov and S. Gluhihs, "Active Control of Structures Using Macro-Fiber Composite (MFC)," *Journal of Physics: Conference Series*, Vol. 93, No. 1, 2007, Article ID: 012034.

[10] G. V. Kumar, S. Raja, K. B. Prasanna and V. Sudha, "Finite Element Analysis and Vibration Control of a Deep Composite Cylindrical Shell Using MFC Actuators," *Smart Materials Research*, Vol. 2012, 2012, Article ID: 513271.

[11] S. John and M. Hariri, "Effect of Shape Memory Alloy Actuation on the Dynamic Response of Polymeric Composite Plates," *Composites Part A*, Vol. 39, No. 5, 2008, pp. 769-776.

[12] K. Yuse and Y. Kikushima, "Development and Experimental Consideration of SMA/CFRP Actuator for Vibration Control," *Sensors and Actuators A*, Vol. 122, No. 1, 2005, pp. 99-107.

[13] R. Zhang, Q. Ni, A. Masuda, T. Yamamura and M. Iwamoto, "Vibration Characteristics of Laminated Composites Plates with Embedded Shape Memory Alloys," *Composite Structures*, Vol. 74, No. 4, 2006, pp. 389-398.

[14] K. Gautier, G. L'Hostis, F. Laurent and B. Durand, "Mechanical Performances of a Thermal Activated Composite," *Composite Science and Technology*, Vol. 69, No. 15-16, 2009, pp. 2633-2639.

23

Use of Vehicle Power-Train Simulation with AMT for Fuel Economy and Performance

Nagwa Ahmed Abdel-Halim
Automotive Engineering Department, Faculty of Engineering, Helwan University, Cairo, Egypt

ABSTRACT

Frequent stops and starts for the traffic lights as well as acceleration/deceleration, rather than at constant speed, are the most common transient operating conditions. Consequently, understanding the dynamic characteristics of the power-train system is vitally important in designing an efficient and fresh vehicle. To investigate the dynamic characteristics of the road vehicle performance, apart from a real road testing, it is necessary to set up a numerical test bed for dynamic performance simulation. The latter is even more convenient, costless and data repeatable. However, the aim of the present paper is to establish a computer simulation package by which the road vehicle power-train system design is developed for a small passenger vehicle based on fuel economy and performance. Furthermore, a proposed professional computer program of Automated Manual Transmission (AMT) system related to organizer algorithm of attractive gear-shift requests is also introduced. The result of the proposed AMT ensures that the drive wheels force balance in order to keep gear shift, to kick down or up not only shift feel but also the fuel consumption, and to alarm by near a specific failure or to get back system performance.

Keywords: Automated Manual Transmission (AMT); Dynamic Characteristics; Dynamic Performance Simulation; Fuel Economy

1. Introduction

To develop the road vehicle power-train system design based on fuel economy and performance a computer simulation package was established for a small passenger vehicle. The developed vehicle power-train system is included automated manual transmission (AMT). AMT has additional advantage rather than the manual transmission (MT) in which the clutch works by an actuator are controlled by number of solenoids instead of manual operation. Also, the transmission select and shift speeds are achieved by actuators and valves. The program is based on calculation difference between vehicle tractive effort F_t and total resistance; also predict by near a specific failure and control it by get back system performance modification.

It is well understood that for the vehicle power train design, the targets are always to achieve minimum fuel consumption, best derivability and acceptable air pollution. On the other hand, the main cause of unwanted emissions and bad fuel economy is the transient operation conditions [1].

In the past, there are some power-train simulation techniques published to aid the design of the vehicle performance predictions and/or experiments. These techniques are normally termed as transient-reconstruction technique because they reconstruct the transient performance from steady state engine performance maps to evaluate the real transient operation. Although this technique is significantly reducing the computational time, the drawback comes from the discrepancy between the simulation condition and the real operation. The main reason is the absence of modeling the dynamic characteristics of the thermal power engine, such as transient effects during acceleration/deceleration [2-5].

The introduction of methods for operating a manual transmission for heavy duty trucks with diesel engine is introduced. The basic logic of the methods is to select the most economical gear position among usable gear positions and to achieve compatibility of fuel economy with drivability. The usable gear positions which mean gear position appropriate for drivability are determined from vehicle acceleration, engine speed, gear position used, etc. While the most economical gear position is determined by comparison of engine specific fuel consumption data between each usable gear position for the same running resistance and vehicle speed [6-8].

In this paper, a professional computer program of AMT system was proposed, which is mainly evolved from a vehicle spark ignition engine. The program uses vehicle speed (or road load) simulation technique. The main aim of the simulation technique is to do the optimum gear change according to the fuel economy based on the data obtained of the vehicle speed, road load and the behavior of the driver. The major input includes the engine/power-train/ vehicle specifications (which may be obtained from its workshop manual); the operation schedule (includes the instantaneous engine speed, vehicle speed, throttle, clutch, and gear positions which are exactly same as those controlled by the driver). The major output consists of the instantaneous vehicle resistances, acceleration or deceleration and the fuel flow rate, and those accumulated results such as accumulated fuel consumption, traveling distance, etc and alarm by near a specific failure, which are regarded as the basic vehicle performance. This model is derived from the engine, the power-train and the vehicle dynamics.

2. Vehicle Designed Parameters

2.1. Engine Power at Full Open Throttle Valve

Accurate engine performance information giving the specific fuel consumption at different loads and speeds is essential if fuel consumption over the test cycle has to be accurately computed. Such performance is commonly presented, either in a power/speed or torque/speed or fuel consumption/speed or engine efficiency/speed form. The power/speed and torque/speed characteristics can be constructed from an empirical formula, if the maximum power, $P_{e\max}$ (in kW) and the angular velocity, ω_N (rad/s) are known. One of the empirical formulas suggested to be used is illustrated below [2]:

$$P_e = P_{e\max}\left[\left(\frac{\omega_e}{\omega_p}\right)+\left(\frac{\omega_e}{\omega_p}\right)^2-\left(\frac{\omega_e}{\omega_p}\right)^3\right]\text{KW} \quad (1)$$

Having specified the angular velocities ω_e and estimated their corresponding values of the power P_e, the speed characteristic of the power $P_e = f(\omega_e)$ can be plotted and then the speed characteristic of the torque $M_e = f(\omega_e)$. The values of the torque, M_e, can be found from the formula [2]:

$$M_e = (10)^3 (P_e/\omega_e)\ \text{N}\cdot\text{m} \quad (2)$$

2.2. The Desirable Number of Gear Changes

The most desirable number of gear changes depends on the difference between the highest and the lowest gear ratios. In most automotive transmissions the gear ratios come fairly close to forming a geometrical series. The diagram there shown applies to a five-speed passenger car transmission. The cars may be accelerated in the first gear until the engine turns over at n_p rpm, which corresponding to a little over V_1 km/hr road speed. This is indicated in the diagram by the first sloping line on the left. For acceleration above this speed the gearbox is shifted to the second shift, which pulls down the engine speed until the engine again reached n_m rpm. In second gear the car is allowed to accelerate until the engine again reached n_p rpm, and then the change to third gear is made, which again slows the engine down to n_m rpm. After bringing the engine up to n_p rpm in third gear, the transmission is shifted into higher shift and the process is repeated until the gear shift reached to the highest reduction ratio. This would insure that in each gear change the fuel economy characteristics are similar [2,3]. The reduction ratios come from the following formula:

$$V_{z-1} = n_p / n_m$$

Or

$$\frac{i_{g(z-4)}}{i_{g(z-3)}} = \frac{i_{g(z-3)}}{i_{g(z-2)}} = \frac{i_{g(z-2)}}{i_{g(z-1)}} = \frac{i_{g(z-1)}}{i_{g(z)}} \quad (3)$$

$$n_p = n_m$$

3. Road Dynamic Properties of a Motor Vehicle

3.1. Vehicle Power Balance

The dynamic properties of a passenger motor vehicle can be analysed by comparing the tractive power which is produced from the engine and the power required to overcome the vehicle motion resistance during ascent and acceleration, the equation of power balance is [2].

$$P_t = P_e - P_{tr} = \acute{\eta}_{tr} P_e = P_r + P_g + P_{air} + P_{in} \quad (4)$$

$$P_t = P_{rd} + P_{air} + P_{in} = R_t v$$

$$P_{res} = P_{rd} + P_{air}$$

$$P_r = f \cdot W \cdot \cos\alpha_{rd} \cdot v/1000\ \text{kW}$$

$$P_g = W \cdot \sin\alpha_{rd} \cdot v/1000\ \text{kW}$$

$$P_{air} = \left(W_{air} \cdot (v + vw)^2 + \frac{v}{1000}\right)\text{kW}$$

$$W_{air} = \frac{1}{2}\rho_{air} \cdot C_d \cdot Af\ \text{N}$$

$$P_{in} = \delta_{rot} \cdot W \cdot j \cdot v/1000\,g\ \text{kW}$$

$$\delta_{rot} = 1 + \left(\acute{\eta}_{tr} \cdot I_f \cdot i_{2tr} + I_r\right) g / W \cdot r_2$$

Or can be calculated from

$$\delta_{rot} = 1 + (\delta_1 + \delta_2 i_2 g) W / W_a$$

$$\delta_1 \approx \delta_2 \approx 0.03 - 0.05$$

The dynamic certificate of a motor vehicle is a combination of the dynamic characteristics, load monogram and skidding control graph. The certificate makes it possible to solve the equation of motion with design parameters of a vehicle, basic characteristics of the road and the vehicle load [2].

$$D = (F_t - F_{air}) / W \text{, and } D_a = D \cdot W_a / W \quad (5)$$

3.3. Vehicle Fuel Consumption

The fuel consumption for a unit can be easily determined rather easily, but it does not take into account the transport operation undertaken by the vehicle, and this may result in wrong conclusions when fuel economy is evaluated. Thus, when a vehicle carries a load its fuel consumption per unit run is always greater than with the same vehicle operating in the identical conditions but without load. However, the fuel consumption for a run of 100 km in liters can be estimated based on the following formula [2]:

$$Q_{run} = \left[0.735 g_e \cdot P_e / (36 v \cdot \rho_f)\right] l/100\text{ km} \quad (6)$$

4. Application

4.1. Data Description

In this application, the modeling analysis stated above was applied on a vehicle. The vehicle is equipped by spark ignition internal combustion engine as a propulsion system, where its dynamic parameters and technical specifications are listed in **Table 1**. The engine performance information is in a power/speed and torque/speed form and is shown in **Figure 1**.

4.2. Results and Discussions

The velocity/engine-speed diagram **Figure 2** gives a good overview of appropriate configurations of the transmission ratios and is reference to a geometrical series for number of gearbox reduction ratios. As mentioned in Secction (2.2) the number of gear changes can be designed related to the maximum and minimum allowable vehicle speed which depends on many factors. This design follows the geometric gear index which can be changed slightly in practice. The gear ratios can be adapted to the vehicle by a bench tests, road tests and computer simulation of field condition.

Figure 3 shows the traction versus vehicle speed. These shows explain how the difference gaps between the available traction are reduced in the higher reduction

Table 1. Small passenger car technical specifications.

S/No	Part	Item	Value
1	Body type	Body on frame	
2	Dimensions	Overall length (mm)	4150
		Overall width (mm)	1940
		Overall height (mm)	1417
		Wheelbase (mm)	2472
3	Engine	Type: Gasoline	
		Max. power (HP @ rpm)	75/5500
		Max. torque (Nm @ rpm)	104/3500
4	Weights	Curb weight (kg)	910
		Gross weight (kg)	1430
5	Vehicle Performance (full load)	Maximum speed, km/hr	170
		Fuel consumption (l/100 km)	8
6	Transmission (manual)	4.2, 2.7, 1.72, 1.1, 0.7	
		4.2 (final drive)	

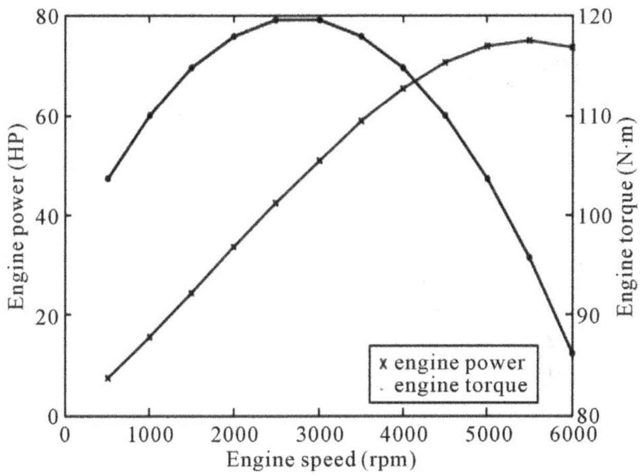

Figure 1. Speed characteristics of the spark ignition engine.

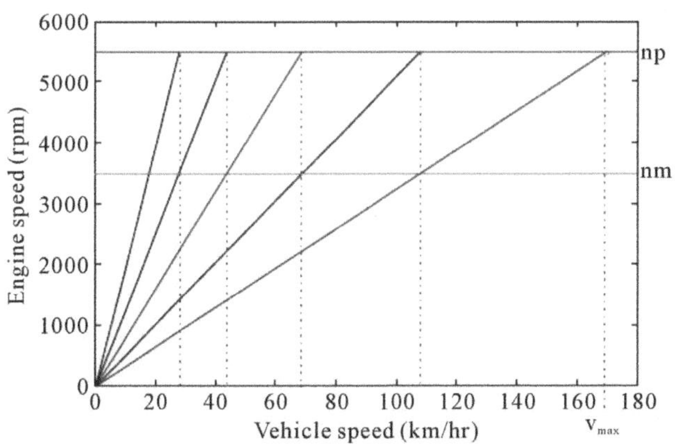

Figure 2. Desirable number of gear changes.

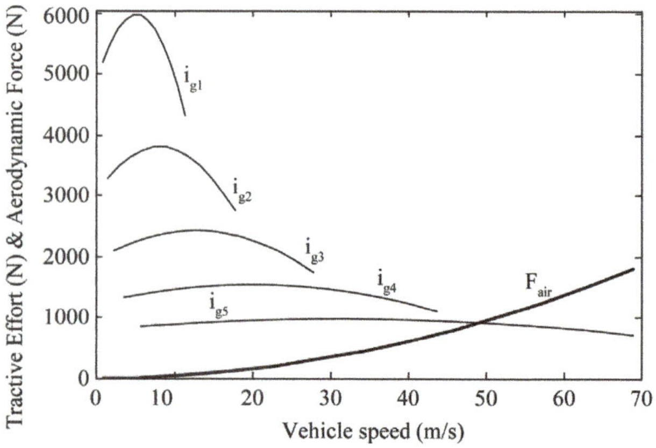

Figure 3. Tractive efforts and resistances versus vehicle speed.

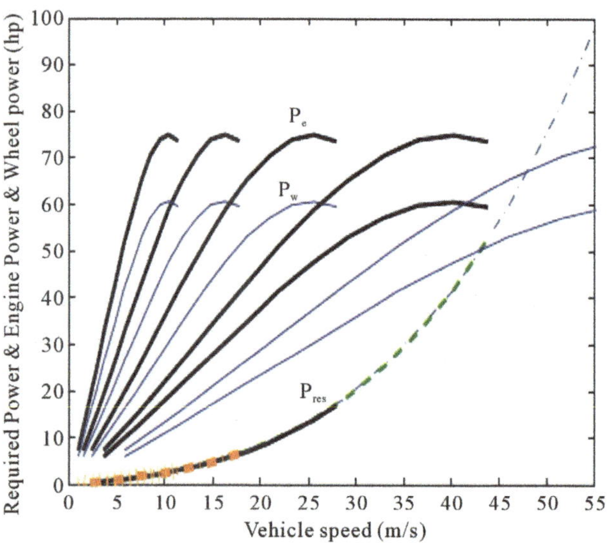

Figure 4. Required, given, and resistance power with vehicle speed.

gear ratios. The high of excess power available in lower speed range of the vehicle means that larger gap in the traction availability.

The power available and power required intersect are shown in **Figure 4**. The difference between the engine given power (P_e) and vehicle wheel sensing power (P_w) is dissipated in the transmission resistance. The graph shows that the required power value is changed according to the vehicle speed range. Fuel consumption is influenced greatly by gear and final ratios specified, and the gearshift point selected shown in **Figure 5**. With geared transmissions the operating points lie at the intersections of the ratio-dependent driving resistance lines and the line showing engine power available.

In geared transmissions, the most favourable consumption area is usually covered by top gear. When the dynamic factor at full load (D) is companied with its value at unlade vehicle weight (D_a); the operation condition, (the coefficient of road rolling-adhesion, the vehicle acceleration, also road gradient) can be defined at any load of the vehicle. This allows the other graphs can be calculated at the definite condition. **Figure 5** shows the dynamic factor of the given data with and without load.

4.3. The Proposed AMT

If the vehicle designer wishes for change the manual synchronizer gearbox to automatic manual transmission as shown in **Figure 6**, in which show that a suggest diagram of clutch and gearbox hydraulic circuit actuators and also the synchronizers motion direction by those actuators. These types of the design give additional advantages compare with conventional automatic gearbox, *i.e.* planetary gearbox. In the proposed AMT the clutch actuator is controlled by a solenoid intake valve (V_2) and two solenoid relief valves. **Figure 7** is summarized the

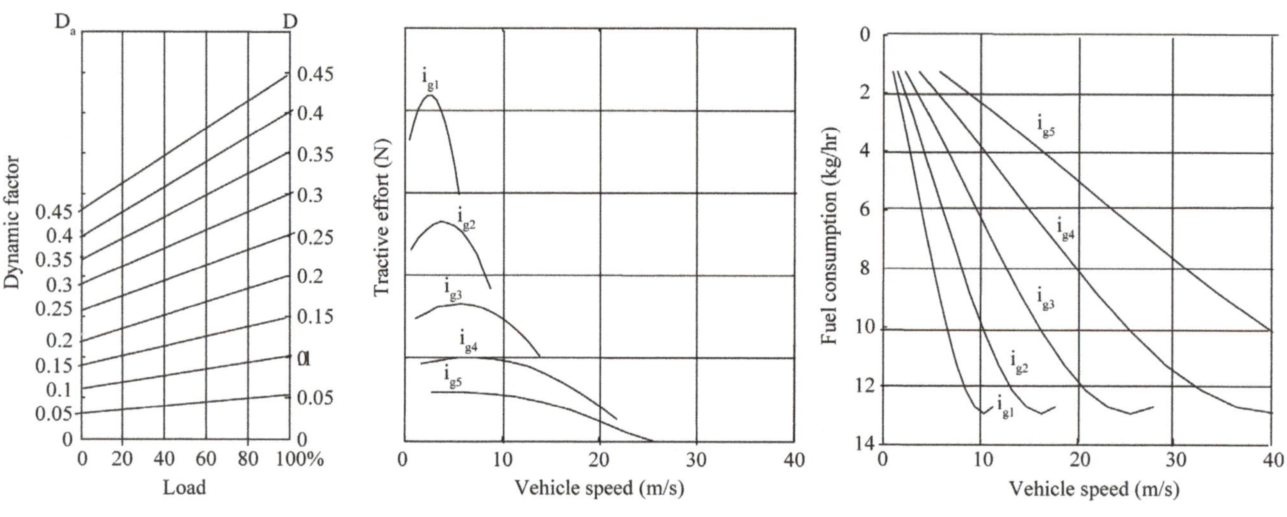

Figure 5. Vehicle dynamic passport in which stand for dynamic factors with different loads and engine fuel consumption at all gear ratios.

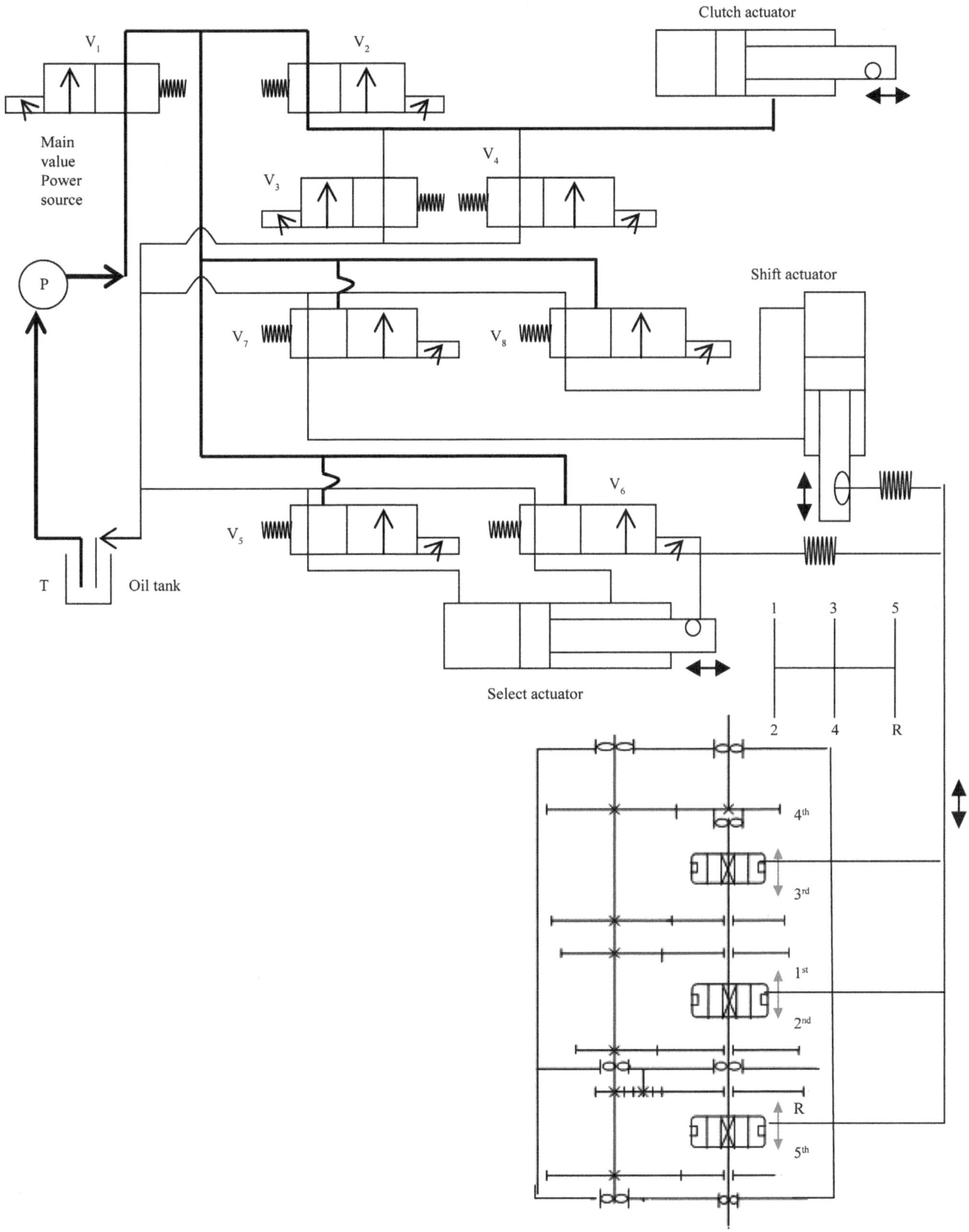

Figure 6. Clutch and gearbox hydraulic circuits, and the actuators with the direction of motion of the synchronizers by the actuators [9,10].

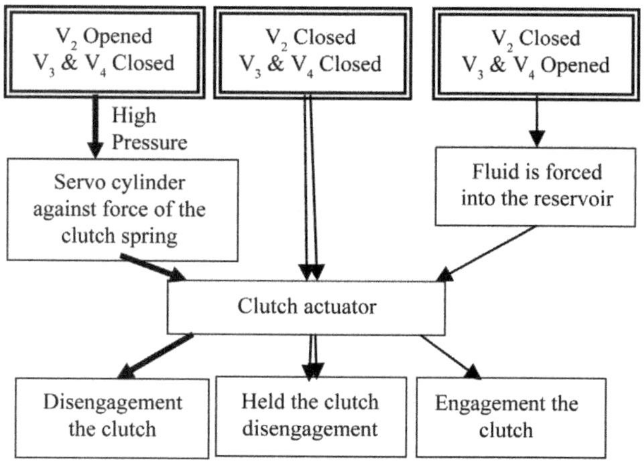

Figure 7. Clutch actuator control [9,10].

clutch actuator control by the intake and relief valves. Also, **Figure 8** is summarized the select and shift actuators control by 3-port solenoid valves. The transmission actuator consists of a shift cylinder and a select cylinder which cross at a right angle. The select cylinder is controlled by the 3-port solenoid valves V_5 and V_6, while the shift cylinder is controlled by the valves V_7 and V_8.

4.4. The Proposed Intelligent Control System

Figures 9-11 show that to expand the capability to simulate the road vehicle performance. **Figure 9** shows a flow chart diagram of the suggested intelligent control system to an automatic manual transmission, (AMT), development based on discover both the eye-catching of gear-shift requests and dreadful conditions measures of AMT,

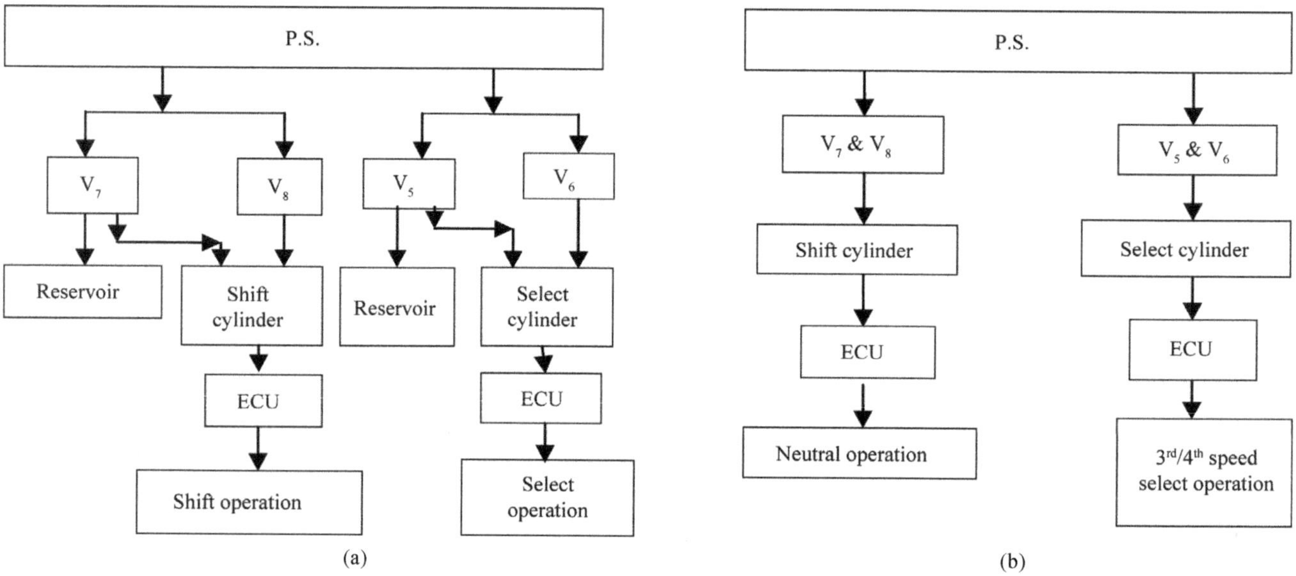

Figure 8. Transmission selection and shift actuators control [9,10]. (a) Shift or select operation; (b) Speed select position 1st/2nd or 3rd/4th.

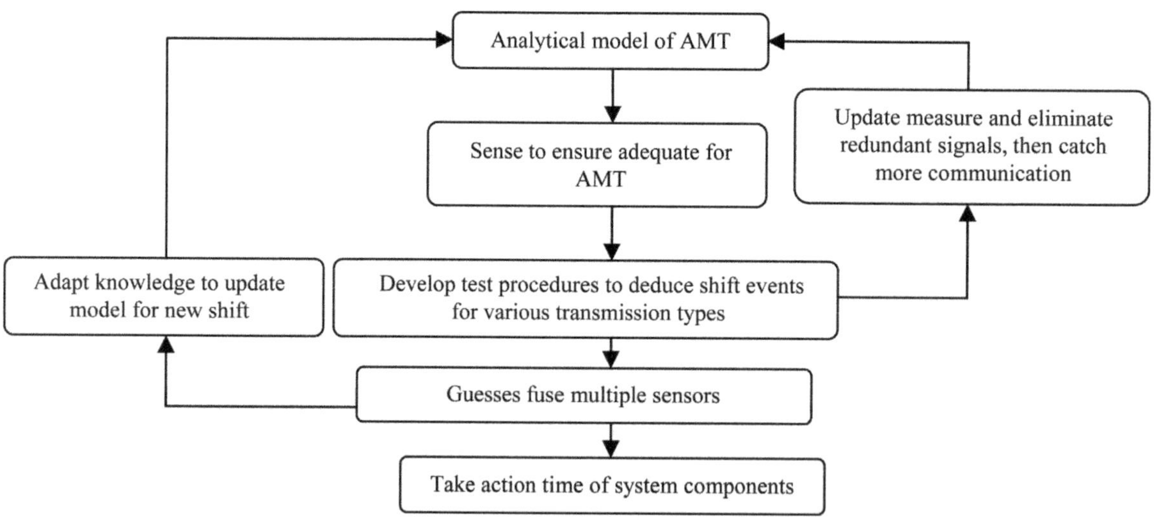

Figure 9. Suggested to an intelligent AMT development [11-18].

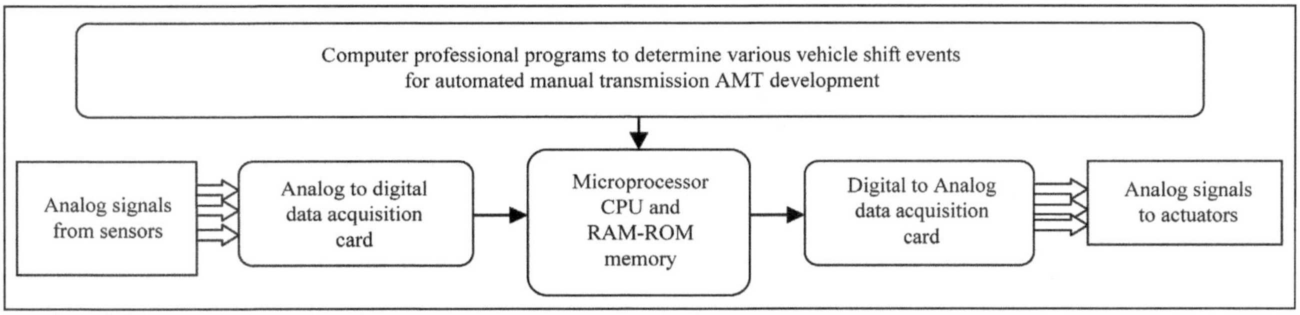

Figure 10. Proposed technologies of program related to AMT system [11-18].

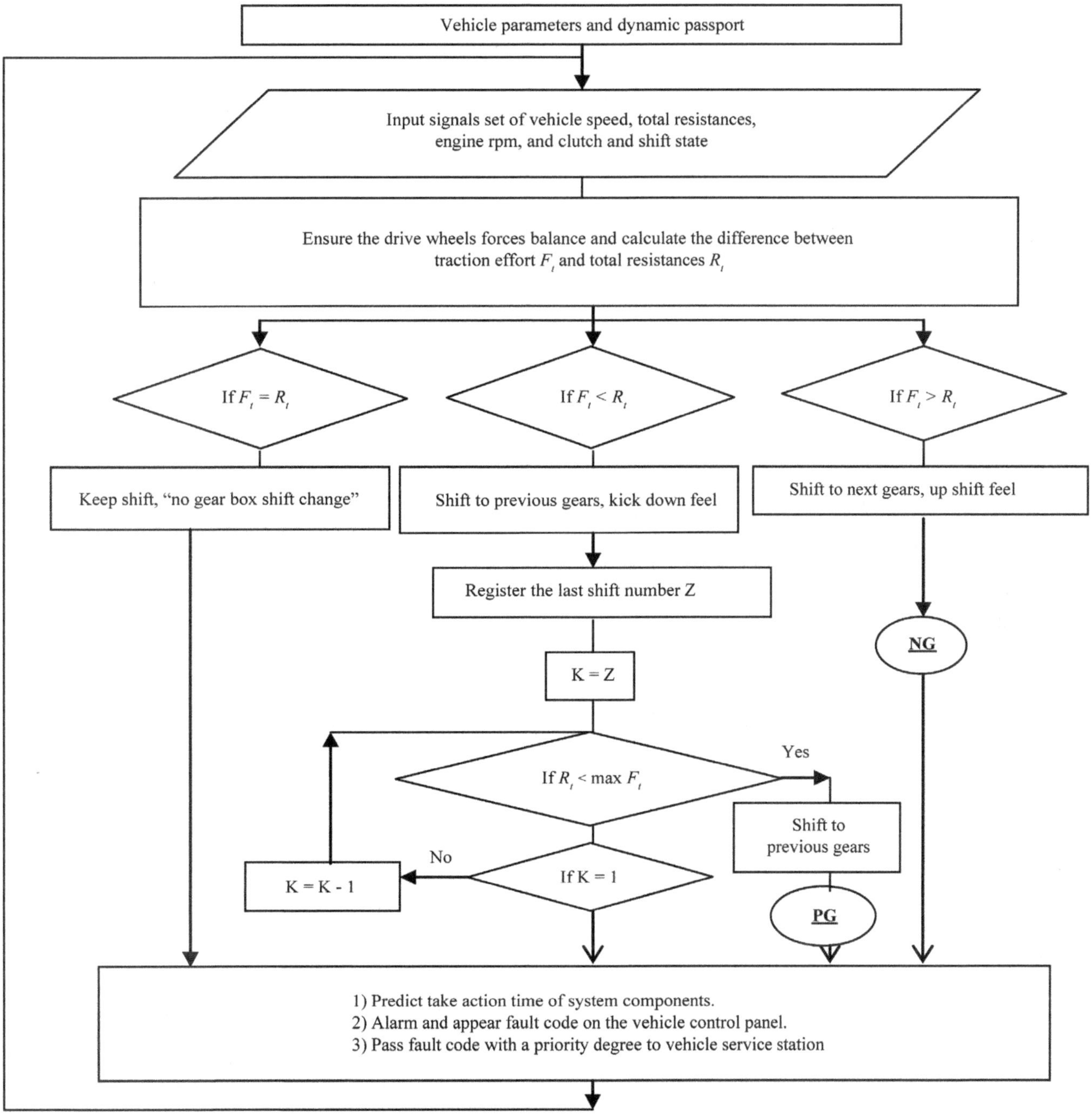

Figure 11. Professional computer program of AMT system [11-18].

those can be deduced from different measures as power-train supply and whole oppositions.

Figure 10 shows the future technologies of program related to AMT and illustrates the conceptual flow data analysis process that starts with sensors data collection at the process control level. The data is subsequently delivered to the control condition assessment algorithm for the system.

The algorithm determines the state of these; firstly, engine performance related to throttle valve position (as: torque, power and petroleum consumption versus engine rpm) secondly, road resistances (as: rolling, air, gradient and acceleration) thirdly, potential-kinetic energy (when a vehicle go away down or reduce its speed) fourthly, gearshift position and finally what is the driver feels like to do using a combination of sensors signals, actuators and dynamic models that represent the process the elements performs. The control algorithm uses the condition assessment information to update or otherwise estimate in service times for a desirable gearshift needs before any event occurs. **Figure 11** shows the flowchart diagram represents suggest a professional computer program of AMT system related to organizer algorithm of attractive gearshift requests. The program ensures the drive wheels forces balance and calculate the difference between traction effort F_t and total resistances R_t in order to keep gear shift, kick down or up shift feel. Also to alarm by near a specific failure or to get back system performances well, a control panel and actuators can be support together with tool that take their signals from suggest intelligent control system.

5. Conclusions

This paper has described theoretical investigation of a different vehicle moving operating condition which can be the basic parameters of the AMT computer designed program. A basic calculation has been done for different mathematical equation. However, the paper introduces a computer simulation program for road vehicle power-train system based on fuel economy, performance and also for AMT system linked to director algorithm of attractive gearshift requests and foretelling by near a specific failure or to get back system performances well.

It was seen that this calculation need more for formulating the complete dynamic behavior of the vehicle on the road. This formulation can help define the suitable minute for gear change. Other improvements can also include the complete mathematical models with inclusion of the gear box, the clutch, and the actuators operating gearbox and clutch.

REFERENCES

[1] M. D. Artamonov, V. A. Ilarionov and M. M. Morin, "Motor Vehicles," Mi Publishers, Union of Soviet Socialist Rrpublics, 1976.

[2] J. Y. Wong, "Theory of Ground Vehicles," John Wley & Sons, New York, 1993.

[3] P. M. Heldt, "Torque Converters or Transmissions," Chilton Company, Philadelphia, 1955.

[4] W. C. Waters, "General Purpose Automotive Vehicle Performance and Economy Simulator," SAE Paper No. 720043, 1972.

[5] K. Post, *et al.*, "Vehicle Characterization and Fuel Consumption Prediction Using Maps and Power Demand Models," *International Journal of Vehicle Design*, Vol. 6, No. 1, pp. 72-92,1985.

[6] S. M. Metwalley, "Hybrid Electric Minivan Performance and Emission Pollution Evaluation," *Engineering Research Journal*, Zagazig University, 2007

[7] S. M. Metwalley, "Comparative Study of Energy Requirement for Conventional and Hybrid Electric Vehicles," *Engineering Research Journal*, Zagazig University, 2007.

[8] A. W. Phillips, *et al.*, "Development and Use of Vehicle Power-Train Simulation for Fuel Economy and Performance Studies," SAE Paper No. 900619, 1990.

[9] A. Watanabe, *et al.*, "Isuzu's New Transmission Control System," *International Journal of Vehicle Design*, Vol. 6, No. 3, 1985, 362-372.

[10] J. Horn, *et al.*, "Flatness-Based Clutch Control for Automated Manual Transmissions," SAE, Paper No. 2004-01-0866, 2004.

[11] C. W. Hong, "A PC Based Computer Simulation Package for Spark Ignition Engine System Design," *Proceedings of the IMechE International Conference on Computers in Engine Technology*, 1991, pp. 127-134.

[12] S. Yamashita, *et al.*, "Development of an Analyzing System for Power-Train Performance," *JSAE*, Vol. 14, No. 4, 1993, pp. 28-34.

[13] R. E. Buck, "A Computer Program (HEVSIM) for Heavy Duty Vehicle Fuel Economy and Performance Simulation," Vol. 2, Users' Manual, US Department of Transportation Document #DOT-HS-805-911, 1981.

[14] M. Shibuya and S. Kubota, "Optimum Control Logic for Manual Transmission," SAE, Paper No. 850297, 1985.

[15] M. D. Schrock and Y. Liu, "Simulation of Control Schemes for Agricultural Tractor Engines and Transmissions," *American Society of Agricultural Engineers*, Vol. 33, No. 6, 1990, 1785-1789.

[16] A. F. Burke and C. B. Somcah, "Computer-Aided Design of Electric and Hybrid Vehicles," *International Journal of Vehicle Design*, 1982, pp. 61-81.

[17] H. W. He, Z. T. Liu, L. M. Zhu and X. L. Liu, "Dynamic Coordinated Shifting Control of Automated Mechanical Transmissions without a Clutch in a Plug-In Hybrid Electric Vehicle," *Energies*, Vol. 5, 2012, pp. 3094-3109.

[18] Z. T. Liu, S. Y. Shi, H. W. He and W. Q. Li, "Study on Dynamic Shift Schedule of Automated Mechanical

Transmission in a Parallel Hybrid Electric Vehicle," *2nd International Conference on Electronic & Mechanical Engineering and Information Technology (EMEIT)*, Shenyang, 7-9 September 2012, pp. 1619-1622.

Nomenclature

AMT	Automatic Manual Transmission or Automated Manual Transmission
A	Vehicle frontal area, (m^2).
C_d	Drag coefficient.
D	Dynamic factor at full load.
D_a	Dynamic factor at no load.
f	Coefficient of rolling resistance.
F_t	Tractive force, (N).
F_{air}	Air resistance force, (N).
g_e	Specific effective fuel consumption = 350, (g/kw·hr).
I_f	The total friction inertia moment for all rotating parts, ($N{\cdot}m{\cdot}s^2$).
i_g	Transmission reduction ratio.
I_r	The total inertia moment for all wheels, ($N{\cdot}m{\cdot}s^2$).
i_{tr}	Total drive train reduction ratio.
J	Acceleration of the wheel centre, (m/s^2).
M_e	Engine torque, (N·m).
n_m	Engine revelation per minute at maximum moment or torque.
n_p	Engine revelation per minute at maximum power.
P_{air}	Air power resistance, (hp).
P_g	Gradient power resistance, (hp).
P_{in}	Inertia power, (hp).
P_{rd}	Road power resistance, (hp).
P_{res}	Total resistance power, (hp).
P_r	Rolling power resistance, (hp).
P_{tr}	Transmission Power, (hp).
Q_{run}	Running fuel consumption, (l/100 km).
R_t	Total resistance
R	Rolling radius of driving wheels, (m).
V	Vehicle speed, (m/s).
W	Total vehicle weight (N).
W_a	Weight of vehicle at a given load, (N).
W_{air}	Streamlining factor, ($N{\cdot}s^2/m^2$).
Z	Nmber of gear change or gear reduction ratios.
ω_e	Angular engine speed, (rad/sec).
ω_P	Angular engine speed at maximum power, (rad/sec).
$\acute{\eta}_{tr}$	Transmission efficiency.
α_{rd}	Steepness of road ascent.
ρ_{air}	Density of air = 1.226 kg/m^3 (at 15°C and 10^5 Pa (1 bar) ambient conditions).
δ_{rot}	The coefficient accounting for rotating masses.
ρ_f	Fuel density, (kg/m^3).
V_1	Main valve.
V_2	Solenoid intake valve.
V_3 & V_4	Solenoid relief valves.
V_5, V_6, V_7 & V_8	3-Port solenoid valves.
NG	Next gear.
PG	Previous gear.
K	Integer number.

Multidisciplinary Constrained Optimization of Power Quality in Doubly Fed Wind Turbine Induction Generator

Seyed Javad Fattahi[1], Abolghasem Zabihollah[2]
[1]Department of Mechanical Engineering, Ottawa University, Ottawa, Canada
[2]School of Science and Engineering, Sharif University of Technology, Inter'l Campus, Kish Island, Iran

ABSTRACT

Shape optimization of turbine blade to maximize the output power usually changes the power factor due to compensate Repower in a wind turbine. This article presents a multidisciplinary optimization technique to maximize the output power in Doubly Fed Induction Generator (DFIG) wind turbine. The most common parameters when operating the turbine, namely, active power, reactive power and power factor, are considered as the problem constraints and the pitch angle grid side variable frequency converter of the turbine blades are optimized to maximize the output power. Numerical simulation has been illustrated to present the performance of the proposed design approach.

Keywords: Optimization; DFIG; WTG; Pitch Angle; RSC; GSC

1. Introduction

Wind Turbine machines are rapidly becoming an economically viable source of renewable energy. Two key elements of wind turbine, performance and availability, serve optimization based on machine design and performance. Wind turbine performance could be characterized by three basis parameters for power, torque and the load of the tower of a wind turbine. The torque size is important to design the mechanical elements like the rotor, gear box and brake. The tower of the wind turbine and foundation are designed to resist against the compression, buckling and the axial loading. The power size is important because it affects the transformed energy by the rotor of wind turbine. Therefore, power quality is a new challenge in wind turbine design [1-3]. Harmonics generated by the interactive wind system were investigated by Giraud and Salameh [4]. Boukhezzar *et al.* [5] proposed a non-linear approach to control a variable-speed turbine to maximize the power of turbine considering the generator torque. Muljadi and Butterfield [6] developed a pitch control strategy to maximize the power and minimize turbine loads for different wind speed scenarios. Chen *et al.* in [7] proposed the reactive power optimization of the distribution network which contains several wind farms by introducing the static VAR compensators, and Bie *et al.* in [8] present a transition-optimized approach based on adaptive load pattern classification for DFIG based on the fluctuation of the output active power and the reactive power capability and this procedure was completed in [9]. Also in [10-12], different multi-objective particle swarm optimization (MOPSO) is proposed where MOPSO considers multi-objective simultaneously rather than a PSO but the output of MOPSO consists of a group of non-dominated instead of PSO that it has only one optimization solution, but these papers focus on active and reactive power of DFIG and grid, on the other hand, Kusiak and Zheng [13] introduced an evolutionary computation approach for optimization of power factor and power output of wind turbines. This article contributes a multidisciplinary optimization for doubly Fed Induction Generator (DFIG) wind turbine subjected to active power, reactive power and power factor mutually, through pitch angle and grid side variable frequency converter. Numerical simulation has been illustrated to present the performance of the proposed design approach.

2. Wind Active Power

Theoretically, using Betz's model, wind energy captured by the rotor of a wind turbine can be expressed by:

$$P_r = 0.5\rho\pi R^2 C_p(\lambda,\beta)v_1^3 \qquad (1)$$

where P_r, ρ, R and v_1 represent, respectively, the wind energy captured by the rotor, the air density, the rotor radius, and wind speed before passing the rotor. The parameter C_p is the power coefficient that depends on the blade pitch angle, β, and the tip-speed ratio, λ, and can be determined from the following relation as:

$$C_p = 4\left(\frac{v_2}{v_1}\right)^3\left(1-\frac{v_2}{v_1}\right) \quad (2)$$

$$v_2 = \frac{v_3+v_1}{2} \quad (3)$$

$$\lambda = \frac{\omega_r R}{v_2} \quad (4)$$

where ω_r is the rotational speed of the rotor and v_1 and v_3 are wind speed in front and behind of turbine blade respectively. The literature on optimization of the power coefficient C_p is quite extensive. The maximum value of the power coefficient C_p is obtained by the derivation of (2) and optimal value of v_2/v_1 is equal to 0.593.

A wind turbine may operate when the tip speed ratio is changing in large limits but a maximum power coefficient, C_p, could be obtained only for an optimal value of λ (tip speed ratio). It results that the maximum efficiency in the wind energy conversion and the rotational speed of the rotor wind turbine must be correlated with the wind speed.

Wind turbines with capability to control the pitch angle may have the power coefficient described by a function that depends on pitch angle and tip speed ratio:

$$C_p(\lambda,\beta) = C_1\left(C_2\lambda_i - C_3\beta - C_4\right)e^{-c_5\lambda_i} + C_6\lambda \quad (5)$$

where:

$$\lambda_i = \frac{1}{\lambda+0.08\beta} - \frac{0.035}{\beta^3+1} \quad (6)$$

The variables C_i are given as: $C_1 = 0.5175$, $C_2 = 116$, $C_3 = 0.4$, $C_5 = 5$, $C_6 = 21$ [14]. **Figure 1** shows the generated power with respect to different wind speed when pitch angle equal to zero in which one may conclude that the generated power has a large sensitivity to pitch angle. The power coefficient C_p for the pitch angle β variation from 0° to 20° is shown in **Figure 2**.

3. Sensitivity Analysis

Design optimization of wind turbine generators using pitch angle does not guarantee that the power quality of the output power to be desirable. Numerically, for rotor speed of 54 rpm, zero pitch angle and the maximum C_p of 0.4296, the tip speed ratio is obtained as 7.9952.

It is worth noting that the variation of the rotor speed is limited by the gear constraint and it is allowed to tolerate ±5%. As it is shown in **Figure 3**, the maximum output power has a large sensitivity to set point of the working rotor speed. **Figure 4** shows the graph of the working plane of a WTG for different rotor speed and wind speed. It is shown that in desired wind speed (around 13.5 - 14.5 m/s) the output power is changed from 82% - 107.5% of nominal (0.95 MW) output power, hence, for delivering a good power quality, multiple variables, including pitch angle, rotor speed limitation or generated reactive power need to be controlled simultaneously.

The power is given as:

$$S^2 = P^2 + Q^2 \quad (7)$$

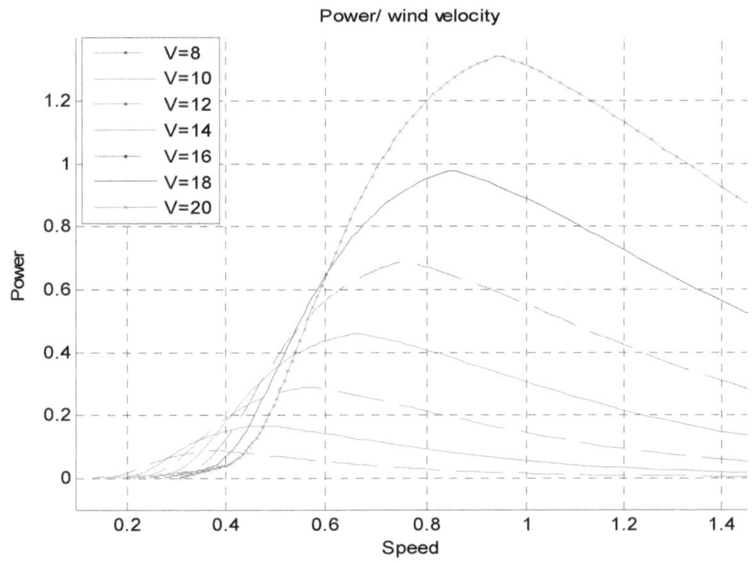

Figure 1. Output power in different wind speed with respect to $\beta = 0°$.

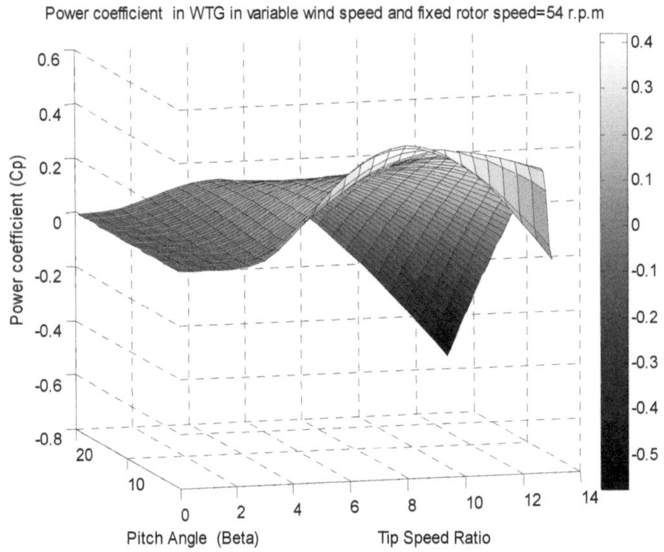

Figure 2. Power coefficient in WTG in variable wind speed and fixed rotor speed 54 r.p.m.

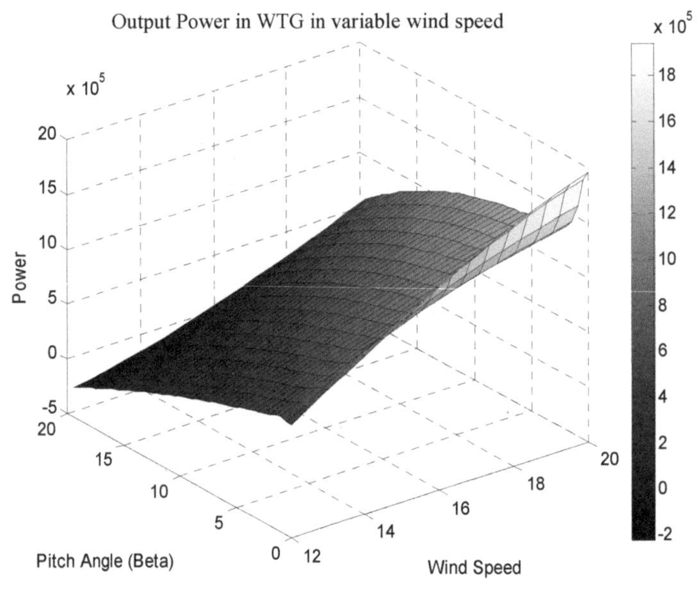

Figure 3. Output power of WTG in when speed of wind and pitch angle varies.

where P is the stator active power measured in Watts (W), S is the apparent power measured in volt-amperes (VA), Q is the reactive power measured in reactive volt-amperes. The Power factor is given by $\cos(\Phi) = P/S$, where Φ is the phase angle between the current and the voltage, measured in degrees.

3.1. Reactive Power Sensitivity

The reactive power flow from the grid to the wind plant at the interconnecting point is simplified and given by [15]:

$$Q_t = Q_g + 3I^2X - V^2(2\pi f)C - Q_c \quad (8)$$

where Q_g is the positive reactive power, consumption, of the turbine which is also negative when WTG is operating at a lagging power factor, X, denotes the equivalent series reactance of cables, lines and transformers, C is the equivalent shunt reactance of cables, and Q_c is the reactive-power injected by any reactive power compensation system at the point of interaction. The second and third terms of Equation (8) are not controllable and approximately, fixed in steady state, thus, control of Q_t has to be done with Q_g and Q_c only. The wind turbine is connected to the Doubly Fed Induction Generators (DFIG) through a mechanical shaft system, which consists of a low-speed and a high-speed shaft which are connected with a gearbox. The wound-rotor induction

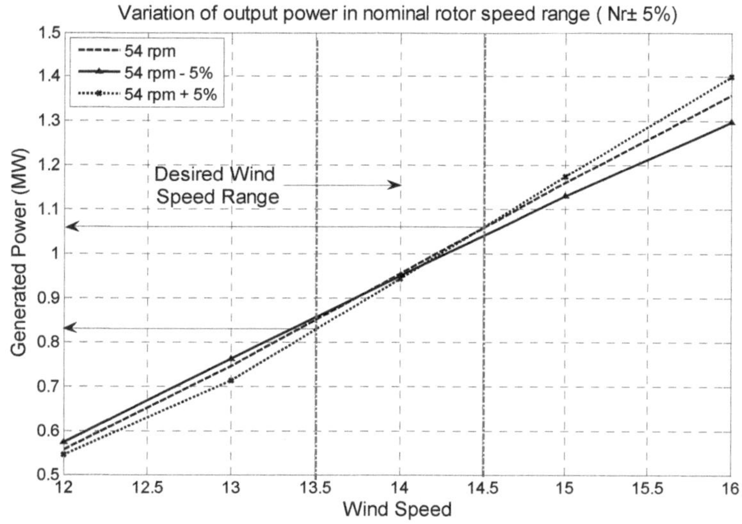

Figure 4. Generated power of WTG with constraint to rotor speed and wind speed.

machine in this configuration is fed from both stator and rotor sides. The stator is directly connected to the grid while the rotor is fed through a variable frequency converter (VFC). In order to produce electrical active power at constant voltage and frequency to the utility grid over a wide operation range from sub-synchronous to super-synchronous speed, the active power flow between the rotor circuit and the grid must be controlled both in magnitude and in direction. In DFIG the variable frequency converter consists of two four-quadrant IGBT PWM converters, rotor-side converter (RSC) and gridside converter (GSC), connected back-to-back by a dc link capacitor. The crow-bar is used to short-circuit the RSC in order to protect the RSC from over-current in the rotor circuit during transient disturbances. The RSC control scheme is expected to achieve the following objectives [16].

1) Regulating the DFIG rotor speed for maximum wind power capture

2) Maintaining the DFIG stator output voltage frequency constant

3) Controlling the DFIG reactive power

Normally, for Doubly Fed Induction Generators (DFIG), the reactive power consumption depends on the terminal voltage and loading according to the following expressions [16,17]:

$$P = -V_s \frac{L_m}{L_s} I_{qr} \tag{9}$$

$$Q_g = \frac{3}{2} \frac{\omega_s L_m^2 I_{ms} \left(I_{ms} - I_{dr} \right)}{L_s} \tag{10}$$

$$I_{ms} = \frac{V_{qs} - R_s I_{qs}}{\omega_s L_m} \tag{11}$$

where, I_{dr} is rotor current in d-axis, L_s and L_m are stator leakage and velocity (rad/s) and V_{qs} is the magnitude of the stator phase voltage in q-axis. The reactive power can be controlled on the terminals of the machine or at the grid side of the turbine transformer within the WTG capability. The voltage variations are much less than the current variation, therefore, the current is the main term that is required for compensation. Thus, by using vector control with d-axis oriented stator flux vector in rotor side converter (Equations (10) and (11)), the stator reactive power Q_g can be controlled by regulating I_{dr}.

On the other hand, grid side converter is presented for keeping DC-link voltage of capacitor constant, regardless of the magnitude and direction of rotor power. Neglecting the power losses in the converter, capacitor current can be described as follow [12]:

$$I_{DC} = \frac{3}{4} k I_{dgc} - I_{rDC} \tag{12}$$

where I_{dgc} stands for the d-axis current flowing between grid and grid side converter, I_{rDC} is the rotor side DC current, C is the DC-link capacitance, and k is the PWM modulation index of the grid side converter. The reactive power flow into the grid from GSC can be expressed as:

$$Q_c = \frac{3}{2} V I_{qgc} \tag{13}$$

where V is the magnitude of grid phase voltage (V) and I_{qgc} is q-axis current of grid side converter (A). Therefore, it is seen from Equations (11) and (12), by adjusting I_{dgc} and I_{qgc}, DC-link voltage and Q_c can be controlled respectively.

Using Equations (10) and (13), the DFIG reactive power can be written as: (See Equation (14)).

The power factor measures the efficiency of electric power utilization. The wind turbines are controlled to maintain a power factor of 1. Substituting Equations (9) and (14) in Equation (8), leads to:

$$S^2 = K\left[\left(I_{qr}\right)^2 + \left(I_{dr} - \frac{\varphi_s}{L_m}\right)^2\right]$$
$$K = \left(-V_s \frac{L_m}{L_s}\right)^2 \qquad (15)$$

However, the power factor of wind turbines is difficult to control, thus, its value is lower than 1 for individual wind turbines and wind farms. **Figure 5** shows the active vs reactive power generation curve, in which one can observe that ideal point $(S = P)$ occurs at "0.99 Pmax". On the other hand, it is known that the generated power is highly sensitive to power coefficient and pitch angle ratio. Therefore, to reach to a desired power factor in grid several variables should be taken into consideration simultaneously, requiring a multidisciplinary optimization technique.

4. The Optimization Problem

The limitations imposed by ambient temperature on the power electronic device, is less restrictive than those fixed by the generator itself. Consequently, the generator must be kept in a safe operation zone. Safe operation zone depends on the type of inverter and generator which is considered as 95% of full power capacity. Optimization of power quality needs solving two optimization problems: 1) maximization of active power and 2) minimization of reactive power, therefore, requiring a multidisciplinary optimization problem to be solved.

4.1. Statement of Optimization Problem

The optimization problems have been cast into a standard format as: (See Equation (19)).
subject to:

$$0 < \beta < 5$$

$$0 < \omega_r R / v_2 < 14$$

$$12 < v_1 < 16 \text{ m/s}$$

Minimizing

$$Q\left(I_{dr}, I_{ms}, V, f\right) = \frac{3}{2}\frac{\omega_s \left(2\pi f x_m\right)^2 I_{ms}\left(I_{ms} - I_{dr}\right)}{L_s} + \frac{3}{2} V I_{qgc} a \qquad (20)$$

subject to:

$$0.95 < s_r < 1.05 \text{ Pu}$$
$$\frac{V_{qs} - R_s I_{qs}}{\omega_s L_m} = I_{dr}$$
$$180 < V_{qs} < 240 \text{ Volt}$$

In general, the pitch angle (λ), tip speed ratio (β), wind speed (v) and rotor speed (S_r) are related constraints of generated active power and I_{dr}, I_{ms}, V and f are related constraints of reactive power.

4.2. Statement of Multidisciplinary Optimization Problem

The objective of Multidisciplinary optimization problem is defined as:
Maximize S through ;

$$S = \sqrt{P^2 + Q^2} \qquad (21)$$

The solution of this problem has been performed as the following:

Solving the maximization problem described at i without any optimizing procedure on reactive power "Q", or vice versa, means solving minimization problem described at ii without any optimizing procedure on active power "P"; both procedures lead to bad power quality output from DFIG. Therefore, it needs to define a Multidisciplinary Optimization (MDO) control in two different domains: 1) MDO is based on active power optimization by using the pitch angle, tip speed ratio, and wind velocity and 2) reactive power optimization by using rotor speed, d-axis of rotor current, q-axis of stator current and voltage simultaneously and priority concern. The priority of solving optimization Problem (1) and Problem (2) depends on the results obtained from each section. This fact has been clearly presented in results provided in **Tables 1-3**. These two problems have been solved using Sequential Quadratic Programming (SQP)

$$Q = \frac{V_s \varphi_s}{L_s} - V_s \frac{L_m}{L_s} I_{dr} = Q_g + Q_c = \frac{3}{2}\frac{\omega_s L_m^2 I_{ms}\left(I_{ms} - I_{dr}\right)}{L_s} + \frac{3}{2} V I_{qgc} \qquad (14)$$

$$P\left(\lambda, \beta, v, S_r\right) = 0.5\rho\pi R^2 \cdots \left\{ C_1\left(C_2\left(\frac{1}{\lambda + 0.08\beta} - \frac{0.035}{\beta^3 + 1}\right) - C_3\beta - C_4\right) * e^{-c_5\left(\frac{1}{\lambda+0.08\beta} - \frac{0.035}{\beta^3+1}\right)\cdots} \cdots + C_6\lambda \right\} v_1^3 \qquad (19)$$

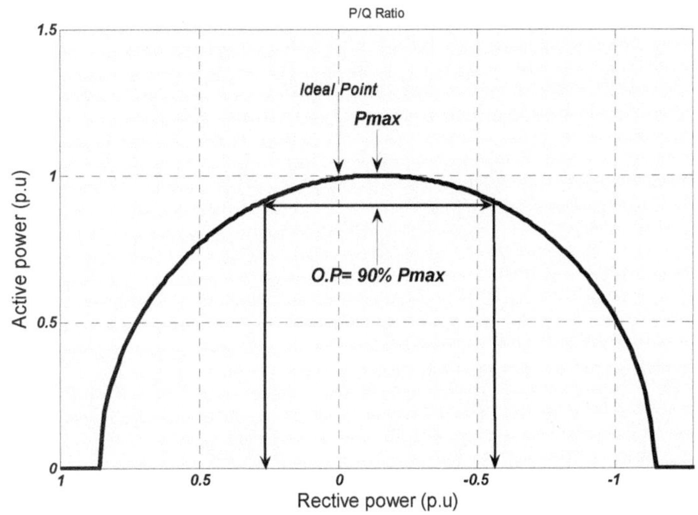

Figure 5. Active vs. reactive power generation curve in DFIGWTG.

approach described in the following section. One may note that due to complexity of the solution procedure and as the (SQP) is a well-defined optimization technique in many commercial packages, including MATLAB®, here, only the basis of the (SQP) is provided. However, to enhance the accuracy and for faster convergence to optimal results, the gradients of the objective and constraints functions have been computed analytically and fed to the optimization toolbox of MATLAB®.

4.3. Sequential Quadratic Programming (SQP)

The optimization problem is formulated as Sequential Quadratic Programming (SQP) which is a widely used method for most complex nonlinear optimization problems owing to its robustness and high efficiency in searching for the optimum point. Considering a general problem of minimizing the objective function $f(x)$, subject to constraint (x), we can write the Lagrangian function as

$$L(\{x_k\},\lambda) = f(\{x\}) + \sum_{i=1}^{m} \lambda_i g_i(\{x_k\}) \tag{22}$$

Then the quadratic programming (QP) sub-problem can be formulated based on a quadratic approximation of the Lagrangian function as:

$$\min_{d \in R^n} \frac{1}{2} d^{\mathrm{T}} H_k d + \delta f(q_k)^{\mathrm{T}} d \tag{23}$$

Subject to

$$\begin{gathered} \nabla g_i(x_k)^{\mathrm{T}} d + g_i(x_k) = 0 \\ \nabla \lambda_i(x_k)^{\mathrm{T}} d + \lambda_i(x_k) \le 0 \end{gathered} \tag{24}$$

Here x_k, g_i and λ_i are design variable vector, constraint functions and Lagrange multipliers, respectively.

Table 1. Maximum output power and recommended priority of control in normal rotor speed (54 r.p.m).

°β	Wind Speed (m/s)	Max Power MW	Reactive Power Control Priority
0	12	0.53943	RSC, GSC, Pitch
0	13	0.74516	RSC, GSC, Pitch
0.07196309	14	0.95	RSC, Pitch, GSC
1.39194002	15	0.95	Pitch, RSC, GSC
3.12291662	16	0.95	Pitch, RSC, GSC

Table 2. Maximum output power and recommended priority of control in rotor sub-speed (54 × 0.95 r.p.m).

°β	Wind Speed (m/s)	Max Power MW	Reactive Power Control Priority
0	12	0.57310	RSC, GSC, Pitch
0	13	0.76210	RSC, GSC, Pitch
0.00351158	14	0.95	RSC, Pitch, GSC
1.10133779	15	0.95	Pitch, RSC, GSC
1.83905968	16	0.95	Pitch, RSC, GSC

Table 3. Maximum output power and recommended priority of control in rotor super-speed (54 × 1.05 r.p.m).

°β	Wind Speed (m/s)	Max Power MW	Reactive Power Control Priority
1.1525	12	0.54649	Pitch, RSC, GSC
0.8616	13	0.71390	Pitch, RSC, GSC
0	14	0.94356	RSC, Pitch,GSC
1.76758958	15	0.95	Pitch, RSC, GSC
4. 39396085	16	0.95	Pitch, RSC, GSC

The matrix H_i is a positive definite approximation of the Hessian matrix of the Lagrangian function and can be updated by using quasi-Newton methods. The solution to the sub-problem is then used to form a new iteration until convergence occurs.

To improve the chances of obtaining a minimum closer to global minimum, a trial-and-error or heuristic approach is used together with the SQP method. A simple heuristic method involves randomly selecting a set of starting points in the hope that one of these starting points is close to the global minimum. While global minimum is not assured, the probability of obtaining better minimum increases with the number of starting points. Examining Equations (23) and (24), one can easily realize that the gradients of objective and constraint functions, $f(\{x\})$ and $g_i(\{x_k\})$ are the most dominant factor in convergence of the optimization procedure. As it mentioned above, in order to reach the global optimum the procedure should be repeated for different starting points, requiring an efficient computations of the function gradients. In the present work, the analytical gradients of objective and constraint functions have been derived and implemented in SQP method for all the optimization problems unless otherwise specified.

Solving the optimization problem described above, leads to the maximum power output for normal, sub-speed and super-speed rotor values. **Table 1** provides the maximum power output for normal speed (54 r.p.m), in which one can conclude that the maximum power obtains at wind speed of 16 m/s and $\beta = 3.12^{\circ}$ **Table 2** provides the maximum power output for rotor sub-speed (54 × 0.95 r.p.m), in which, once again, the maximum power obtains at wind speed of 16 m/s and $\beta = 1.84^{\circ}$ and the priority in control is similar that of normal rotor speed. For super-speed of rotor (54 × 1.05 r.p.m), the maximum power achieves at $\beta = 4.39^{\circ}$ as given in **Table 3**. It is worth noting that to obtain the maximum power; the priority in control for normal, sub-speed and super-speed is always recommended as: Pitch, RSC, GSC.

5. Modeling and Simulation

By simulating the DFIG that it's connected directly to the network through the stator, and controlled by its rotor through a PWM direct torque control converter, based on a vector control approach via stator flux. The simultaneous control methodology is implemented and simulated on DFIG based on pitch angle control, rotor-side converter (RSC) control and grid-side converter (GSC) control. **Figure 6** shows the regulated speed of GFIG rotor,and it is shown that the rotor speed remained fixed to nominal speed (54 r.p.m)when the wind speed is varied in step down and up to 13 m/sand 15 m/srespectively. **Figure 7** shows the rotor current regulated in DFIG by using PWM converter on rotor side. Finally, the delivered current (stator current) to grid side is presented in **Figure 8**, in which one can see that after 0.2 second the delivered power (current) to the grid is remained constant, and it's calculated RMSE value is 0.43% after steady state time (after 500 ms) by concurrent using of RSC, GSC, Pitch and PWM control together.

6. Conclusion

A multidisciplinary optimization technique has been

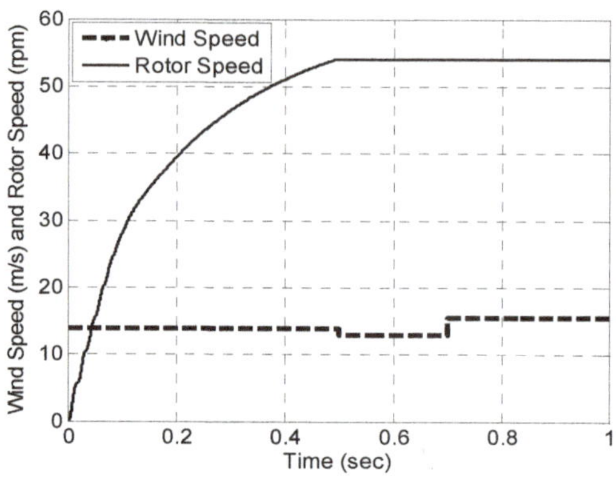

Figure 6. Rotor speed of DFIG in variable wind speed.

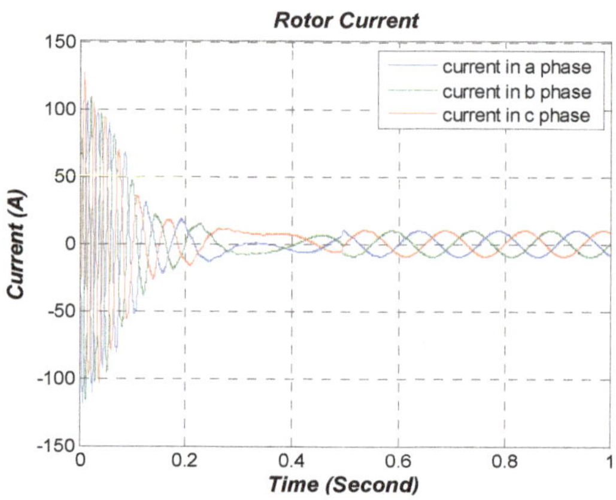

Figure 7. Regulated rotor current in DFIG in variable wind speed.

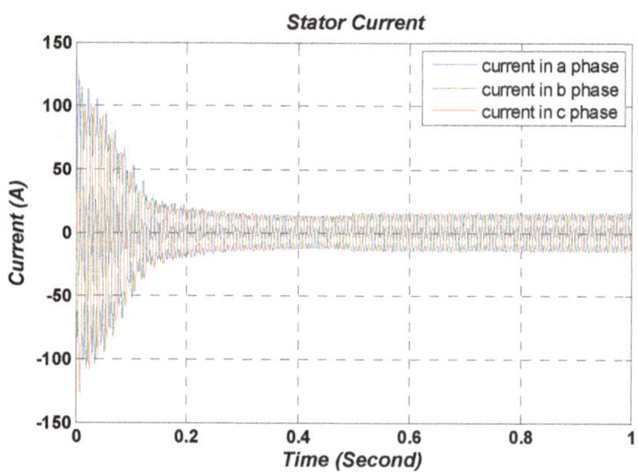

Figure 8. Stator current in DFIG in variable wind speed.

presented to maximize the power generated of Doubly Fed Induction Generator (DFIG) wind turbine. The most common constraints, namely, active power, reactive power, power factor, pitch angle and grid side variable frequency converter has been taken into consideration. The design and operation optimization problems have been considered and solved simultaneously to determine the optimal pitch angle and power control. It is observed that optimizing the pitch angle is not sufficient to maximize the generated power. It may cause an undesired power quality in grid, requiring a multi-disciplinary optimization approach to obtain the optimum power quality.

REFERENCES

[1] A. Tapia, G. Tapia and J. X. Ostolaza, "Reactive Power Control of Wind Farms for Voltage Control Applications," *Renewable Energy*, Vol. 29, No. 3, 2004, pp. 377-392.b

[2] A. Dittrich, W. Hofmann, A. Stoev and A. Thieme, "Design and Control of a Wind Power Station with Double Fed Generator," *7th European Conference on Power Electronics and Application*, Trondheim, 8-10 September 1997, pp. 723-728.

[3] S. H. Jangamshetti and V. Guruprasada Rau, "Normalized Power Curves as a Tool for Identification of Optimum Wind Turbine Generator Parameters," *IEEE Transactions on Energy Conversion*, Vol. 16, No. 3, 2001, pp. 283-288.

[4] F. Giraud and Z. M. Salameh, "Measurement of Harmonics Generated by an Interactive Wind/Photovoltaic Hybrid Power System," *Electric Power Components and Systems*, Vol. 35, No. 7, 2007, pp. 757-768.

[5] B. Boukhezzar, H. Siguerdidjane and M. Maureen Hand, "Nonlinear Control of Variable-Speed Wind Turbines for Generator Torque Limiting and Power Optimization," *ASME Transactions*: *Journal of Solar Energy Engineering*, Vol. 128, No. 4, 2006, pp. 516-530.
2006

[6] E. Muljadi and C. P. Butterfield, "Pitch-Controlled Variable-Speed Wind Turbine Generation," *IEEE Transactions on Industry Applications*, Vol. 37, No. 1, 2003, pp. 240-246.

[7] L. Chen, J. Zhong, Y. X. Ni, D. Q. Gan, *et al*., "Reactive Power Optimization in Distribution Network Including Distributed Generators," *Automation of Electric Power Systems*, Vol. 30, No. 14, 2006, pp. 20-24.

[8] Z. H. Bie, Y. H. Song, X. F. Wang, G. A. Taylor and M. R. Irving, "A Transition-Optimized Approach to Reactive Power and Voltage Control," *IEEE Power Engineering Society General Meeting*, Denver, 6-10 June 2004, pp. 226-232.

[9] Z. Jiang, J. Chen, N. Li, M. Yao and H. Li, "Dynamic Optimization of Reactive Power and Voltage Control in Distribution Network Considering the Connection of DFIG," *Power Engineering and Automation Conference* (*PEAM*), Wuhan, 8-9 September 2011, pp. 30-34.

[10] C. A. C. Coello and M. S. Lechuga, "MOPSO: A Proposal for Multiple Objective Particle Swarm Optimization," *Proceedings of the* 2002 *Congress on Evolutionary Computation*, Honolulu, 12-17 May 2002, pp. 1051-1056.

[11] H. Radwan, A. Hamid Amr, M. A. Amin Refaat, S. Ahmed Adel and A. A. El-Gammal, "Optimal Operation of Induction Motors Based on Multi-Objective Particle Swarm Optimization (MOPSO)," *The* 33*rd Annual Conference of the IEEE Industrial Electronics Society* (*IECON*), Taipei 5-8 November 2007, pp. 1079-1084.

[12] A. Laifa and B. Mohamed, "FACTS Allocation for Power Systems Voltage Stability Enhancement Using MOPSO," 5*th International Multi-Conference on Systems*, *Signals and Devices*, Amman, 20-22 July 2008, pp. 1-6.

[13] A. Kusiak and H. Y. Zheng, "Optimization of Wind Turbine Energy and Power Factor with an volutionary Computation Algorithm," *Energy*, Vol. 35, No. 3, 2010, pp. 1324-1332.

[14] S. Heier, "Grid Integration of Wind Energy Conversion Systems," John Wiley & Sons Ltd., Hoboken, 1998.

[15] R. Pena, J. C. Clare and G. M. Asher, "Doubly Fed Induction Generator Using Back-to-Back PWM Converters and Its Application to Variable Speed Wind-Energy Generation," *Proceedings of the Institute of Electrical and Electronics Engineers* (*IEEE*), Vol. 143, No. 3, 1996, pp. 231-241.

[16] A. Petersson, L. Harnefors and T. Thiringer, "Evaluation of Current Control Methods for Wind Turbines Using Doubly-Fed Induction Machines," *IEEE Transactions on Power Electronics*, Vol. 20, No. 1, 2005, pp. 227-235.

[17] Q. Wei, "Wind Speed Estimation Based Sensor Less Output Maximization Control for a Wind Turbine Driving a DFIG," *IEEE Transactions on Power Electronics*, Vol. 23, No. 3, 2008.

Permissions

The contributors of this book come from diverse backgrounds, making this book a truly international effort. This book will bring forth new frontiers with its revolutionizing research information and detailed analysis of the nascent developments around the world.

We would like to thank all the contributing authors for lending their expertise to make the book truly unique. They have played a crucial role in the development of this book. Without their invaluable contributions this book wouldn't have been possible. They have made vital efforts to compile up to date information on the varied aspects of this subject to make this book a valuable addition to the collection of many professionals and students.

This book was conceptualized with the vision of imparting up-to-date information and advanced data in this field. To ensure the same, a matchless editorial board was set up. Every individual on the board went through rigorous rounds of assessment to prove their worth. After which they invested a large part of their time researching and compiling the most relevant data for our readers. Conferences and sessions were held from time to time between the editorial board and the contributing authors to present the data in the most comprehensible form. The editorial team has worked tirelessly to provide valuable and valid information to help people across the globe.

Every chapter published in this book has been scrutinized by our experts. Their significance has been extensively debated. The topics covered herein carry significant findings which will fuel the growth of the discipline. They may even be implemented as practical applications or may be referred to as a beginning point for another development. Chapters in this book were first published by Scientific Research Publishing Inc.; hereby published with permission under the Creative Commons Attribution License or equivalent.

The editorial board has been involved in producing this book since its inception. They have spent rigorous hours researching and exploring the diverse topics which have resulted in the successful publishing of this book. They have passed on their knowledge of decades through this book. To expedite this challenging task, the publisher supported the team at every step. A small team of assistant editors was also appointed to further simplify the editing procedure and attain best results for the readers.

Our editorial team has been hand-picked from every corner of the world. Their multi-ethnicity adds dynamic inputs to the discussions which result in innovative outcomes. These outcomes are then further discussed with the researchers and contributors who give their valuable feedback and opinion regarding the same. The feedback is then collaborated with the researches and they are edited in a comprehensive manner to aid the understanding of the subject.

Apart from the editorial board, the designing team has also invested a significant amount of their time in understanding the subject and creating the most relevant covers. They scrutinized every image to scout for the most suitable representation of the subject and create an appropriate cover for the book.

The publishing team has been involved in this book since its early stages. They were actively engaged in every process, be it collecting the data, connecting with the contributors or procuring relevant information. The team has been an ardent support to the editorial, designing and production team. Their endless efforts to recruit the best for this project, has resulted in the accomplishment of this book. They are a veteran in the field of academics and their pool of knowledge is as vast as their experience in printing. Their expertise and guidance has proved useful at every step. Their uncompromising quality standards have made this book an exceptional effort. Their encouragement from time to time has been an inspiration for everyone.

The publisher and the editorial board hope that this book will prove to be a valuable piece of knowledge for researchers, students, practitioners and scholars across the globe.

List of Contributors

N. A. Abdel-Halim
Automotive Engineering Department, Faculty of Engineering, Helwan University, Cairo, Egypt

Ben Zion Sandler and Vladimir Chapsky
Mechanical Engineering Department, Ben Gurion University, Beer Sheva, Israel

Frank Domroes, Carsten Krewet and Bernd Kuhlenkoetter
Institute of Production Systems, TU Dortmund University, Dortmund, Germany

Ebrahim Ebrahimi, Mohammad Hadi Jalili, Nasrolah Astan and Mojtaba Bavandpour
Department of Mechanical Engineering of Agricultural Machinery, Islamic Azad University, Kermanshah Branch, Kermanshah, Iran

Payam Javadikia
Department of Mechanical Engineering of Agricultural Machinery, Razi University, Kermanshah, Iran

Majid Haidari
Department of Mechanical Engineering of Agricultural Machinery, Islamic Azad University, Takestan Branch, Takestan, Iran

Ali Zarei
Department of Mechanical Engineering of Agricultural Machinery, East Azarbaijan Sience and Research Branch, Islamic Azad University, Tabriz, Iran

Chi-Chang Hsieh
Department of Mechanical and Automation Engineering, National Kaohsiung First University of Science and Technology, Taiwan

Yan-Huei Li
Instiute of Engineering Science and Technology, National Kaohsiung First University of Science and Technology, Taiwan

Yubin Lu
Key Laboratory of Testing Technology for Manufacturing Process (Ministry of Education), Southwest University of Science and Technology, Mianyang, China

Songyan Zhang
Student Office of School of Arts and Literature, Southwest University of Science and Technology, Mianyang, China

Thin-Lin Horng
Department of Mechanical Engineering, Kun-Shan University, Tainan, Chinese Taipei

Marcos Daniel Souza dos Santos, Daniel Valença Fiuza Lima, Jorge Eliécer Campuzano Carmona, Suzana Moreira Avila and Graciela Nora Doz de Carvalho
Departamento de Engenharia Civil e Ambiental, Faculdade de Tecnologia, Universidade de Brasilia, Brasilia, Brasil

Shiping Liu and Shujuan Zheng
North China Institute of Water Conservancy and Hydro-Electric Power, Zhengzhou, China

Chi-Chang Hsieh
Department of Mechanical and Automation Engineering, National Kaohsiung First University of Science and Technology, Taiwan

Chih-Ching Hung
Department of Product Design, Shu-Te University, Taiwan

Yan-Huei Li
Department of Mechanical and Automation Engineering, National Kaohsiung First University of Science and Technology, Taiwan

Jingyang Peng and Xiongbiao Chen
Department of Mechanical Engineering, University of Saskatchewan, Saskatoon, Canada

Naziha Zerari and Abdelouaheb Benretem
Laboratory of the Electromechanicaly Engineering Badji Mokhtar-Annaba University, Annaba, Algeria

Tarik Saidouni
Laboratory of the Mechanical Systems, Path, Engineering of the Mechanical Systems, Option Structure and Production Ministry of the National Defense, School Military Polytechnic EMP, Alger, Algeria

D. Koulocheris, A. Stathis, Th. Costopoulos and G. Gyparakis
School of Mechanical Engineering, National Technical University of Athens, Athens, Greece

Yuntao Wang
College of Mechanical Engineering, Liaoning Technical University, Fuxin, China

Yunhong Cheng
Department of Civil Engineering, Northeastern University, Shenyang, China

Nianchun Lü
School of Material Science and Engineering, Shenyang Ligong University, Shenyang, China

Jin Cheng
Department of Astronautics and Mechanics, Harbin Institute of Technology, Harbin, China

Frank Wardle
UPM Ltd., Swindon, England

Timothy Minton and Saiful Bin Che Ghani
Brunel University, London, England

Paul Förstmann, Martin Roeder, Sebastian Richarz and Fiona Sammler
Technische Universität, Berlin, Germany

P. N. Jiki
Civil Engineering Department, University of Agriculture, Makurdi, Nigeria

Tao Jiang
Military Representative Office, Dalian, China

Wu Wei
China Ship Development and Design Center, Wuhan, China

Yuan-Ping Luh and Yin-Chang Liu
Department of Mechanical and Electrical Engineering, National Taipei University of Technology, Chinese Taipei

Hélio Dorneles Etchepare and Wilson Kindlein Júnior
PPG3M, Federal University of Rio Grande do Sul, Porto Alegre, Brazil

Sharad P. Jagtap
Mechanical Engineering Department, Government Polytechnic, Murtizapur, India

Anand N. Pawar
Mechanical Engineering Department, Government Polytechnic, Amravati, India

Axel Imbert
Laboratoire de Physiques et Mécanique, Ecole Nationale d'Ingénieur de Sud Alsace, Mulhouse, France
Centre Technique des Industrie Mécanique, Centre Recherche Matériaux Textiles, Mulhouse, France

Gildas L'Hostis, Fabrice Laurent and Bernard Durand
Laboratoire de Physiques et Mécanique, Ecole Nationale d'Ingénieur de Sud Alsace, Mulhouse, France

David Rigel
Centre Technique des Industrie Mécanique, Centre Recherche Matériaux Textiles, Mulhouse, France

Nagwa Ahmed Abdel-Halim
Automotive Engineering Department, Faculty of Engineering, Helwan University, Cairo, Egypt

Seyed Javad Fattahi
Department of Mechanical Engineering, Ottawa University, Ottawa, Canada

Abolghasem Zabihollah
School of Science and Engineering, Sharif University of Technology, Inter'l Campus, Kish Island, Iran